Metal Halide Perovskite Crystals

Metal Halide Perovskite Crystals

Growth Techniques, Properties and Emerging Applications

Special Issue Editor

Wei Zhang

MDPI • Basel • Beijing • Wuhan • Barcelona • Belgrade

MDPI

Special Issue Editor
Wei Zhang
University of Surrey
UK

Editorial Office
MDPI
St. Alban-Anlage 66
4052 Basel, Switzerland

This is a reprint of articles from the Special Issue published online in the open access journal *Crystals* (ISSN 2073-4352) from 2017 to 2018 (available at: https://www.mdpi.com/journal/crystals/special_issues/emerging_applications)

For citation purposes, cite each article independently as indicated on the article page online and as indicated below:

LastName, A.A.; LastName, B.B.; LastName, C.C. Article Title. *Journal Name* **Year**, *Article Number*, Page Range.

ISBN 978-3-03897-558-8 (Pbk)
ISBN 978-3-03897-559-5 (PDF)

Contents

About the Special Issue Editor

Wei Zhang is a Lecturer (Assistant Professor) in Energy Technology at the Advanced Technology Institute, University of Surrey. He obtained his PhD at the National University of Singapore, working on solid state sensitized solar cells. He then moved to the University of Oxford, where he began working on perovskite solar cells. His current research interests include halide perovskites for photovoltaic and light-emission applications and low-dimensional nanomaterials (semiconducting metal oxides, plasmonic metal nanoparticles, photonic crystals, etc.) for energy conversion and storage. He has published over 50 peer-reviewed journal articles (6 on Science and Nature Family journals and 18 have been selected as ESI highly cited papers), with a total citation of over 6800 and Google H-index of 34. He is the Associate Editor of *Energy and Environmental Materials* (since 2018), *Science and Technology of Advanced Materials* (since 2017), and a Guest Editor for the Special Issue of Crystals on Nanoscale Research Letters.

Preface to "Metal Halide Perovskite Crystals"

In recent years, metal halide perovskites have emerged as a rising star among semiconductor materials owing to their low cost, solution processability, and fascinating combination of material properties enabling a broad range of energy applications. Accompanied by the unprecedented success in the photovoltaic community, which has witnessed a certified power conversion efficiency of 23.7%, rapid advancement has also been achieved in the areas of light-emitting diodes, lasers, photodetectors, and solar-to-fuel energy conversion devices. Beyond the dominant format of polycrystalline perovskite thin films for solar cell applications, recent progress in metal halide perovskite crystals, ranging from nanocrystals to macroscopic single-crystals, has spurred a great deal of both scientific and industrial interest. Great research efforts have endeavored to develop new techniques for crystal growth and investigate the physical and chemical properties of the materials and explore their emerging applications. These exciting achievements call for a rationalization of the different forms of perovskite semiconductors beyond the widely used polycrystalline thin films. In the current Special Issue, "Metal Halide Perovskite Crystals: Growth Techniques, Properties and Emerging Applications", we aim to provide a forum for the discussion and presentation of recent advances in the fields of research related to metal halide perovskite crystals.

Wei Zhang
Special Issue Editor

crystals

MDPI

Article

Metal Halide Perovskite Single Crystals: From Growth Process to Application

Shuigen Li [1,*], Chen Zhang [2], Jiao-Jiao Song [2], Xiaohu Xie [1], Jian-Qiao Meng [2,3,*] and Shunjian Xu [1,*]

1   School of New Energy Science and Engineering, Xinyu University, Xinyu 338004, China; xie19930723@163.com
2   Hunan Key Laboratory for Super-Microstructure and Ultrafast Process, School of Physics and Electronics, Central South University, Changsha 410083, China; zhangchen@csu.edu.cn (C.Z.); 162211012@csu.edu.cn (J.-J.S.)
3   Synergetic Innovation Center for Quantum Effects and Applications (SICQEA), Hunan Normal University, Changsha 410081, China
*   Correspondence: sgli76@163.com (S.L.); jqmeng@csu.edu.cn (J.-Q.M.); xushunjian@126.com (S.X.)

Received: 21 April 2018; Accepted: 9 May 2018; Published: 17 May 2018

Abstract: As a strong competitor in the field of optoelectronic applications, organic-inorganic metal hybrid perovskites have been paid much attention because of their superior characteristics, which include broad absorption from visible to near-infrared region, tunable optical and electronic properties, high charge mobility, long exciton diffusion length and carrier recombination lifetime, etc. It is noted that perovskite single crystals show remarkably low trap-state densities and long carrier diffusion lengths, which are even comparable with the best photovoltaic-quality silicon, and thus are expected to provide better optoelectronic performance. This paper reviews the recent development of crystal growth in single-, mixed-organic-cation and fully inorganic halide perovskite single crystals, in particular the solution approach. Furthermore, the application of metal hybrid perovskite single crystals and future perspectives are also highlighted.

Keywords: perovskite single crystals; growth process; application; solar cell; photodetector

1. Introduction

Recently, organic-inorganic metal hybrid perovskites have shown great applied potential because of their impressive optical and electrical properties [1–5], which can be represented by the structure ABX_3, where A is $CH_3NH_3^+$, $CH(NH_2)_2^+$ or Cs^+, B is Pb^{2+} or Sn^{2+}, and X is I^-, Br^- or Cl^- [6–27]. The ideal ABX_3 structure is cubic symmetry, where A and B ions are located at the eight corners and center of a cubic unit, respectively. The symmetry of ABX_3 structures is based on the atomic species of the A and B sites. In a typical perovskite crystal structure, the A, B, and X ionic radii, e.g., R_A, R_B, and R_X, should correspond to a specific geometric relationship, known as the Tolerance factor [28–30]: $t = (R_A + R_X)/\sqrt{2}(R_B + R_X)$. The ideal value of t should be 1 for cubic structures; otherwise, the structure tends to be distorted, or even destroyed [28,30,31]. For lead hybrid perovskite, the large organic cation at the A position, e.g., methylammonium (MA^+) or formamidinium (FA^+), is able to match the large radius of the Pb^{2+} ion at the B position and meet the tolerance factor t, while the halogen anions or their mixtures occupy the C positions, resulting in the formation of a 3D perovskite structure [32]. These perovskite-based materials, when used in the photovoltaic field, can provide remarkable properties, such as broad absorption from the visible to the near-infrared region, tunable optical and electronic properties [15,33–36], high charge mobility, and long exciton diffusion length and carrier recombination lifetime [32,37–48]. Within a few years, they have revolutionized the photovoltaic field; an efficiency of 22.1% from solution-processable perovskite-based solar cells has been reported [49]. In addition, lead hybrid perovskites have also been used in some other fields, such as laser [50], photodetector [51], light-emitting-diodes [52], thermoelectricity [53], and catalysis [54], demonstrating their potential application prospects.

Until now, many intensive investigations have been based on polycrystalline thin films, one of the existing forms of perovskite, and most of the results have been focused on the perovskite polycrystalline film. With in-depth research, single crystals—another form of perovskite—have been found with low defect density. The carrier diffusion length of perovskite is sensitive to defects. When expanding the grain size, the carrier diffusion length of polycrystalline forms can increase to up to 1 μm, while large single crystals are able to provide even longer carrier diffusion lengths. Dong et al. prepared millimeter-sized MAPbI$_3$ single crystals via a low-temperature solution approach, in which a carrier diffusion length of over 175 μm was obtained under 1 sun illumination, and a longer carrier diffusion length exceeding 3 mm could be produced under a weaker illumination with 0.003% sun illumination [55]. Shi et al. reported low trap-state density of states with an order of 10^9–10^{10} cm^{-3} and carrier diffusion length > 10 mm in MAPbX$_3$ single crystals [40]. The longer carrier diffusion length in single crystals with low trap-state density derives from their better extraction and transport of photogenerated charge carriers, resulting in a performance boost for optoelectronic devices. These meaningful findings will contribute to the development of perovskite-based materials, and will be extremely beneficial to further fundamentally investigate the intrinsic properties of perovskites single crystals. To date, single-organic-cation, mixed-organic-cation, and all-inorganic kinds of metal halide perovskite single crystals have been demonstrated. In this review, we will summarize the advances in the growth and application of the above perovskite single crystals.

2. Growth of Organic-Inorganic Hybrid Halide Perovskite Single Crystals

Since organic-inorganic metal hybrid perovskite solar cells (PSCs) were studied for the first time [5], they have attracted particular attention due to their extraordinary performance. Since then, in-depth study on perovskite-based materials and devices has been carried out, and a series of research results have been obtained. Meanwhile, perovskite single crystals, which were reported about forty years ago [40,56,57], are studied again.

2.1. Growth of Single-Organic-Cation Halide Perovskite Single Crystals

Solution temperature lowering (STL) is a traditional single crystal growth process. In 1987, Poglitsch et al. gained MA-based perovskite single crystals via a temperature-lowering method [58] in which they heated the mixed solution to 100 °C, and perovskite single crystals were grown by cooling the solution to room temperature. In general, minimizing the number of nuclei is crucial to growing large single crystals. As an improved technology, seed-assisted growth is often adopted for the purpose of growing large-sized and high-quality single crystals, i.e., small crystals are firstly put into a single crystal precursor, followed the temperature-lowering process. Using a slow cooling rate of 0.1–0.2 °C/h, Su et al. obtained large MAPbI$_3$ single crystals with sizes of up to 1 cm [59] by a process in which small crystals were firstly obtained by spontaneous nucleation (Figure 1a), then high-quality as-grown crystals were selected as seeds and dropped back into the mother liquid to grow large single crystals (Figure 1b). Similarly, MAPbBr$_3$ single crystals with perfect cubic structure were formed. Dang et al. grew tetragonal MAPbI$_3$ bulk single crystals with dimensions of 10 mm × 10 mm × 8 mm by a seed-assisted growth method [60] in which the seeded crystal was fixed in the middle of the solution. The solution was saturated gradually with a decrease of temperature from 65 °C to 40 °C, resulting in the formation of high-quality single crystals over the following few days.

Huang's [36] group grew large bulk MAPbI$_3$ single crystals with a size of 10 mm × 3.3 mm via a temperature-lowering method, as presented in Figure 1c, in which the seed crystal was fixed in the top half of the solution. Importantly, they dissolved the seed crystals in the bottom of the precursor solution, and the super saturation of the top solution was readily induced because of the temperature gradient between the top and bottom, leading to fast-growing of single crystals with a rate of about 2 mm per day.

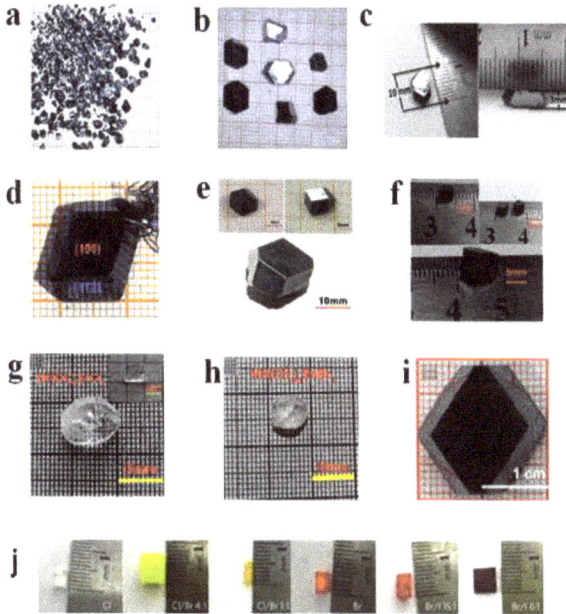

Figure 1. (a) MAPbI$_3$ crystal grains obtained by spontaneous nucleation; and (b) grown large MAPbI$_3$ crystals obtained using seeds. Reprinted with permission from [59], Copyright 2015, Elsevier B.V.; (c) large bulk MAPbI$_3$ single crystals with sizes of 10 mm × 3.3 mm. Reprinted with permission from [36], Copyright 2015, American Association for the Advancement of Science; (d) MAPbI$_3$ single crystal grown using the BSSG method. Reprinted with permission from [61], Copyright 2015, Springer; (e) CH$_3$NH$_3$SnI$_3$ and (f) CH(NH$_2$)$_2$SnI$_3$ single crystals grown via the TSSG method. Reprinted with permission from [62], Copyright 2016, Wiley; (g) NH(CH$_3$)$_3$SnCl$_3$ and (h) NH(CH$_3$)$_3$SnBr$_3$ single crystals prepared via the BSSG method. Reprinted with permission from [63], Copyright 2016, American Chemical Society; (i) MAPbI$_3$(Cl) bulk single crystals grown by rapid solution temperature-lowering method. Reprinted with permission from [64], Copyright 2016, American Chemical Society; (j) mixed-halide perovskite single crystals with different halide compositions. Reprinted with permission from [65], Copyright 2015, Springer.

Using the bottom-seeded solution growth (BSSG) method, Lian et al. [61] prepared centimeter-sized bulk MAPbI$_3$ single crystals. To eliminate the negative effect of multiple nuclei, a seed crystal was fixed by platinum wire to segregate the seed crystal from the bottom of the flask, and the desired single crystal, 12 mm × 12 mm × 7 mm in size, was obtained by lowering the temperature of the growth solution from 373 K to 330 K, as shown in Figure 1d.

Tin-based perovskite single crystals have also been harvested by the temperature-lowering method. Tao and co-worker reported bulk cubic CH$_3$NH$_3$SnI$_3$ and CH(NH$_2$)$_2$SnI$_3$ single crystals grown via a top-seeded solution growth (TSSG) method (Figure 1e,f) [62]. Similarly, NH(CH$_3$)$_3$SnX$_3$ (X = Cl, Br) single crystals were prepared via the TSSG method [63], as shown in Figure 1g,h.

Lian et al. reported a rapid solution temperature-lowering method to prepare Mixed-halide perovskite single crystals based on the addition of chlorine [64]. With the addition of chlorine, the surface free energy and the edge free energy were changed. The resulting edge free energy is expressed by: $\rho_{Chlorine} = \rho - k_B T \ln C_{Chlorine}$, given that the edge free energy $\rho_{Chlorine}$ decreases with $C_{Chlorine}$, the growth rate of $CH_3NH_3PbI_3(Cl)$ crystal face will increase. They harvested large $CH_3NH_3PbI_3(Cl)$ with sizes of 20 mm $\times$ 18 mm $\times$ 6 mm within only 5 days (Figure 1i) that possessed excellent properties, i.e., a high carrier mobility of 167 ± 35 $cm^2 \cdot V^{-1} \cdot s^{-1}$, a low trap-state density of 7.6×10^8 cm^{-3}, and a transient carrier lifetime as long as 449 ± 76 μs. Mixed-halide perovskite single crystals can also be grown via solw temperature-lowering method [65]. At 100 °C, a super-saturated aqueous solution, including single or mixed haloid acid of different halide ratios, mixing methylamine and lead (II) acetate, was prepared. By gradually lowering the precursor solution temperature, a series of perovskite single crystals, depending on the different halide ratios, could readily be formed, as presented in Figure 1j. As is shown, for both $MAPbBr_{3-x}Cl_x$ and $MAPbI_{3-x}Br_x$ single crystals, the color varies with the different halide ratios of Br/(Cl + Br) and I/(I + Br).

The STL method provides a simple and effective approach for the growth of $MAPbX_3$ and $FAPbX_3$, in which the crystals are formed with the decrease in the temperature of the precursor solution. However, the drawback of its being highly time-consuming (typically more than two weeks to gain one-centimeter-sized crystals [61,62]) limits its extensive use. On the other hand, the solute solubility in solvents decreases with increasing temperature for a few materials [66], i.e., inverse temperature solubility, which is also present for organic-inorganic metal hybrid perovskites. For example, $MAPbI_3$, $MAPbBr_3$ and $MAPbCl_3$ show inverse temperature solubility in certain solvents-gamma-butyrolactone (GBL), *N,N*-dimethylformamide (DMF), and dimethylsulfoxide (DMSO), respectively. Based on the above characteristics, inverse temperature crystallization (ITC) was first introduced to grow $MAPbI_3$ single crystals. During the ITC, the precursor was dissolved in GBL and the temperature was increased until single crystals formed at about 190 °C [67]. At the same time, Saidaminov et al. designed inverse temperature crystallization (ITC) for $MAPbX_3$ perovskites due to their inverse temperature solubility behavior in some solvents [68]. They grew size- and shape-controlled high-quality $MAPbI_3$ and $MAPbBr_3$ single crystals within several hours (Figure 2a).

To grow large single crystals and understand their growth mechanism, Liu's group developed the seed-repeated method and harvested the largest $MAPbI_3$ single crystal, with a size of 71 mm $\times$ 54 mm $\times$ 39 mm, as well as inch-sized $MAPbBr_3$ and $MAPbCl_3$ crystals (Figure 2g) [69]. In the process of crystal growth, small perovskite particulates were harvested as seed crystals to keep in precursor solution at 100 °C for 24 h. A seed crystal was placed in precursor solution to keep at 100 °C for 48 h, resulting in the formation of a larger crystal. By repeating the above process, the final large crystals would be produced. Figure 2j shows the absorbance spectra of $MAPbX_3$ (X = Cl, Br, I) perovskites. A clear band edge without excitonic signature or absorption is shown, indicating high-quality single crystals with low defect concentration. Furthermore, all of the PL spectra of $MAPbCl_3$, $MAPbBr_3$, and $MAPbI_3$ perovskites exhibited narrow PL peaks at $\approx$402, $\approx$537, and $\approx$784 nm, respectively (Figure 2k), and the PL peak values were smaller than the corresponding absorption onsets (431, 574, and 836 nm), indicating their advantageous application in solar cells. The X-ray Diffraction (XRD) measurement displayed that the (200) diffraction peak of $MAPbI_3$ single crystal showed a FWHM of 0.3718°, which indicated that the single crystal held a respectable crystalline quality. Electric characterization showed that the electron trap was 1.1×10^{11} cm^{-3} for $MAPbBr_3$ and 4.8×10^{10} cm^{-3} for $MAPbI_3$. The hole trap densities for $MAPbBr_3$ and $MAPbI_3$ were determined to be 2.6×10^{10} cm^{-3} and 1.8×10^9 cm^{-3}, respectively. Furthermore, the crystalline $MAPbX_3$ (X = I, Cl, Br) gave a high carrier mobility of 34 $cm^2 \cdot V^{-1} \cdot s^{-1}$, 179 $cm^2 \cdot V^{-1} \cdot s^{-1}$ and 4.36 $cm^2 \cdot V^{-1} \cdot s^{-1}$. It is expected that such wafer-sized single-crystalline $MAPbX_3$ (X = I, Cl, Br) with superior properties in terms of defect state and carrier density are promising materials for high-performance optoelectronic devices.

The ITC method is a highly effective approach for growing metal hybrid halide perovskite crystals that possesses a much faster growth rate than that of the typical STL method and has been used extensively to grow single or mixed halide perovskite crystalline materials [67,70–74]. Additionally, the ITC method meets the requirements of $FAPbX_3$ single crystals. Bakr's group reported the retrograde solubility of FA-based perovskites and grew high-quality crack-free $FAPbI_3$ and $FAPbBr_3$ single crystals (Figure 2h,i) [75]. By improving the onset of crystallization temperature, they obtained grain boundary-free FAPbI$_3$ crystals within 3 h. Yang's group obtained 5-mm-sized FAPbI$_3$ single crystals for the first time via a modified ITC method [74]. They first grew the FAPbI$_3$ seed crystal via a cooling solution method, followed by growing larger single crystals by placing the small seed crystals into the ITC precursor and keeping at 100 °C.

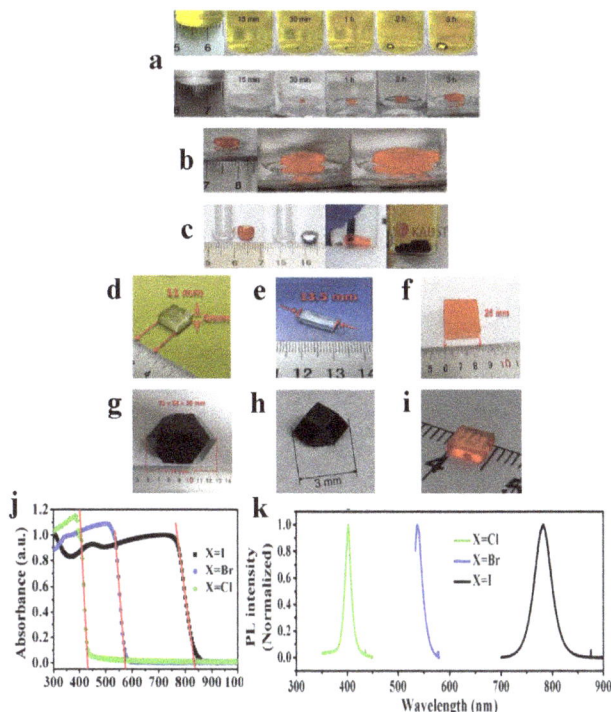

Figure 2. (a) MAPbI$_3$ and MAPbBr$_3$ single crystal growth at different time intervals, (b) continuous growth of MAPbBr$_3$, and (c) crystal shape control of MAPbBr$_3$ (red) and MAPbI$_3$ (black). Reprinted with permission from [68], Copyright 2015, Springer; Photographs taken from the as-grown MAPbX$_3$ crystals: (d,e) MAPbCl$_3$ (f) MAPbBr$_3$ (g) MAPbI$_3$. Reprinted with permission from [69], Copyright 2015, Wiley; (h) FAPbI$_3$ (black) and (i) FAPbBr$_3$ (red) single crystal grown by the ITC method. Reprinted with permission from [75], Copyright 2016, Royal Society of Chemistry; UV-vis-NIR absorption spectrum and photoluminescence (PL) properties of CH$_3$NH$_3$PbX$_3$ (X = Cl, Br, I): (j) absorption spectrum and (k) photoluminescence spectrum Reprinted with permission from [69], Copyright 2015, Wiley.

In addition to the above growth method based upon temperature, the antisolvent vapor-assisted crystallization (AVC) method, a temperature-independent process, was also developed [40]. In AVC, a proper anti-solvent slowly diffuses into the crystal precursor solution, resulting in formation of sizable MAPbY$_3$. Using the AVC method, Bakr's group [40] gained high-quality MAPbI$_3$ and MAPbBr$_3$ single crystals, and implemented a solvent with high solubility for MAX and PbX$_2$, i.e., N,N-Dimethylformamide (DMF) or γ-butyrolactone (GBA). Dichloromethane (DCM) acted as the antisolvent to avoid the formation of hydrogen bonds due to its poor solubility for both PbX$_2$ and MAX, thus minimizing asymmetric interactions with the ions during their assembly into crystal form. This approach created the conditions for the coprecipitation of the ionic building blocks of perovskite. When DCM diffused into DMF or GBA at a slow and controlled rate (Figure 3a), millimeter-sized MAPbBr$_3$ and MAPbI$_3$ single crystals were grown. As shown in Figure 3a, to obtain MAPbBr$_3$ single crystals, PbBr$_2$/MABr molar ratio of 1:1 was chosen and dissolved in DMF to form the precursor with

PbI_2 of 0.2 mol·L^{-1}. For MAPbI$_3$ single crystals, a precursor with PbI$_2$/MAI molar ratio of 1:3 was dissolved in GBA, and PbI$_2$ of 0.5 mol·L^{-1} was prepared. In this case, high-quality MAPbX$_3$ single crystals were obtained, with super-excellent performance. The absorbance of MAPbX$_3$ (X = Br$^-$ and I$^-$) (Figure 3e) exhibited a clear band edge cutoff without excitonic signature that showed a minimal number of in-gap defect states. This confirmed a carrier (holes) concentration of 5×10^9–5×10^{10} cm^{-3}. The time-dependent PL signals of MAPbI$_3$ and MAPbBr$_3$ single crystals were obtained in order to quantify the carrier dynamics (Figure 3f), and showed a superposition of surface components (fast) $\tau \approx 41$ ns and bulk components (slow) $\tau \approx 357$ ns for MAPbBr$_3$, and fast $\tau \approx 22$ ns and slow $\tau \approx 1032$ ns for MAPbI$_3$. For MAPbBr$_3$ single crystals, the carrier lifetime τ was also estimated by transient absorption (TA). It was shown that the fast component amounts to only 3.6% of the total TA signal in MAPbBr$_3$, and to 7% and 12% of the total PL signal in MAPbI$_3$ and MAPbBr$_3$, respectively. As an alternative antisolvent, toluene can also effectively induce formation of MAPbBr$_3$ single crystals (Figure 3b) with low surface recombination velocity ($\sim(3.4 \pm 0.1) \times 10^3$ cm·s^{-1}) [76] through a process in which the crystal precursor is derived from dissolution of PbBr$_2$ and MABr (1/1 by molar, 0.1 mol·L^{-1}) in DMF. In another study, diethyl ether was reported as the antisolvent [77], in which HI served as the good solvent instead of organic solvents. Where PbI$_2$ was firstly dissolved in HI solution upon heating to 120 °C, and formed a hot bright yellow solution; to the hot solution was added MAI, which dissolved immediately, leading to the formation of the crystal precursor. X-ray Diffraction (XRD) analysis revealed that an intermediate product of H$_x$PbI$_{2+x}$·xH$_2$O was created because of the coordination between HI and PbI$_2$. With inflow of diethyl ether into the precursor, MAPbI$_3$ single crystals emerged developmentally. It's worth mentioning that the growth mechanism of halide perovskite single crystals was studied. Recently, Chen et al. reported the growth mechanism of MAPbBr$_3$ single crystal, which was synthesized by the antisolvent method [78]. The assembly model is shown in Figure 3c. CH$_3$NH$_3$Br·PbBr$_2$·DMF adduct complex was first formed in the precursor solution, and crystallization occurred when the solution was supersaturated. The MAPbBr$_3$ molecule in the saturated solution condensed into numerous small nuclei with the coalescence of the nuclei into bigger particles. The perovskite particles were gradually self-assembled into a hollow structure. The crystals were twisted, and their faces were peculiarly inclined toward each other. Subsequently, MAPbBr$_3$ crystals exhibited a layered stacked structure, and continued to grow until the final single crystal was formed. Li and co-workers investigated the crystallization of MAPbI$_{3-x}$Br$_x$ by adjusting the molar ratio of I/Br in precursor solution [79]. It was found that the crystallization and perovskite morphology were heavily affected by the composition of precursor solutions (Figure 3d). It was reported that Br has a smaller ionic radius and lower solubility in organic solvents because of the stronger bond strength [80]. As a comparison with PbBr$_2$, PbI$_2$ has a stronger electron-accepting ability and Lewis acidity [81]. Therefore, it was more likely to form needle crystals of MAPbI$_{3-x}$Br$_x$·DMF with high iodine concentration, which might lead to morphology evolution as the molar ratio of I/Br.

Bulk perovskite single crystals can show advanced properties, such as higher carrier mobility, longer carrier lifetime and diffusion length. However, bulk single crystals may cause degradation of device performance, because a thick active layer will increase the charge recombination. In this case, it is desirable to achieve the fabrication of perovskite single-crystal thin films, and thus to enhance the performance of the device. Bakr's group grew MAPbBr$_3$ monocrystalline film successfully via a cavitation-triggered asymmetrical (CTAC) strategy [82] in which a very short ultrasonic pulse was introduced to a low supersaturation level solution with antisolvent vapor diffusion; perovskite monocrystalline films were able to grow within several hours under the ultrasonic pulse. These obtained films were free of grain boundaries and were homogeneous, with the films having thicknesses ranging from one up to several tens of micrometers, and lateral dimensions varying from hundreds of microns to three millimeters, as shown in Figure 4a.

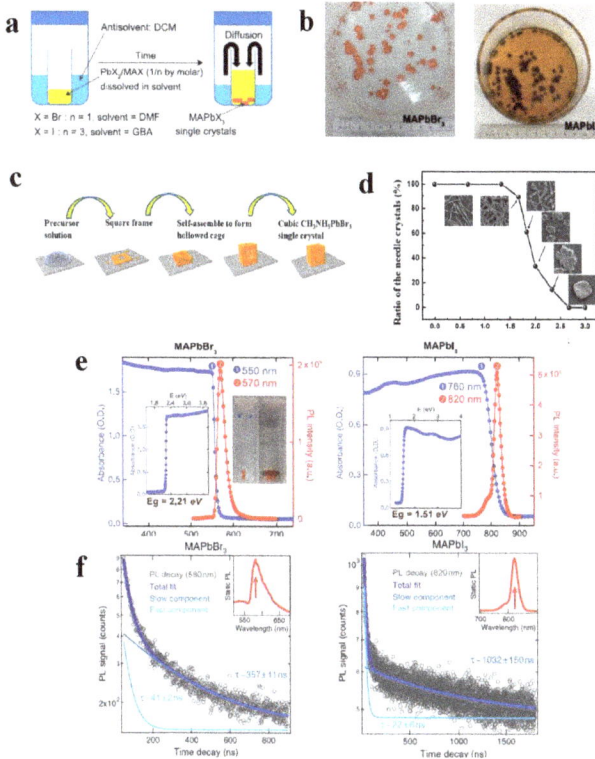

Figure 3. (**a**) Chemotic diagram of crystallization process of AVC method. Reprinted with permission from [40], Copyright 2015, AAAS; (**b**) MAPbBr₃ (red) and MAPbI₃ (black) single crystals grown by AVC. Reprinted with permission from [76], Copyright 2015, American Association for the Advancement of Science; (**c**) Schematic illustration of MAPbBr₃ single crystal crystallization. Reprinted with permission from [78], Copyright 2018, American Chemical Society; (**d**) Dependence of ratio of the needle crystals on the bromine concentration of *x*. Reprinted with permission from [79], Copyright 2018, American Chemical Society; (**e**) Steady-state absorbance and photoluminescence and (**f**) PL time decay trace on MAPbBr₃ and MAPbI₃ crystal. Reprinted with permission from [40], Copyright 2015, AAAS.

To obtain MAPbI₃ perovskite single-crystalline wafer, Liu's group designed an ultrathin geometry-defined dynamic-flow reaction system (Figure 4c) to obtain single crystals with different thicknesses and shapes (Figure 4d–m) [83]. It was shown that the two glass slides of the reaction system were separated and aligned in parallel by two spacers, leading to the single-crystalline wafer thickness and shape being defined by the spacers and slit channel design. Using the thickness-controllable reaction system, wafer as thin as about 150 μm with high crystallinity and a low trap state density of 6×10^8 cm^{-3} was prepared.

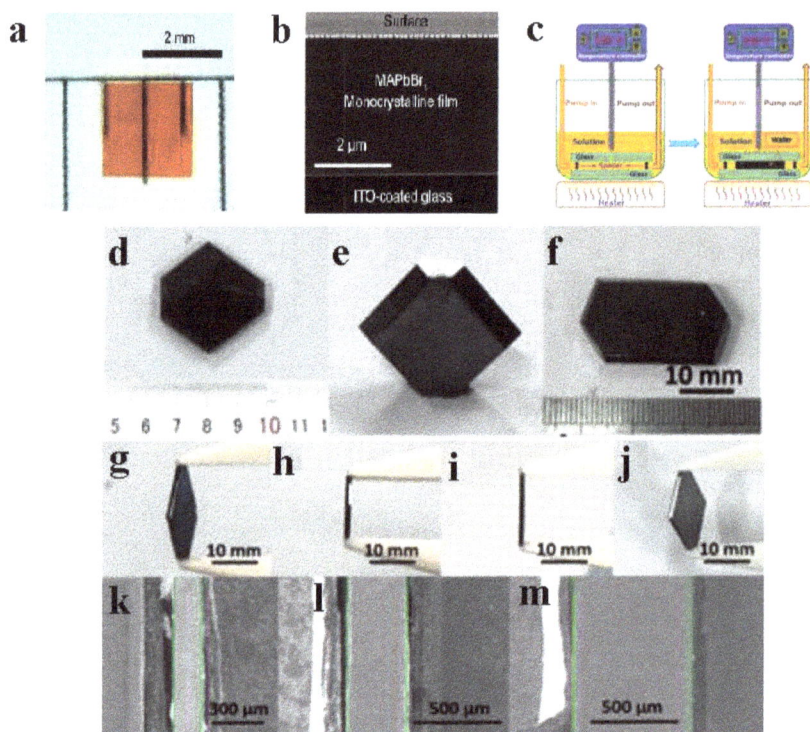

Figure 4. (**a**) Optical image and (**b**) cross-section SEM image of MAPbBr$_3$ monocrystalline film. Reprinted with permission from [82], Copyright 2016, Wiley; (**c**) Schematic illustration for the ultrathin single crystal wafer preparation, (**d–j**) Photos of the single crystal wafers with different thicknesses and shapes, (**k–m**) Cross-sectional view of single crystal perovskite wafers showing different thicknesses. Reprinted with permission from [83], Copyright 2016, Wiley.

Chen et al. reported the controllable fabrication of air-stable, sub-millimeter-size perovskite single-crystalline thin films (SCTFs) [84]. For the preparation process, two flat substrates were clipped together and vertically immersed in perovskite precursor (Figure 5a), the thickness of the solution film could be easily tuned using clipping force, and the resulting SCTF thickness could be adjusted with an aspect ratio of up to 10^5 from nano- to micrometers (Figure 5b). The prepared SCTFs exhibited outstanding air stability and comparable quality to bulk single crystals with trap density (n$_{trap}$) of 4.8 × 10^{10} cm^{-3}, carrier mobility (μ) of 15.7 cm$^2 \cdot$V$^{-1} \cdot$s^{-1}, and a carrier lifetime (τ$_r$) of 84 μs. In addition, perovskite SCTF growth is a substrate-independent strategy, which would offer appealing potentials, such as SCTF/ITO for PSCs, SCTF/PET for flexible devices, SCTF/quartz for optical devices and SCTF/Si for electronic devices, etc.

Recently, Rao et al. developed a space-limited inverse temperature crystallization (SLITC) method, in which the limited spatial module has a tripartite structure: a FTO glass, a U-style thin PTFE, and a PTFE board (Figure 5c). For the preparation of continuous and dense MAPbBr$_3$ crystal film, the precursor solution was injected into the module and a decreased temperature gradient was applied. As a result, MAPbBr$_3$ crystal film with a super-large area of 120 cm^2 and a controllable thickness of 0.1–0.8 μm was prepared [85] (Figure 5d).

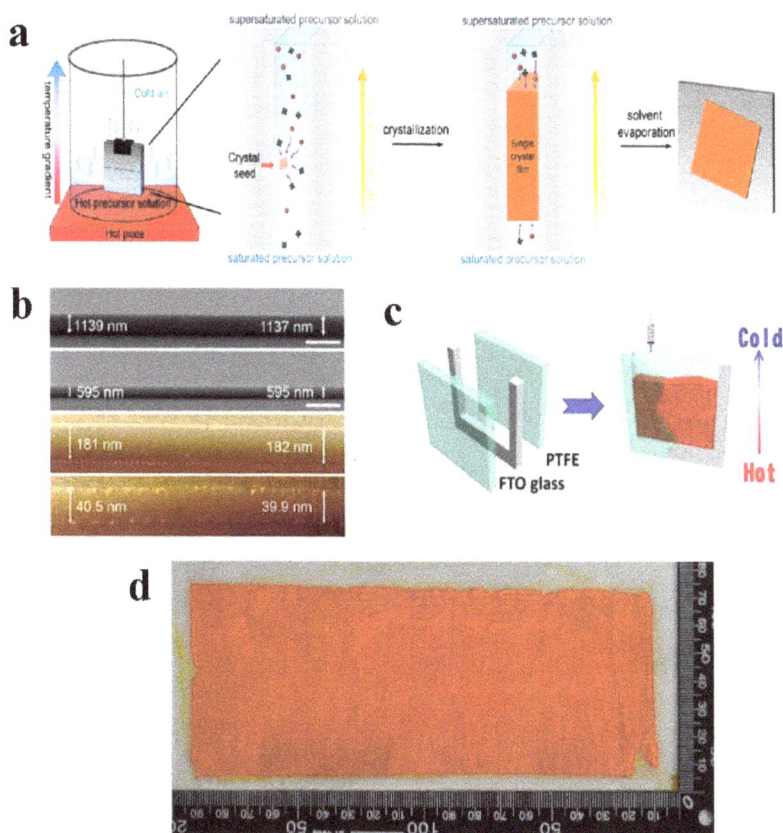

Figure 5. (**a**) Scheme for growth of perovskite single-crystalline thin films; and (**b**) Cross-section SEM images and AFM images of MAPbBr$_3$ single-crystalline thin films with varied thicknesses. Reprinted with permission from [84], Copyright 2016, American Chemical Society; (**c**) Schematic diagram of the module for growing MAPbBr$_3$ crystal films; (**d**) MAPbBr$_3$ crystal films with a thickness of 0.4 mm and an area about 120 cm^2. Reprinted with permission from [85], Copyright 2017, Wiley.

Owing to the high photoluminescence quantum yields, metal halide perovskite nanocrystals (Ncs) have also attracted great attention. Using a solvent-induced reprecipitation approach [86], MAPbBr$_3$ NCs were first synthesized in a process in which octylammonium bromide and octadecylammonium bromide acted as surfactants to stabilize the nanocrystals for up to 3 months. Urban's group reported dilution-induced formation of hybrid perovskite nanoplatelets (NPls) [87]. This proceeded by fragmentation of the NCs into NPls, with an excess of organic ligands stabilizing the newly formed surfaces. Such fragmentation was in excellent agreement with the effects of cation intercalation and increased solvent osmotic pressure, resulting in the formation of small nanoplatelets. Vybornyi et al. reported hot injection-based synthesis [88], which was basically an ionic metathesis approach. Varying the amounts of surfactants (octylamine (OAm)/oleic acid (OA) mixture), MAPbI$_3$ NCs and MAPbBr$_3$ NPLs, NWs were successfully obtained. It was shown that the resultant NCs had poorer optical properties than those of NCs synthesized by ligand-assisted reprecipitation. Similarly, FA-based nanocrystals have received considerable interest. Protesescu et al. prepared FAPbX$_3$ (X = Br, I) NCs via a three-step polar solvent-free hot-injection method [89]. NCs with cubic morphology and a high photoluminescence quantum yield (PLQY) of 85% were obtained; however, its potential versatility was limited by the formation of phase impurities ($NH_4Pb_2Br_5$). Manna's group synthesized $FAPbX_3$ NCs with excellent phase purity via a modified three-precursor hot-injection technique [71] in which benzoyl halide acted as a halide precursor; thus, the metal cation sources and halide ions were not delivered together, making it possible to work with the ideal stoichiometry of ions.

2.2. Growth of Mixed-Organic-Cation Halide Perovskite Single Crystals

It has been reported that the mixed-organic-cation based on FA and MA halide perovskite integrates nearly all of their advantages, which include extended absorption, decreased trap-assisted recombination and enhanced ambient stability [90,91]. Li et al. fabricated the $MA_{0.45}FA_{0.55}PbI_3$ single crystal using a modified ITC method [92]. That is, to prepare $MAPbI_3/FAPbI_3$ (1.0 M) precursor solutions, equimolar PbI_2 and MAI/FAI were dissolved in γ-butyrolactone at 60 °C overnight. Subsequently, the $MAPbI_3$ and $FAPbI_3$ solutions were blended in a certain ratio to form the mixed-cation precursor solution. Seed crystals with sizes of 0.5–1 mm were obtained by placing 2 mL mixed-cation perovskite precursor solution in an oil bath at 160 °C for 30 min. To obtain mixed-cation alloy perovskite crystal, a fresh precursor solution containing the corresponding seed crystal was kept at 120 °C for 3 h. Thus, the final $MA_{0.45}FA_{0.55}PbI_3$ single crystals maintained an impressive stability, and were still able to maintain their original black color after exposure in ambient air for more than 14 months (Figure 6a); however, the α-$FAPbI_3$ single crystal changed from black to yellow and $MAPbI_3$ exhibited pale yellow spots on the surface. The carrier lifetime τ of $MA_{0.45}FA_{0.55}PbI_3$ was characterized by transient photovoltaic (TPV); the TPV curves and corresponding τ are shown in Figure 6d,e. Compared to $FAPbI_3$ and $MAPbI_3$, a longer carrier lifetime of 93 µs for $MA_{0.45}FA_{0.55}PbI_3$ was obtained. In another work, an ITC method assisted by hydroiodic acid for the mixed-organic-cation perovskites ($APbI_3$, A = MA^+ or FA^+) was developed [93]. The given PbI_2, MAI and FAI were dissolved in GBL with the introduction of appropriate amount of HI, leading to a change in chemical environment in the precursor solution and the formation of $APbX_3$-GBL, and further to H^+, MA^+, FA^+ and GBL molecules being inserted in the $PbI_{2+x}{}^{x-}$ layer. Based on the HI-assisted ITC method, $FA_{(1-x)}MA_xPbI_3$ single crystals of millimeter size with different compositions were obtained (Figure 6b). By using time-resolved photoluminescence (TRPL) measurements, the carrier lifetimes of fresh $FA_{(1-x)}MA_xPbI_3$ single crystals were investigated. Figure 6f,g shows the TRPL spectra of $FA_{(1-x)}MA_xPbI_3$ crystals with different x values, and the corresponding lifetimes are summarized in Table 1. It was found that the carrier lifetime of $FA_{(1-x)}MA_xPbI_3$ (x = 0.8–0.95) was significantly improved compared to the counterpart $MAPbI_3$ single crystal, which disclosed the effective FA^+ cation doping in $MAPbI_3$ crystal with respect to carrier behavior.

Table 1. Lifetimes extracted from the PL spectra of $FA_{(1-x)}MA_xPbI_3$ single crystals with x = 0, 0.05, 0.1, 0.15, 0.2, 0.8, 0.85, 0.9, 0.95, 1.

$FA_{(1-x)}MA_xPbI_3$	x = 0	x = 0.05	x = 0.1	x = 0.15	x = 0.2
τ_1 (ns)	91.26	61.43	52.11	27.26	31.58
τ_1 (ns)	839.31	689.92	381.86	579.75	236.74
$FA_{(1-x)}MA_xPbI_3$	x = 1	x = 0.95	x = 0.9	x = 0.85	x = 0.8
τ_1 (ns)	7	32.45	122	88.05	105.3
τ_1 (ns)	145.65	557.5	1074.78	926.39	956.8

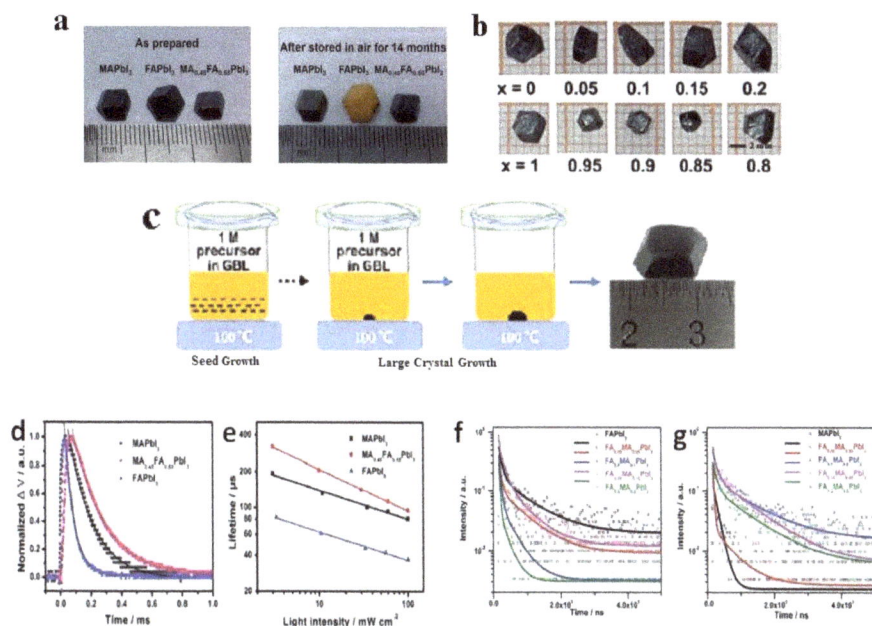

Figure 6. (**a**) Photographs of as-prepared MAPbI$_3$, FAPbI$_3$ and MA$_{0.45}$FA$_{0.55}$PbI$_3$ single crystals and the same crystals after being stored in air for 14 months. Reprinted with permission from [92], copyright 2017, The Royal Society of Chemistry; (**b**) FA$_{(1-x)}$MA$_x$PbI$_3$ single crystals with different compositions grown via HI-assisted ITC method. Reprinted with permission from [93], Copyright 2016, Royal Society of Chemisty; (**c**) Schematic illustration of (FAPbI$_3$)$_{1-x}$(MAPbBr$_3$)$_x$ crystal growth process. Reprinted with permission from [94], Copyright 2017 American Chemical Society; (**d**) Transient photovoltaic curves and (**e**) the extracted charge lifetime from TPV measurement of MAPbI$_3$, FAPbI$_3$ and MA$_{0.45}$FA$_{0.55}$PbI$_3$ single crystals. Reprinted with permission from [92], copyright 2017, The Royal Society of Chemistry; Time-resolved photoluminescence spectra of FA$_{(1-x)}$MA$_x$PbI$_3$ single crystals (**f**) with $x = 0, 0.05, 0.1, 0.15, 0.2$ and (**g**) with $x = 0.8, 0.85, 0.9, 0.95$. Reprinted with permission from [93], Copyright 2016, Society of Chemisty.

Similarly, high-quality mixed-cation and -halide perovskite single crystals, with the formula (FAPbI$_3$)$_{1-x}$(MAPbBr$_3$)$_x$ ($x = 0, 0.05, 0.1, 0.15, 0.2$), were also successfully grown via the ITC method [94]. The small seeds, grown at 100 °C in the precursor solution, were used to grow crystals, inducing the growth of centimeter sized single crystals, as shown in Figure 6c.

3. Growth of Fully Inorganic Halide Perovskite Single Crystals

Recently, CsPbX$_3$ perovskites have rightfully been receiving attention because of their promising potential in photovoltaics [95] and bright light emission [96]. Perhaps such fully inorganic halide perovskite can overcome the chemical instability of organic-inorganic hybrid halide perovskite. It was once reported that CsPbX$_3$ perovskite single crystals could be obtained via the Bridgmann method, a melt crystallization method, which was carried out at a high temperature and with highly pure starting reagents [97–100]. In 2008, for the first time, CsPbCl$_3$ single crystals were grown with the Bridgmann method using a process in which the precursors of the PbI$_2$ and CsCl powders were sealed in a quartz crucible under vacuum [98]. In addition, CsPbBr$_3$ single crystals were also grown using the melt crystallization method [97]. Other than the above, the fast and simple route via solution growth of hybrid halide perovskite single crystals was also used to grow inorganic CsPbX$_3$ single crystals using the ITC method. Dirin et al. presented the growth of $CsPbBr_3$ single crystals under ambient atmosphere using the ITC method [101]. The optimal solvent for the growth of $CsPbBr_3$ was reported to be DMSO. In particular, a solution of CsBr and $PbBr_2$ (1/2 by molar), dissolved in a mixed solvent of DMSO with cyclohexanol and DMF, could grow 1–3 nuclei at 90 °C, and further crystal growth could take place without additional nucleation until the temperature increased to 110 °C. As a result, a flat and orange-colored $CsPbBr_3$ single crystal of ~8 mm in length was obtained (Figure 7).

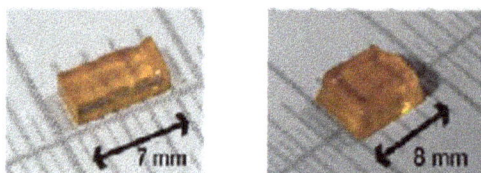

Figure 7. Photographs of the obtained $CsPbBr_3$ single crystals. Reprinted with permission from [101], Copyright 2016, American Chemical Society.

The AVC method can be used to prepared inorganic halide perovskite, too. Rakita et al. grew $CsPbBr_3$ single crystals from a DMSO precursor solution with CsBr and $PbBr_2$ (1/1 by molar) [102]. The precursor solution was titrated by MeCN or MeOH until a saturated system was achieved. The saturated solution was filtered to use for crystal growth. During the AVC process, MeCN or MeOH was developed as the antisolvent. A balanced antisolvent atmosphere was created, and crystal formation occurred at room temperature; the crystal growth could be accelerated by heating the antisolvent bath. In addition, crystal growth using H_2O as an antisolvent was investigated, and it was found that the orange crystals were inclined to blench. $CsPbBr_3$ crystal growth via the ITC method was also studied by the same group. To eliminate the formation of undesirable precipitants, they developed a two-step heating cycle. The precursor solution was firstly heated to the desired temperature and allowed return to room temperature under continuous stirring. After being filtered, the final crystalline precursor solution was obtained, and orange $CsPbBr_3$ crystals were shown after the second heating cycle. It is noted that $CsPbBr_3$ crystals begin to appear at above 120 °C in the MeCN-saturated solution; however, this can occur at about 40 °C in the MeOH-saturated solution.

Tong et al. reported high-quality colloidal $CsPbX_3$ (X = I, Br, and Cl) perovskite nanocrystals (NCs) [103]. The synthesis was based on direct tip sonication of precursor mixtures under ambient atmospheric conditions. This method was based on the formation of a metal-ligand complex, which was then further reduced into metal nanoparticles. Urban and Co-workers developed a single-step ligand-mediated synthesis of single-crystalline $CsPbBr_3$ nanowires (NWs) directly from the precursor powders [104]. Through an oriented-attachment mechanism, the initially formed $CsPbBr_3$ nanocubes were transformed into NWs, which exhibited strongly polarized emission and could self-assemble at an air/liquid interface.

In addition, $Cs_2AgBiBr_6$, a bismuth-halide double perovskite single crystal was also reported by Karunadasa's group [105]. To obtain the large single crystal, the precursor solution was kept at 110 °C for 2 h, then cooled to room temperature at a cooling rate of 1 °C/h. The results indicated that $Cs_2AgBiBr_6$ crystal had an indirect bandgap of 1.95 eV and a long room-temperature PL lifetime of ca. 660 ns.

The growth methods and properties of as-grown perovskite single crystals are summarized in Table 2.

Table 2. Summary of perovskite single crystal growth methods and properties of as-grown crystals.

Single Crystal	Growth Method	Size (mm)	Growth Period	Properties		Ref.
				Carrier Mobility ($cm^2 \cdot V^{-1} \cdot s^{-1}$)	Trap State Density (cm^{-3})	
MAPbI$_3$		2–3	48 h			[60]
CH$_3$NH$_3$PbBr$_3$		5	More than 10 days			[60]
MAPbI$_3$	STL	10 × 10 × 8				[61]
MAPbI$_3$		12 × 12 × 7	2–4 weeks	105 ± 35	10^{10}	[62]
MAPbI$_3$		20 × 18 × 6	5 days			[63]
CH$_3$NH$_3$ PbI$_3$ (Cl)		20 × 18 × 6	5 days	167 ± 35	7.6×10^8	[66]
MAPbI$_3$		71 × 54 × 39		34	1.4×10^{10}	[71]
MAPbCl$_3$		7	3 days	179	1.8×10^9	[71]
MAPbBr$_3$	ITC	11 × 11 × 4		4.36	2.6×10^{10}	[71]
FPbI$_3$		5		4.4	1.5×10^{11}	[76]
MA$_{0.45}$FA$_{0.55}$PbI$_3$		8	4 h	271 ± 60	2.6×10^9	[86]
CH$_3$NH$_3$PbBr$_3$	AVC	1.4 × 1.4 × 0.7			10^9–10^{10}	[78]

4. Application of Halide Perovskite Single Crystals

Through the CATC method, Bakr's group grew hybrid perovskite monocrystalline films and carried out explorative study for the first time on perovskite monocrystalline solar cells with two simple device structures, i.e., ITO/MAPbBr$_3$/Au and FTO/TiO$_2$/MAPbBr$_3$/Au [82]. Figure 8 shows the photoelectric characteristics of those monocrystalline solar cells. Without any HTLs and ETLs, the optimized solar cell of ITO/MAPbBr$_3$ (4 μm)/Au offers an ultra-stable photoelectric conversion efficiency (PCE) of over 5% and close to 100% IQE (Figure 8a,c). Furthermore, for the p-n-junction-based architecture, near 100% IQE and higher efficiency (6.5%) than the best HTL-free MAPbBr$_3$ solar cells was achieved (Figure 8b,d). These significant works made clear that the optoelectronic properties of monocrystalline-film-based devices are superior to their polycrystalline counterparts. Interestingly, based on the bulk single crystal, Huang's group fabricated a lateral-structured device with a maximum PCE of 5.36% at 170 K and a comparable J_{SC} to the best thin-film solar cells [106].

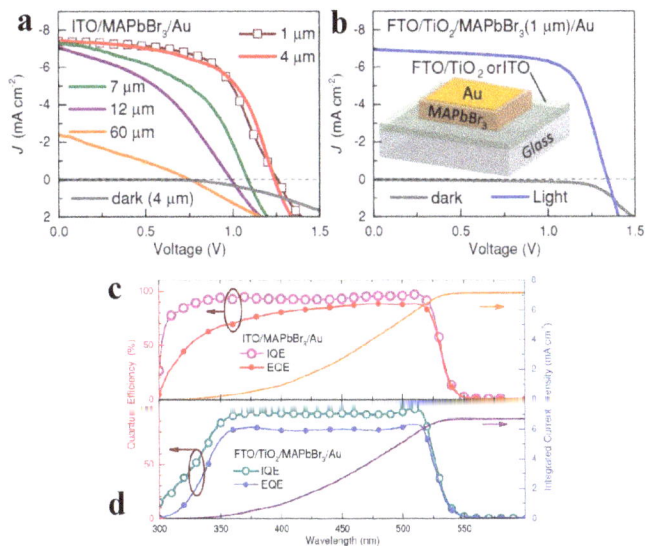

Figure 8. Illuminated and dark *J-V* curves, wavelength-dependent *IQE* and *EQE*: (**a,c**) ITO-based and (**b,d**) FTO/TiO$_2$-based monocrystalline film solar cells. Reprinted with permission from [82], Copyright 2016, Wiley.

Photodetectors are another device for realizing photoelectric conversion, and have attracted much attention because of their extensive applications, which include biological sensing, camera imaging, missile warning, and communication [107–110]. Organic–inorganic hybrid perovskite single crystals have also been utilized to fabricate photodetectors [111–113]. Lian et al. firstly fabricated high-performance planar-type photodetector on the (100) facet of MAPbI$_3$ single crystal [61]. The spectral responsivity (R) and EQE, the crucial parameters for photodetectors, were measured and calculated. For MSCP (MAPbI$_3$ single crystal photodetector) and MPFP (MAPbI$_3$ polycrystalline film photodetector), the R values were estimated to be 2.55 A W^{-1} and 0.0197 A W^{-1}, respectively, with the corresponding EQE values calculated to be 5.95 $\times$ 10^2% and 4.59%, respectively (Figure 9a), which are over two orders of magnitude higher than the R and EQE values obtained for MSCP. Furthermore, the photoresponse times of MSCP (74 μs and 58 μs for the rise time and decay time, respectively) were about three orders of magnitude faster than those for MPFP (52 ms and 36 ms for the rise time and decay time, respectively) (Figure 9b).

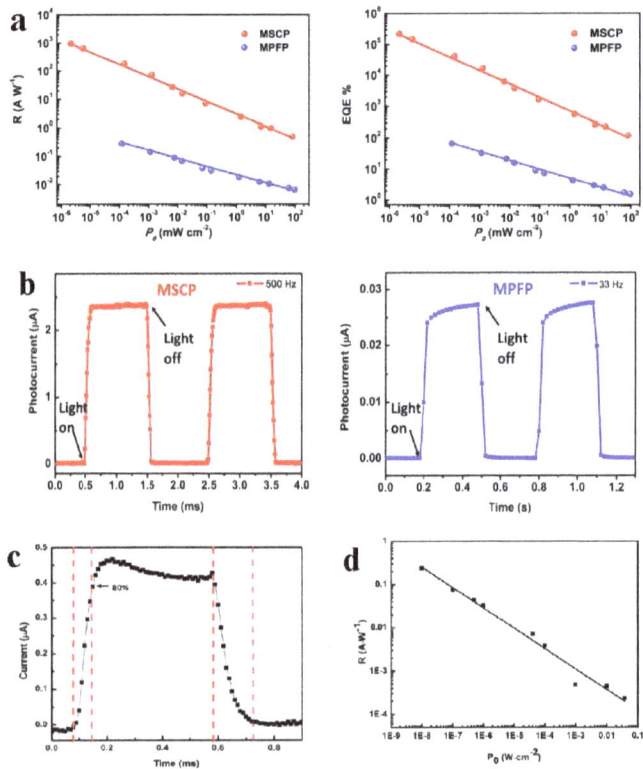

Figure 9. (a) Responsivity, external quantum efficiency EQE; and (b) Transient photocurrent response for as-fabricated MSCP and MPFP. Reprinted with permission from [61], Copyright 2015, Springer; (c) Time response of the MAPbI$_3$ perovskite photodetector; and (d) Responsivity of MAPbI$_3$ single crystal photodetector at different radiance power. Reprinted with permission from [114], copyright 2016, The Royal Society of Chemistry.

Ding et al. fabricated a self-powered MAPbI$_3$ perovskite single crystal photodetector with Au-Al electrodes [114], which exhibited a fast rise and decay time of 71 μs and 112 μs (Figure 9c). A good R value of 0.24 A W^{-1} at the lowest incident power density of 1 $\times$ 10^{-8} W cm^{-2} was demonstrated (Figure 9d).

Liu's group reported a MAPbI$_3$ single-crystalline wafer photodetector [83] that showed a photocurrent response about 350 times higher than that of the microcrystalline thin film detector. Moreover, nearly 100 photodetectors were fabricated on a piece of single-crystalline perovskite wafer (Figure 10a), highlighting the feasibility of batch-processing integrated circuits on ultrathin single-crystalline wafers. In addition, they fabricated FAPbI$_3$-wafer-based photodetectors, which exhibited a photoresponse 90 times higher than its thin-film perovskite counterpart. Furthermore, an array of more than 150 photodetectors were also designed on a piece of thin wafer. Using MAPbBr$_3$ single crystal, Shaikh et al. constructed Schottky-type photodetectors [115]. These devices exhibited response times on the scale of 100 µs and a photodetectivity of 1.4 × 10^{10} Jones at zero bias. Recently, narrowband photodetector devices based on large-area MAPbBr$_3$ crystal films have been studied [85], which has enabled high narrow response under a low bias of −1 V, a broad linear response range of 10^{-4}–10^2 mW cm^{-2} and 3 dB cutoff frequency (f$_{3dB}$) of ~110 kHz. Different from one-component single-crystal perovskite photodetectors, a core-shell heterojunction photodetector based on MAPbBr$_3$ single crystal was developed [116]. It was found that the photodetector offered the feature of self-power and exhibited a peak R of 11.5 mA W^{-1} at zero bias under 450 nm, which was one order of magnitude higher than that of MAPbBr$_3$ single crystal. The *EQE* of 3.17% was also much higher than the reported MAPbBr$_3$ single crystal (0.2%). The high-quality MAPbCl$_3$ crystals for UV photodetection were grown using DMSO-DMF solution. This demonstrated that the MAPbCl$_3$ single crystal-based UV-photodetector possessed an on-off ratio as high as 1.1 × 10^3 and a calculated detectivity of 1.2 × 10^{10} Jones. Li et al. reported that the mixed cation MA$_{0.45}$FA$_{0.55}$PbI$_3$ perovskite single crystal photodetector [88], which showed high on-off ratio of about 1000, a low detection limit of ~1 nW cm^{-2}, and a short response time of less than 200 µs. It is noted that the photodetector showed stable characteristics for a long period at both zero and −1 V bias, as shown in Figure 10d.

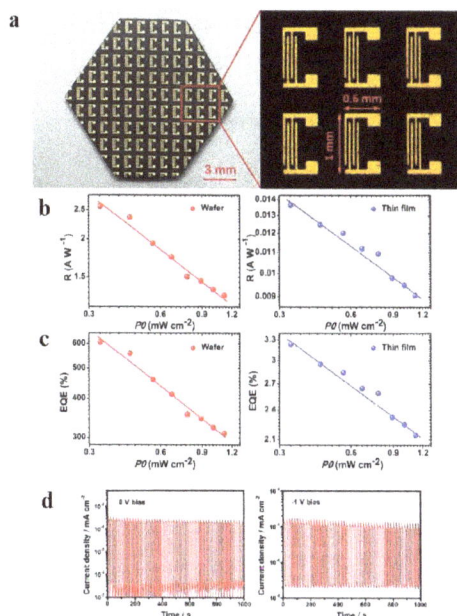

Figure 10. (a) Photograph of ≈100 photodetectors fabricated on a single-crystalline wafer; (**b**) *R* and (**c**) *EQE* of a detector made of single-crystalline perovskite wafer and microcrystalline perovskite thin film. Reprinted with permission from [83], Copyright 2017, Wiley; (**d**) Stability test of MA$_{0.45}$FA$_{0.55}$PbI$_3$ perovskite single crystal photodetector at 0 V and −1 V bias. Reprinted with permission from [88], copyright 2017, The Royal Society of Chemistry.

Further, owing to the excellent carrier transport properties, metal halide perovskite single crystals can also be used in high-energy ray detectors, which are able to convert some high-energy rays like X-ray and γ-ray photons into charges. Using the $MAPbI_3$ films, an X-ray detector with responsivity of 1.9×10^{-4} carriers/photon was first reported [117]. Based on bulk $MAPbBr_3$ single crystal, through structural optimization and surface passivation, Wei et al. fabricated the X-ray detector and obtained a detector with a sensitivity of 80 $\mu C \cdot mGy_{air}^{-1} \cdot cm^{-2}$ [118].

5. Summary and Future Perspectives

Metal hybrid perovskites have been extensively studied for solar cells, photodetectors, lasing, light-emitting diodes, etc., owing mainly to their excellent semiconductor properties, low cost and facile deposition techniques. It is noted that metal hybrid perovskite single crystals show remarkably low trap-state densities and long carrier diffusion lengths, which can even compare with the best photovoltaic-quality silicon. These properties enable metal hybrid perovskite single crystals to act as desirable semiconductors for optoelectronic applications. This review focuses on the recent progress in the growth and application of different metal hybrid perovskite single crystals. Single- and mixed-organic-cation halide perovskite single crystals can be prepared in solution, with the STL process having been demonstrated to be an effective method. To grow large-sized and high-quality single crystals, the seed-assisted growth method was developed, and centimeter-sized single crystals were obtained. However, STL is a time-consuming process. The ITC and AVC methods are improved processes that have been widely investigated for growth of high-quality single crystals. It is noted that the fully inorganic halide perovskite single crystals can be grown by solution process. The preparation of the above metal hybrid halide perovskite single crystals may contribute to their further application research.

To date, optoelectronic devices based on perovskite single crystals are gradually being considered. The brilliant performance of single-crystal-based photodetectors is further testimony to the benefits of metal hybrid halide perovskite materials. However, the application in solar cells based on single crystals is still limited because of the lack of high-quality monocrystal films with appropriate thickness for perovskite solar cells. In the future, the preparation of high-quality perovskite single crystals with controlled thickness and orientation able to meet the requirements of this application is urgently needed.

Author Contributions: Shuigen Li, Jian-Qiao Meng and Shunjian Xu conceived and designed the analyzed, Shuigen Li, Chen Zhang, Jiao-Jiao Song and Xiaohu Xie collected the data. Shuigen Li, Jian-Qiao Meng and Shunjian Xu wrote the paper.

Acknowledgments: This work was supported by China 1000-Young Talents Plan, the National Natural Science Foundation of China (51203192, 61172047, 51673214), the Jiangxi Provincial Natural Science Foundation of China (GJJ171063, GJJ171060, GJJ171066).

Conflicts of Interest: The authors declare no conflict of interest.

References

1. Burschka, J.; Pellet, N.; Moon, S.-J.; Humphry-Baker, R.; Gao, P.; Nazeeruddin, M.K.; Grätzel, M. Sequential deposition as a route to high-performance perovskite-sensitized solar cells. *Nature* **2013**, *499*, 316–319. [CrossRef] [PubMed]
2. Heo, J.H.; Im, S.H.; Noh, J.H.; Mandal, T.N.; Lim, C.-S.; Chang, J.A.; Lee, Y.H.; Kim, H.J.; Sarkar, A.; Nazeeruddin, M.K.; et al. Efficient inorganic organic hybrid heterojunction solar cells containing perovskite compound and polymeric hole conductors. *Nat. Photonics* **2013**, *7*, 486–491. [CrossRef]
3. Im, J.-H.; Lee, C.-R.; Lee, J.-W.; Park, S.-W.; Park, N.-G. 6.5% Efficient Perovskite Quantum-Dot-Sensitized Solar Cell. *Nanoscale* **2011**, *3*, 4088–4093. [CrossRef] [PubMed]
4. Kim, H.-S.; Lee, C.-R.; Im, J.-H.; Lee, K.-B.; Moehl, T.; Marchioro, A.; Moon, S.-J.; Humphry-Baker, R.; Yum, J.-H.; Moser, J.E.; et al. Lead iodide perovskite sensitized all-solid-state submicron thin film mesoscopic solar cell with efficiency exceeding 9%. *Sci. Rep.* **2012**, *2*, 591. [CrossRef] [PubMed]

5.   Kojima, A.; Teshima, K.; Shirai, Y.; Miyasaka, T. Organometal halide perovskites as visible-light sensitizers for photovoltaic cells. *J. Am. Chem. Soc.* **2009**, *131*, 6050–6051. [CrossRef] [PubMed]

6.   Zhao, Y.; Zhu, K.C. Organic–inorganic hybrid lead halide perovskites for optoelectronic and electronic applications. *Chem. Soc. Rev.* **2016**, *45*, 655–689. [CrossRef] [PubMed]

7.   Tiep, N.H.; Ku, Z.; Fan, H.J. Recent advances in improving the stability of perovskite solar cells. *Adv. Energy Mater.* **2016**, *6*, 1501420. [CrossRef]

8.   Lü, X.; Wang, Y.; Stoumpos, C.C.; Hu, Q.; Guo, X.; Chen, H.; Yang, L.; Smith, J.S.; Yang, W.; Zhao, Y.; et al. Enhanced Structural Stability and Photo Responsiveness of $CH_3NH_3SnI_3$ Perovskite via Pressure-Induced Amorphization and Recrystallization. *Adv. Mater.* **2016**, *28*, 8663–8668. [CrossRef] [PubMed]

9.   Jeon, N.J.; Noh, J.H.; Yang, W.S.; Kim, Y.C.; Ryu, S.; Seo, J.; Seok, S.I. Compositional Engineering of Perovskite Materials for High-Performance Solar Cells. *Nature* **2015**, *517*, 476–480. [CrossRef] [PubMed]

10.   Albero, J.; Asiri, A.M.; Garcia, H.; Mater, J. Influence of the composition of hybrid perovskites on their performance in solar cells. *J. Mater. Chem. A* **2016**, *4*, 4353–4364. [CrossRef]

11.   Chen, J.; Zhou, S.; Jin, S.; Li, H.; Zhai, T. Crystal organometal halide perovskites with promising optoelectronic applications. *Mater. Chem. C* **2016**, *4*, 11–27. [CrossRef]

12.   Prochowicz, D.; Franckevičius, M.; Cieślak, A.M.; Zakeeruddin, S.M.; Grätzel, M.; Lewiński, J. Mechanosynthesis of the hybrid perovskite $CH_3NH_3PbI_3$: Characterization and the corresponding solar cell efficiency. *Mater. J. Chem. A* **2015**, *3*, 20772–20777. [CrossRef]

13.   Harikesh, P.C.; Mulmudi, H.K.; Ghosh, B.; Goh, T.W.; Teng, T.; Thirumal, K.; Lockrey, M.; Weber, K.; Koh, T.M.; Li, S.; et al. Rb as an Alternative Cation for Templating Inorganic Lead-Free Perovskites for Solution Processed Photovoltaics. *Chem. Mater.* **2016**, *28*, 7496–7504. [CrossRef]

14.   Cheng, Y.; Li, H.-W.; Zhang, J.; Yang, Q.-D.; Liu, T.; Guan, Z.; Qing, J.; Lee, C.; Tsang, S.-W. Spectroscopic study on the impact of methyl ammonium iodide loading time on the electronic properties in perovskite thin films. *Mater. J. Chem. A* **2016**, *4*, 561–567. [CrossRef]

15.   Eperon, G.E.; Stranks, S.D.; Menelaou, C.; Johnston, M.B.; Herz, L.M.; Snaith, H.J. Formamidinium lead trihalide: A broadly tunable perovskite for efficient planar heterojunction solar cells. *Energy Environ. Sci.* **2014**, *7*, 982–988. [CrossRef]

16.   Leyden, M.R.; Lee, M.V.; Raga, S.R.; Qi, Y. Large formamidinium lead traled perovskite solar cells using chemical vapor deposition with high reproducibility and tunable chlorine concentrations. *Mater. J. Chem. A* **2015**, *3*, 16097–16103. [CrossRef]

17.   Eperon, G.E.; Beck, C.E.; Snaith, H.J. Cation exchange for thin film lead iodide perovskite interconversion. *Mater. Horiz.* **2016**, *3*, 63–71. [CrossRef]

18.   Pellet, N.; Gao, P.; Gregori, G.; Yang, T.-Y.; Nazeeruddin, M.K.; Maier, J.; Grätzel, M. Mixed-Organic-Cation Perovskite Photovoltaics for Enhanced Solar-Light Harvesting. *Angew. Chem. Int. Ed.* **2014**, *53*, 3151–3157. [CrossRef] [PubMed]

19.   Starr, D.E.; Sadoughi, E.; Handick, E.; Wilks, G.; Alsmeier, J.H.; Köhler, L.; Gorgoi, M.; Snaith, H.J.; Bär, M. Direct observation of an inhomogeneous chlorine distribution in $CH_3NH_3PbI_{3-x}Cl_x$ layers: Surface depletion and interface enrichment. *Energy Environ. Sci.* **2015**, *8*, 1609–1615. [CrossRef]

20.   Huang, Z.; Hu, Z.; Yue, G.; Liu, J.; Cui, X.; Zhang, J.; Zhu, Y. $CH_3NH_3PbI_{3-x}Cl_x$ films with coverage approaching 100% and with highly oriented crystal domains for reproducible and efficient planar heterojunction perovskite solar cells. *Phys. Chem. Chem. Phys.* **2015**, *17*, 22015–22022. [CrossRef] [PubMed]

21.   Brenner, T.M.; Egger, D.A.; Kronik, L.; Hodes, G.; Cahen, D. Hybrid organic-inorganic perovskites: Low-cost semiconductors with intriguing charge-transport properties. *Nat. Rev. Mater.* **2016**, *1*, 15007. [CrossRef]

22.   Heo, J.H.; Im, S.H. Highly reproducible, efficient hysteresis-less $CH_3NH_3PbI_{3-x}Cl_x$ planar hybrid solar cells without requiring heat-treatment. *Nanoscale* **2016**, *8*, 2554–2560. [CrossRef] [PubMed]

23.   Kim, T.G.; Seo, S.W.; Kwon, H.; Hahn, J.; Kim, J.W. Influence of halide precursor type and its composition on the electronic properties of vacuum deposited perovskite films. *Phys. Chem. Chem. Phys.* **2015**, *17*, 24342–24348. [CrossRef] [PubMed]

24.   Jiang, M.; Wu, J.; Lan, F.; Tao, Q.; Gao, D.; Li, G. Enhancing the performance of planar organo-lead halide perovskite solar cells by using a mixed halide source. *Mater. J. Chem. A* **2014**, *3*, 963–967. [CrossRef]

25.   Qing, J.; Chandran, H.-T.; Cheng, Y.-H.; Liu, X.-K.; Li, H.-W.; Tsang, S.-W.; Lo, M.-F.; Lee, C.-S. Chlorine Incorporation for Enhanced Performance of Planar Perovskite Solar Cell Based on Lead Acetate Precursor. *ACS Appl. Mater. Interfaces* **2015**, *7*, 23110–23116. [CrossRef] [PubMed]

26. Zhang, W.; Saliba, M.; Moore, D.T.; Pathak, S.K.; Hörantner, M.T.; Stergiopoulos, T.; Stranks, S.D.; Eperon, G.E.; Alexander-Webber, J.A.; Abate, A.; et al. Ultrasmooth organic-inorganic perovskite thin-film formation and crystallization for efficient planar heterojunction solar cells. *Nat. Commun.* **2015**, *6*, 6142. [CrossRef] [PubMed]

27. Dharani, S.; Dewi, H.A.; Prabhakar, R.R.; Baikie, T.; Shi, C.; Yonghua, D.; Mathews, N.; Boix, P.P.; Mhaisalkar, S.G. Incorporation of Cl into sequentially deposited lead halide perovskite films for highly efficient mesoporous solar cells. *Nanoscale* **2014**, *6*, 13854–13860. [CrossRef] [PubMed]

28. Bhalla, A.S.; Guo, R.; Roy, R. The perovskite structure-a review of its role in ceramic science and technology. *Mater. Res. Innov.* **2000**, *4*, 3–26. [CrossRef]

29. Borriello, I.; Cantele, G.; Ninno, D. Ab initio investigation of hybrid organic-inorganic perovskites based on tin halides. *Phys. Rev. B* **2008**, *77*, 235214. [CrossRef]

30. Søndenå, R.; Ravindran, P.; Stølen, S.; Grande, T.; Hanfland, M. Electronic structure and magnetic properties of cubic and hexagonal $SrMnO_3$. *Phys. Rev. B* **2006**, *74*, 144102. [CrossRef]

31. Mitzi, D.B. Organic−Inorganic Perovskites Containing Trivalent Metal Halide Layers: The Templating Influence of the Organic Cation Layer. *Inorg. Chem.* **2000**, *39*, 6107–6113. [CrossRef] [PubMed]

32. Johnston, M.B.; Herz, L.M. Hybrid Perovskites for Photovoltaics: Charge-Carrier Recombination, Diffusion, and Radiative Efficiencies. *Acc. Chem. Res.* **2016**, *49*, 146–154. [CrossRef] [PubMed]

33. Noel, N.K.; Stranks, S.D.; Abate, A.; Wehrenfennig, C.; Guarnera, S.; Haghighirad, A.-A.; Sadhanala, A.; Eperon, G.E.; Pathak, S.K.; Johnston, M.B.; et al. Lead-free organic–inorganic tin halide perovskites for photovoltaic applications. *Energy Environ. Sci.* **2014**, *7*, 3061–3068. [CrossRef]

34. Hao, F.; Stoumpos, C.C.; Cao, D.H.; Chang, R.P.H.; Kanatzidis, M.G. Lead-Free Solid-State Organic-Inorganic Halide Perovskite Solar Cells. *Nat. Photonics* **2014**, *8*, 489–494. [CrossRef]

35. Hoke, E.T.; Slotcavage, D.J.; Dohner, E.R.; Bowring, A.R.; Karunadasa, H.I.; McGehee, M.D. Reversible photo-induced trap formation in mixed- halide hybrid perovskites for photovoltaics. *Chem. Sci.* **2015**, *6*, 613–617. [CrossRef] [PubMed]

36. Suarez, B.; Gonzalez-Pedro, V.; Ripolles, T.S.; Sanchez, R.S.; Otero, L.; Mora-Sero, I. Recombination Study of Combined Halides (Cl, Br, I) Perovskite Solar Cells. *Phys. J. Chem. Lett.* **2014**, *5*, 1628–1635. [CrossRef] [PubMed]

37. Park, N.-G. Perovskite solar cells: An emerging photovoltaic technology. *Mater. Today* **2015**, *18*, 65–72. [CrossRef]

38. Shen, Q.; Ogomi, Y.; Chang, J.; Toyoda, T.; Fujiwara, K.; Yoshino, K.; Sato, K.; Yamazaki, K.; Akimoto, M.; Kuga, Y.; et al. Optical absorption, charge separation and recombination dynamics in Sn/Pb cocktail perovskite solar cells and their relationships to photovoltaic performances. *Mater. J. Chem. A* **2015**, *3*, 9308–9316. [CrossRef]

39. Koren, E.; Lortscher, E.; Rawlings, C.; Knoll, A.W.; Duerig, U. Adhesion and friction in mesoscopic graphite contacts. *Science* **2015**, *348*, 679–683. [CrossRef] [PubMed]

40. Shi, D.; Adinolfi, V.; Comin, R.; Yuan, M.; Alarousu, E.; Buin, A.; Chen, Y.; Hoogl, S.; Rothenberger, A.; Katsiev, K.; et al. Low trap-state density and long carrier diffusion in organolead trihalide perovskite single crystals. *Science* **2015**, *347*, 519–522. [CrossRef] [PubMed]

41. Li, C.; Wang, F.; Xu, J.; Yao, J.; Zhang, B.; Zhang, C.; Xiao, M.; Dai, S.; Li, Y.; Tan, Z. Efficient perovskite/fullerene planar heterojunction solar cells with enhanced charge extraction and suppressed charge recombination. *Nanoscale* **2015**, *7*, 9771–9778. [CrossRef] [PubMed]

42. Wang, Y.; Wang, H.; Yu, M.; Fu, F.; Qin, Y.; Zhang, J.; Ai, X. Trap-limited charge recombination in intrinsic perovskite film and meso-superstructured perovskite solar cells and the passivation effect of the hole-transport material on trap states. *Phys. Chem. Chem. Phys.* **2015**, *17*, 29501–29506. [CrossRef] [PubMed]

43. Troughton, J.; Carnie, M.J.; Davies, M.L.; Charbonneau, C.; Jewell, E.H.; Worsley, D.A.; Watson, T.M.; Mater, J.; Watson, T. Photonic flash-annealing of lead halide perovskite solar cells in 1 ms. *Mater. J. Chem. A* **2016**, *4*, 3471–3476. [CrossRef]

44. Seetharaman, S.M.; Nagarjuna, P.; Kumar, P.N.; Singh, S.P.; Deepa, M.; Namboothiry, M.A. Efficient organic-inorganic hybrid perovskite solar cells processed in air. *Phys. Chem. Chem. Phys.* **2014**, *16*, 24691–24696. [CrossRef] [PubMed]

45. Wang, H.-Y.; Wang, Y.; Yu, M.; Han, J.; Guo, Z.-X.; Ai, X.-C.; Zhang, J.; Qin, Y. Mechanism of biphasic charge recombination and accumulation in TiO_2 mesoporous structured perovskite solar cells. *Phys. Chem. Chem. Phys.* **2016**, *18*, 12128–12134. [CrossRef] [PubMed]

46. Zhao, J.; Wang, P.; Wei, L.; Liu, Z.; Fang, X.; Liu, X.; Ren, D.; Mai, Y. Efficient charge-transport in hybrid lead iodide perovskite solar cells. *Dalton Trans.* **2015**, *44*, 16914–16922. [CrossRef] [PubMed]

47. Bi, C.; Shao, Y.; Yuan, Y.; Xiao, Z.; Wang, C.; Gao, Y.; Huang, J. Understanding the formation and evolution of interdiffusion grown organolead halide perovskite thin films by thermal annealing. *Mater. J. Chem. A* **2014**, *2*, 18508–18514. [CrossRef]

48. Wang, B.; Wong, K.Y.; Yang, S.; Chen, T. Crystallinity and defect state engineering in organo-lead halide perovskite for high-efficiency solar cells. *Mater. J. Chem. A* **2016**, *4*, 3806–3812. [CrossRef]

49. Yang, S.Y.; Park, B.-W.; Jung, E.H.; Jeon, N.J.; Kim, Y.C.; Lee, D.U.; Shin, S.S.; Seo, J.; Kim, E.K.; Noh, J.H.; et al. Iodide management in formamidinium-lead-halide-based perovskite layers for efficient solar cells. *Science* **2017**, *356*, 1376–1379. [CrossRef] [PubMed]

50. Xing, G.; Mathews, N.; Lim, S.S.; Yantara, N.; Liu, X.; Sabba, D.; Grätzel, M.; Mhaisalkar, S.; Sum, T.C. Low-temperature solution-processed wavelength-tunable perovskites for lasing. *Nat. Mater.* **2014**, *13*, 476–480. [CrossRef] [PubMed]

51. Guo, Y.; Liu, C.; Tanaka, H.; Nakamura, E. Air-stable and solution-processable perovskite photodetectors for solar-blind UV and visible light. *J. Phys. Chem. Lett.* **2015**, *6*, 535–539. [CrossRef] [PubMed]

52. Tan, Z.-K.; Moghaddam, R.S.; Lai, M.L.; Docampo, P.; Higler, R.; Deschler, F.; Price, M.; Sadhanala, A.; Pazos, L.M.; Credgington, D.; et al. Bright light-emitting diodes based on organometal halide perovskite. *Nat. Nanotechnol.* **2014**, *9*, 687–692. [CrossRef] [PubMed]

53. He, Y.; Galli, G. Perovskites for solar thermoelectric applications: A first principle study of $CH_3NH_3AI_3$ (A = Pb and Sn). *Chem. Mater.* **2014**, *26*, 5394–5400. [CrossRef]

54. Da, P.; Cha, M.; Sun, L.; Wu, Y.; Wang, Z.S.; Zheng, G. High-performance perovskite photoanode enabled by Ni passivation and catalysis. *Nano Lett.* **2015**, *15*, 3452–3457. [CrossRef] [PubMed]

55. Dong, Q.; Fang, Y.; Shao, Y.; Mulligan, P.; Qiu, J.; Cao, L.; Huang, J. Electron-hole diffusion lengths >175 m in solution-grown $CH_3NH_3PbI_3$ single crystals. *Science* **2015**, *347*, 967–970. [CrossRef] [PubMed]

56. Weber, D. $CH_3NH_3PbX_3$, ein Pb(II)-system mit kubischer perowskitstruktur/$CH_3NH_3PbX_3$, a Pb(II)-system with cubic perovskite structure. *Z. Naturforsch. B* **1978**, *33*, 1443–1445. [CrossRef]

57. Weber, D. $CH_3NH_3SnBrxI_{3-x}$ (x = 0–3), ein Sn(II)-system mit kubischer perowskitstruktur/$CH_3NH_3SnBr_xI_{3-x}$ (x = 0–3), a Sn(II)-system with cubic perovskite structure. *Z. Naturforsch. B* **1978**, *33*, 862–865. [CrossRef]

58. Niu, G.; Li, W.; Li, J.; Wang, L. Progress of interface engineering in perovskite solar cells. *Sci. China Mater.* **2016**, *59*, 728–742. [CrossRef]

59. Su, J.; Chen, D.; Lin, C. Growth of large $CH_3NH_3PbX_3$ (X = I, Br) single crystals in solution. *J. Cryst. Growth* **2015**, *422*, 75–79. [CrossRef]

60. Dang, Y.; Liu, Y.; Sun, Y.; Yuan, D.; Liu, X.; Lu, W.; Liu, G.; Xia, H.; Tao, X. Bulk crystal growth of hybrid perovskite material $CH_3NH_3PbI_3$. *CrystEngComm* **2015**, *17*, 665–670. [CrossRef]

61. Lian, Z.; Yan, Q.; Lv, Q.; Wang, Y.; Liu, L.; Zhang, L.; Pan, S.; Li, Q.; Wang, L.; Sun, J.-L. High-performance planar-type photodetector on (100) facet of MAPbI3 single crystal. *Sci. Rep.* **2015**, *5*, 16563. [CrossRef] [PubMed]

62. Dang, Y.; Zhou, Y.; Liu, X.; Ju, D.; Xia, S.; Xia, H.; Tao, X. Formation of hybrid perovskite tin iodide single crystals by top-seeded solution growth. *Angew. Chem. Int. Ed.* **2016**, *55*, 3447–3450. [CrossRef] [PubMed]

63. Dang, Y.; Zhong, C.; Zhang, G.; Ju, D.; Wang, L.; Xia, S.; Xia, H.; Tao, X. Crystallographic investigations into properties of acentric hybrid perovskite single crystals $NH(CH_3)_3SnX_3$ (X = Cl, Br). *Chem. Mater.* **2016**, *28*, 6968–6974. [CrossRef]

64. Lian, Z.; Yan, Q.; Gao, T.; Jie, D.; Lv, Q.; Ning, C.; Li, Q.; Sun, J. Perovskite $CH_3NH_3PbI_3(Cl)$ single crystals: Rapid solution growth, unparalleled crystalline quality, and low trap density toward 108 cm^{-3}. *J. Am. Chem. Soc.* **2016**, *138*, 9409–9412. [CrossRef] [PubMed]

65. Fang, Y.; Dong, Q.; Shao, Y.; Yuan, Y.; Huang, J. Highly narrowband perovskite single-crystal photodetectors enabled by surface-charge recombination. *Nat. Photonics* **2015**, *9*, 679–686. [CrossRef]

66. Söhnel, O.; Novotný, P.; Solc, Z. Densities of Aqueous Solutions of Inorganic Substances. *J. Chem. Eng. Data* **1985**, *29*, 379–382. [CrossRef]

67. Kadro, J.M.; Nonomura, K.; Gachet, D.; Hagfeldt, A. Facile route to freestanding $CH_3NH_3PbI_3$ crystals using inverse solubility. *Sci. Rep.* **2015**, *5*, 11654. [CrossRef] [PubMed]

68. Saidaminov, M.I.; Abdelhady, A.L.; Murali, B.; Alarousu, E.; Burlakov, V.M.; Peng, W.; Dursun, I.; Wang, L.; He, Y.; Maculan, G.; et al. High-quality bulk hybrid perovskite single crystals within minutes by inverse temperature crystallization. *Nat. Commun.* **2015**, *6*, 7586. [CrossRef] [PubMed]

69. Liu, Y.; Yang, Z.; Cui, D.; Ren, X.; Sun, J.; Liu, X.; Zhang, J.; Wei, Q.; Fan, H.; Yu, F.; et al. Two-inch-sized perovskite CH$_3$NH$_3$PbX$_3$ (X = Cl, Br, I) crystals: Growth and characterization. *Adv. Mater.* **2015**, *27*, 5176–5183. [CrossRef] [PubMed]

70. Maculan, G.; Sheikh, A.D.; Abdelhady, A.L.; Saidaminov, M.I.; Haque, M.A.; Murali, B.; Alarousu, E.; Mohammed, O.F.; Wu, T.; Bakr, O.M. CH$_3$NH$_3$PbCl$_3$ single crystals: Inverse temperature crystallization and visible-blind UV-photodetector. *J. Phys. Chem. Lett.* **2015**, *6*, 3781–3786. [CrossRef] [PubMed]

71. Zhumekenov, A.A.; Saidaminov, M.I.; Haque, M.A.; Alarousu, E.; Sarmah, S.P.; Murali, B.; Dursun, I.; Miao, X.-H.; Abdelhady, A.L.; Wu, T.; et al. Formamidinium lead halide perovskite crystals with unprecedented long carrier dynamics and diffusion length. *ACS Energy Lett.* **2016**, *1*, 32–37. [CrossRef]

72. Abdelhady, A.L.; Saidaminov, M.I.; Murali, B.; Adinolfi, V.; Voznyy, O.; Katsiev, K.; Alarousu, E.; Comin, R.; Dursun, I.; Sinatra, L.; et al. Heterovalent dopant incorporation for bandgap and type engineering of perovskite crystals. *J. Phys. Chem. Lett.* **2016**, *7*, 295–301. [CrossRef] [PubMed]

73. Han, Q.; Bae, S.H.; Sun, P.; Hsieh, Y.T.; Yang, Y.; Rim, Y.S.; Zhao, H.; Chen, Q.; Shi, W.; Li, G. Single crystal formamidinium lead iodide (FAPbI$_3$): Insight into the structural, optical, and electrical properties. *Adv. Mater.* **2016**, *28*, 2253–2258. [CrossRef] [PubMed]

74. Zhang, T.; Yang, M.; Benson, E.E.; Li, Z.; Lagemaat, J.; Luther, J.M.; Yan, Y.; Zhu, K.; Zhao, Y. A facile solvothermal growth of single crystal mixed halide perovskite CH$_3$NH$_3$Pb(Br$_{1-x}$Cl$_x$)$_3$. *Chem. Commun.* **2015**, *51*, 7820–7823. [CrossRef] [PubMed]

75. Saidaminov, M.I.; Abdelhady, A.L.; Maculan, G.; Bakr, O.M. Retrograde solubility of formamidinium and methylammonium lead halide perovskites enabling rapid single crystal growth. *Chem. Commun.* **2015**, *51*, 17658–17661. [CrossRef] [PubMed]

76. Yang, Y.; Yan, Y.; Yang, M.; Choi, S.; Zhu, K.; Luther, J.M.; Beard, M.C. Low surface recombination velocity in solution-grown CH$_3$NH$_3$PbBr$_3$ perovskite single crystal. *Nat. Commun.* **2015**, *6*, 7961. [CrossRef] [PubMed]

77. Zhou, H.; Nie, Z.; Yin, J.; Sun, Y.; Zhou, H.; Li, D.; Dou, J.; Zhang, X.; Ma, T. Antisolvent diffusion-induced growth, equilibrium behaviours in aqueous solution and optical properties of CH$_3$NH$_3$PbI$_3$ single crystals for photovoltaic applications. *RSC Adv.* **2015**, *5*, 85344–85349. [CrossRef]

78. Chen, F.; Xu, C.; Xu, Q.; Zhu, Y.; Zhu, Z.; Liu, W.; Dong, X.; Qin, F.; Shi, Z. Structure Evolution of CH$_3$NH$_3$PbBr$_3$ Single Crystal Grown in N,N-Dimethylformamide Solution. *Cryst. Growth Des.* **2018**, *18*, 3132–3137. [CrossRef]

79. Li, Y.; Zhang, Y.; Zhao, Z.; Zhi, L.; Cao, X.; Jia, Y.; Lin, F.; Zhang, L.; Cui, X.; Wei, J. In Situ Investigation of the Growth of Methylammonium Lead Halide (MAPbI$_{3-x}$Br$_x$) Perovskite from Microdroplets. *Cryst. Growth Des.* **2018**. [CrossRef]

80. Tidhar, Y.; Edri, E.; Weissman, H.; Zohar, D.; Hodes, G.; Cahen, D.; Rybtchinski, B.; Kirmayer, S. Crystallization of methyl ammonium lead halide perovskites: Implications for photovoltaic applications. *J. Am. Chem. Soc.* **2014**, *136*, 13249–13256. [CrossRef] [PubMed]

81. Wharf, I.; Gramstad, T.; Makhija, R.; Onyszchuk, M. Synthesis and vibrational spectra of some lead(II) halide adducts with O-, S-, and N-donor atom ligand. *Can. J. Chem.* **1976**, *54*, 3430–3438. [CrossRef]

82. Peng, W.; Wang, L.; Murali, B. Solution-grown monocrystalline hybrid perovskite films for hole-transporter-free solar cells. *Adv. Mater.* **2016**, *28*, 3383–3390. [CrossRef] [PubMed]

83. Liu, Y.; Zhang, Y.; Yang, Z.; Yang, D.; Ren, X.; Pang, L.; Liu, S. Thinness and shape-controlled growth for ultrathin single-crystalline perovskite wafers for mass production of superior photoelectronic devices. *Adv. Mater.* **2016**, *28*, 9204–9209. [CrossRef] [PubMed]

84. Chen, Y.X.; Ge, Q.Q.; Shi, Y.; Liu, J.; Xue, D.; Ma, J.; Ding, J.; Yan, H.; Hu, J.; Wan, L. General space-confined on-substrate fabrication of thickness-adjustable hybrid perovskite single-crystalline thin films. *J. Am. Chem. Soc.* **2016**, *138*, 16196–16199. [CrossRef] [PubMed]

85. Rao, H.S.; Li, W.G.; Chen, B.X.; Kuang, D.; Su, C. In situ growth of 120 cm^2 CH$_3$NH$_3$PbBr$_3$ perovskite crystal film on FTO glass for narrow band-photodetectors. *Adv. Mater.* **2017**, *29*, 1602639. [CrossRef] [PubMed]

86. Schmidt, L.C.; Pertegás, A.; González-Carrero, S.; Malinkiewicz, O.; Agouram, S.; Mínguez Espallargas, G.; Bolink, H.J.; Galian, R.E.; Pérez-Prieto, J. Nontemplate Synthesis of CH$_3$NH$_3$PbBr$_3$ Perovskite Nanoparticles. *J. Am. Chem. Soc.* **2014**, *136*, 850–853. [CrossRef] [PubMed]

87. Tong, Y.; Ehrat, F.; Vanderlinden, W.; Cardenasdaw, C.; Stolarczyk, J.K.; Polavarapu, L.; Urban, A.S. Dilution-Induced Formation of Hybrid Perovskite Nanoplatelets. *ACS Nano* **2016**, *10*, 10936–10944. [CrossRef] [PubMed]

88. Imran, M.; Caligiuri, V.; Wang, M.; Goldoni, L.; Prato, M.; Krahne, R.; De Trizio, L.; Manna, L. Benzoyl Halides as Alternative Precursors for the Colloidal Synthesis of Lead-Based Halide Perovskite Nanocrystals. *J. Am. Chem. Soc.* **2018**, *140*, 2656–2664. [CrossRef] [PubMed]

89. Protesescu, L.; Yakunin, S.; Kumar, S.; Bär, J.; Bertolotti, F.; Masciocchi, N.; Guagliardi, A.; Grotevent, M.; Shorubalko, I.; Bodnarchuk, M.I.; et al. Dismantling the "Red Wall" of Colloidal Perovskites: Highly Luminescent Formamidinium and Formamidinium–Cesium Lead Iodide Nanocrystals. *ACS Nano* **2017**, *11*, 3119–3134. [CrossRef] [PubMed]

90. Bi, D.; Tress, W.; Dar, M.I.; Gao, P.; Luo, J.; Renevier, C.; Schenk, K.; Abate, A.; Giordano, F.; Baena, J.P.C.; et al. Efficient luminescent solar cells based on tailored mixed-cation perovskites. *Sci. Adv.* **2016**, *2*, 1501170. [CrossRef] [PubMed]

91. Jacobsson, T.J.; Correa-Baena, J.; Pazoki, M.; Saliba, M.; Schenk, K.; Gratzel, M.; Hagfeldt, A. Exploration of the compositional space for mixed lead halogen perovskites for high efficiency solar cells. *Energy Environ. Sci.* **2016**, *9*, 1706–1724. [CrossRef]

92. Li, W.G.; Rao, H.S.; Chen, B.X.; Wang, X.D.; Kuang, D.B. A formamidinium–methylammonium lead iodide perovskite single crystal exhibiting exceptional optoelectronic properties and long-term stability. *J. Mater. Chem. A* **2017**, *5*, 19431–19438. [CrossRef]

93. Huang, Y.; Li, L.; Liu, Z.; Jiao, H.; Jiao, H.Y.; Wang, X.; Zhu, R.; Wang, D.; Sun, J.; Chen, Q.; et al. The Intrinsic Properties of FA $_{(1-x)}$MA$_x$PbI$_3$ Perovskite Single Crystals. *J. Mater. Chem. A* **2017**, *5*, 8537–8544. [CrossRef]

94. Xie, L.Q.; Chen, L.; Nan, Z.A.; Lin, H.X.; Wang, T.; Zhan, D.P. Understanding the cubic phase stabilization and crystallization kinetics in mixed cations and halides perovskite single crystals. *J. Am. Chem. Soc.* **2017**, *139*, 3320–3323. [CrossRef] [PubMed]

95. Kulbak, M.; Cahen, D.; Hodes, G. How Important Is the Organic Part of Lead Halide Perovskite Photovoltaic Cells? Efficient CsPbBr$_3$ Cells. *J. Phys. Chem. Lett.* **2015**, *6*, 2452–2456. [CrossRef] [PubMed]

96. Protesescu, L.; Yakunin, S.; Bodnarchuk, M.I.; Krieg, F.; Caputo, R.; Hendon, C.H.; Yang, R.X.; Walsh, A.; Kovalenko, M.V. Nanocrystals of Cesium Lead Halide Perovskites (CsPbX$_3$, X = Cl, Br, and I): Novel Optoelectronic Materials Showing Bright Emission with Wide Color Gamut. *Nano Lett.* **2015**, *15*, 3692–3696. [CrossRef] [PubMed]

97. Stoumpos, C.C.; Malliakas, C.D.; Peters, J.A.; Liu, Z.; Sebastian, M.; Im, J.; Chasapis, T.C.; Wibowo, A.C.; Chung, D.Y.; Freeman, A.J. Crystal Growth of the Perovskite Semiconductor CsPbBr$_3$: A New Material for High-Energy Radiation Detection. *Cryst. Growth Des.* **2013**, *13*, 2722–2727. [CrossRef]

98. Kobayashi, M.; Omata, K.; Sugimoto, S.; Tamagawa, Y.; Kuroiwa, T.; Asada, H.; Takeuchi, H.; Kondo, S. Scintillation Characteristics of CsPbCl$_3$ Single Crystals. Nucl. Instrum. *Methods Phys. Res. Sect. A* **2008**, *592*, 369–373.

99. Clark, D.J.; Stoumpos, C.C.; Saouma, F.O.; Kanatzidis, M.G.; Jang, J.I. Polarization-Selective Three-Photon Absorption and Subsequent Photoluminescence in CsPbBr$_3$ Single Crystal at Room Temperature. *Phys. Rev. B* **2016**, *93*, 195202. [CrossRef]

100. Nitsch, K.; Hamplová, V.; Nikl, M.; Polák, K.; Rodová, M. Lead Bromide and Ternary Alkali Lead Bromide Single Crystals-Growth and Emission Properties. *Chem. Phys. Lett.* **1996**, *258*, 518–522. [CrossRef]

101. Dirin, D.N.; Cherniukh, I.; Yakunin, S.; Shynkarenko, Y.; Kovalenko, M.V. Solution-Grown CsPbBr$_3$ Perovskite Single Crystals for Photon Detection. *Chem. Mater.* **2016**, *28*, 8470–8474. [CrossRef] [PubMed]

102. Rakita, Y.; Kedem, N.; Gupta, S.; Sadhanala, A.; Kalchenko, V.; Bohm, M.L.; Kulbak, M.; Friend, R.H.; Cahen, D.; Hodes, G. Low-temperature solution-grown CsPbBr$_3$ single crystals and their characterization. *Cryst. Growth Des.* **2016**, *16*, 5717–5725. [CrossRef]

103. Tong, Y.; Bladt, E.; Ayguler, M.F.; Manzi, A.; Milowska, K.Z.; Hintermayr, V.A.; Hintermayr, P.D.; Sara, B.; Alexander, S.U.; Lakshminarayana, P.; et al. Highly Luminescent Cesium Lead Halide Perovskite Nanocrystals with Tunable Composition and Thickness by Ultrasonication. *Angew. Chem. Int. Ed.* **2016**, *55*, 13887–13892. [CrossRef] [PubMed]

104. Tong, Y.; Bohn, B.J.; Bladt, E.; Wang, K.; Peter, M.-B.; Bals, S.; Urban, A.S.; Polavarapu, L.; Feldmann, J. Feldmann From Precursor Powders to CsPbX$_3$ Perovskite Nanowires: One-Pot Synthesis, Growth Mechanism, and Oriented Self-Assembly. *Angew. Chem. Int. Ed.* **2017**, *56*, 13887–13892. [CrossRef] [PubMed]

105. Slavney, A.H.; Hu, T.; Lindenberg, A.M.; Karunadasa, H.I. A Bismuth-Halide Double Perovskite with Long Carrier Recombination Lifetime for Photovoltaic Applications. *J. Am. Chem. Soc.* **2016**, *138*, 2138–2141. [CrossRef] [PubMed]

106. Dong, Q.; Song, J.; Fang, Y.; Shao, Y.; Ducharme, S.; Huang, J. Lateral-structure single-crystal hybrid perovskite solar cells via piezoelectric poling. *Adv. Mater.* **2016**, *28*, 2816–2821. [CrossRef] [PubMed]
107. Monroy, E.; Omnès, F.; Calle, F. Topical review: Wide-band gap semiconductor ultraviolet photodetectors. *Semicond. Sci. Technol.* **2003**, *18*, R33–R51. [CrossRef]
108. Ghezzi, D.; Antognazza, M.R.; Dal, M.M.; Lanzarini, E.; Benfenati, F.; Lanzani, G. A hybrid bioorganic interface for neuronal photoactivation. *Nat. Commun.* **2011**, *2*, 166. [CrossRef] [PubMed]
109. Razeghi, M.; Rogalski, A.J. Semiconductor ultraviolet detectors. *J. Appl. Phys.* **1996**, *79*, 7433–7473. [CrossRef]
110. Manga, K.K.; Wang, J.; Lin, M.; Zhang, J.; Nesladek, M.; Nalla, V.; Ji, W.; Loh, K.P. High-Performance Broadband Photodetector Using Solution-Processible PbSe-TiO$_2$-Graphene Hybrids. *Adv. Mater.* **2012**, *24*, 1697–1702. [CrossRef] [PubMed]
111. Li, S.; Tong, S.; Meng, J.; Zhang, C.; Zhang, C.; Shen, J.; Xiao, S.; Sun, J.; He, J.; Gao, Y.; et al. Fast-response and high-responsivity FA$_x$MA$_{(1-x)}$PbI$_3$ photodetectors fabricated via doctor-blading deposition in ambient condition. *Org. Electron.* **2018**, *52*, 190–194. [CrossRef]
112. Tong, S.; Wu, H.; Zhang, C.; Li, S.; Wang, C.; Shen, J.; Xiao, S.; He, J.; Yang, J.; Sun, J.; et al. Large-area and high-performance CH$_3$NH$_3$PbI$_3$ perovskite photodetectors fabricated via doctor blading in ambient condition. *Org. Electron.* **2017**, *49*, 347–354. [CrossRef]
113. Tong, S.; Sun, J.; Wang, C.; Huang, Y.; Zhang, C.; Shen, J.; Xie, H.; Niu, D.; Xiao, S.; Yuan, Y.; et al. High-Performance Broad band Perovskite Photodetectors Based on CH$_3$NH$_3$PbI$_3$/C8BTBT Heterojunction. *Adv. Electron. Mater.* **2017**, *3*, 1700058. [CrossRef]
114. Ding, J.; Fang, H.; Lian, Z.; Li, J.; Lv, Q.; Wang, L.; Sun, J.; Yan, Q. A self-powered photodetector based on a CH$_3$NH$_3$PbI$_3$ single crystal with asymmetric electrodes. *CrystEngComm* **2016**, *18*, 4405–4411. [CrossRef]
115. Shaikh, P.; Shi, D.; Retamal, J.; Sheikh, A.D.; Haque, M.A. Schottky junctions on perovskite single crystals: Light-modulated dielectric constant and self-biased photodetection. *J. Mater. Chem. C* **2016**, *4*, 8304–8312. [CrossRef]
116. Cao, M.; Tian, J.; Cai, Z.; Peng, L.; Yang, L.; Wei, D. Perovskite heterojunction based on CH$_3$NH$_3$PbBr$_3$ single crystal for high-sensitive self-powered photodetector. *Appl. Phys. Lett.* **2016**, *109*, 233303. [CrossRef]
117. Yakunin, S.; Sytnyk, M.; Kriegner, D.; Shrestha, S.; Richter, M.; Matt, G.J.; Azimi, H.; Brabec, C.J.; Stangl, J.; Kovalenko, M.V.; et al. Detection of X-ray photons by solution-processed lead halide perovskites. *Nat. Photonics* **2015**, *9*, 444–449. [CrossRef] [PubMed]
118. Wei, H.; Fang, Y.; Mulligan, P.; Chuirazzi, W.; Fang, H.-H.; Wang, C.; Ecker, B.R.; Gao, Y.; Loi, M.A.; Cao, L.; et al. Sensitive X-ray detectors made of Methylammonium lead tribromide perovskite single crystals. *Nat. Photonics* **2016**, *10*, 333–339. [CrossRef]

crystals

MDPI

Article

Effects of Iodine Doping on Carrier Behavior at the Interface of Perovskite Crystals: Efficiency and Stability

Guilin Liu [1], Lang Liu [2], Xiuxiu Niu [2], Huanping Zhou [3] and Qi Chen [2,*]

[1]   School of Science, Jiangnan University, Wuxi 214122, China; guilinliu@jiangnan.edu.cn
[2]   School of Materials Science & Engineering, Beijing Institute of Technology, Beijing 100081, China;
       amn716@163.com (L.L.); 15650785083@163.com (X.N.)
[3]   College of Engineering, Peking University, Beijing 100871, China; hpzhou@pku.edu.cn
*   Correspondence: qic@bit.edu.cn

Received: 22 March 2018; Accepted: 21 April 2018; Published: 25 April 2018

Abstract: The interface related to the polycrystalline hybrid perovskite thin film plays an essential role in the resulting device performance. Iodine was employed as an additive to modify the interface between perovskite and spiro-OMeTAD hole transport layer. The oxidation ability of iodine significantly improved the efficiency of charge extraction for perovskite solar cells. It reveals that the Open Circuit Voltage (V_{oc}) and Fill Factor (FF) of perovskite solar cells were improved substantially due to the dopant, which is mainly attributed to the interfacial improvement. It was found that the best efficiency of the devices was achieved when the dopant of iodine was in equivalent mole concentration with that of spiro-OMeTAD. Moreover, the long-term stability of the corresponding device was investigated.

Keywords: perovskite solar cells; iodine doping; interfacial interaction

1. Introduction

Organometal halide perovskite solar cells (PSCs) have attracted much attention because of their advantages such as high-power efficiency [1–3], flexibility [4–6], and long-term stability [7–9]. Compared with silicon solar cells [10–12], perovskite solar cells have simplified manufacture processing, and can be rapidly fabricated by spin-coating with precursor solutions of lead (II) iodide and methylamine hydroiodide (MAI). After only a few years of development, perovskite solar cells have reached a comparable efficiency to that of crystalline silicon solar cells [13]. To further improve the high efficiency, issues regarding crystallinity [14,15], film morphology [16,17], and grain size [18,19] have been widely investigated. The photon absorption [20] was a key purpose in those abovementioned investigations. Meanwhile, the charge extraction is equally important for perovskite crystals. Perovskite crystal surface and interfaces play an essential role in determining device performance, which includes different functional components such as the electron transport layer (ETL), hole transport layer (HTL), counter electrode, and so forth. In contrast to electrons, the transportation for holes within devices is challenging because of the heavier effective mass of holes [21] and the lack of HTLs with satisfactory hole conductivity. Moreover, carrier accumulation would be disastrous if carriers are not balanced in the photovoltaic devices [22]. Therefore, in the charge extraction step, the interfaces between the perovskite crystal and HTLs should be carefully studied with the emphasis on charge extraction and interface interactions [23]. To improve the hole conductivity in the organic HTL, oxidization is a feasible approach to increase the carrier concentration and thus the depletion length. However, the application of oxidant may deteriorate the perovskite crystals and its interface. Coincidentally, spiro-OMeTAD aging takes a long time in oxygen [24].

Considering HTL aging, we propose iodine as a dopant in the spiro-OMeTAD layer in order to increase the efficiency of charge extraction without damaging the polycrystalline perovskite thin films. Seok [25] recently found that additional iodide ions could decrease the concentration of deep-level defects in formamidinium-lead-halide-based perovskite solar cells. This research indicated that using the dopant of iodine was a feasible method for methylamine-lead-halide-based perovskite solar cells as well.

2. Experimental

2.1. Fabrication

The Indium Tin Oxides (ITO) substrates were cleaned with isopropanol, acetone, ethanol, and distilled water in a sonicator and then were blown with nitrogen gas. All substrates were sent into an ozonizer for a 20-min UV treatment. SnOx solution (15% in H_2O, Alfa Aesar, Haverhill, MA, USA) were then spin-coated on substrates as the electron transport layer (ETL) and annealed at 150 °C for 30 min in ambient. The thickness was around 40 nm. Methylamine iodide (MAI, >99%, Sigma Aldrich, St. Louis, MO, USA) and lead (II) iodide (PbI_2, 99%, Sigma Aldrich) were predissolved in N, N-dimethylformamide (DMF, >99%, Sigma Aldrich) as precursor solution before spin-coating process. After SnOx annealing, methylamine-lead(II)-iodide ($MAPbI_3$) precursor solution was then spin-coated on ETL as an intrinsic layer at 3000 rpm for 20 s; 500 μL of diethyl ether (analytical reagent, Sinopharm Chemical Reagent Co., Ltd., Shanghai, China) was dripped (before end of the 14 s) on perovskite films during the spin-coating operation. Substrates were immediately moved to hot plate for 10 min annealing at 75 °C to exclude solvents. The thickness of perovskite film was around 300 nm. 2,2′,7,7′–tetrakis(N, N-dimethoxyphenylamine)-9,9′-spirobiflurorene (Spiro-OMeTAD, 99%, Sigma Aldrich) solution was prepared by dissolving 80 mg of spiro-OMeTAD, 30 μL of 4-*tert*-butylpyridine (TBP, 99.9%, Sigma Aldrich), and 35 μL of a stock solution of 260 mg/mL lithium *bis*(trifluoromethylsulfonyl)-imide (Li-TFSI, 99.9%, Sigma Aldrich) in acetonitrile (99.9%, Sigma Aldrich) in 1 mL of chlorobenzene (99.9%, Sigma Aldrich). The iodine (99.8%, Sigma Aldrich) was doped into the spiro-OMeTAD solution in different concentrations. The dopant was varied from 0 mg/mL to 15 mg/mL in a gradient of 3 mg/mL. Before the metal electrode evaporation, samples were sent into an oxygenated, sealed cabin for 12-h of aging. Finally, the 100-nm counter electrode was deposited by thermal evaporation of gold under a pressure of 5×10^{-5} Pa to form five functional cells. The effective area was 0.102 cm^2.

2.2. Characterization

The X-ray diffraction (XRD) spectrum of the ITO/perovskite film was measured using a PANalytical X-ray diffractometer with Cu k_α radiation at a scanning rate (2θ) of 3° min^{-1}. The accelerating voltage and current were 40 kV and 40 mA, respectively. Ultraviolet/visible (UV/vis) absorption spectrum was recorded by a Hitachi (model UH4150) spectrophotometer at a rate of 0.5 nm/s. Photocurrent–voltage (J–V) measurements for the perovskite solar cells were characterized under both dark and AM 1.5 G irradiation (100 mW/cm^2) by using a xenon lamp simulator (Enlitech, model-SS-F7-3A, Kaohsiung City, Taiwan) after sample fabrication. The fluorescence (FL) and time-resolved photoluminescence were characterized by Oxford FLS 920 and FLS 980, respectively. Cross-sectional images were measured by field emission scanning electron microscope (FE-SEM, Hitachi S-4800FESEM, Hitachi, Tokyo, Japan) operated at 5 kV. The transient photovoltage and transient photocurrent measurements were carried out by a home-built system. The Nd:YAG laser irradiated with a 1064-nm pulse with 200-ns bandwidth and then generated 532-nm green light after passing through a second harmonic generation (SHG) crystal. The light spot was incident on the optical attenuator and then excited samples. Before the laser output, the laser respectively travels through a polarizer, a half-wave plate, and the shutter, generating a p-polarized pulse. The transient voltage and current were then traced and recorded by a Tektronix MDO 400 oscilloscope.

3. Results and Discussion

The employment of iodine aims to improve the carrier extraction efficiency and avoid other negative effects. XRD analysis, shown in Figure 1a, indicated that the hexagonal phase of the perovskite crystal was stable following doped HTL deposition. The perovskite thin film has strong diffraction patterns at 14.1°, and 28.3°, assigned to 110 and 220 of MAPbI3, respectively. We did not observe the diffraction peak often shown at 12.7° with perovskite, which belongs to excess PbI2. The XRD diffraction pattern strongly supports the crystal stability. Similar results were shown in Figure S1: the spiro-OMeTAD HTL as well as the dopant gathered and transferred extracted charge carriers without damaging perovskite crystal quality. The absorption spectrum was slightly changed in the wavelength range from 600 to 800 nm, indicating that the dopant has a small influence on the perovskite crystal as well. Meanwhile, spiro-OMeTAD was rapidly oxidized as shown in the absorption spectrum (Figure 1b). Commonly, spiro-OMeTAD [26] absorbs photons in the wavelength around 300–420 nm, and oxidized spiro-OMeTAD$^+$ has a polaronic band at 500 nm. Hence, we zoomed into the absorption spectrum in the 350–550 nm and 730–800 nm regions for comparison. The comparison was summarized in Figure 2.

The tail bandgap was in accordance with the XRD patterns in the perovskite samples with spiro-OMeTAD on top. In Figure 2b, the cutoff wavelengths were all around 1.59 eV, without shifts, which also strongly supports the stability of the perovskite crystal. Interestingly, in the 400 to 500 nm range, the absorption intensity gradient was increased when the doping concentration was below 9 mg/mL, mainly attributed to the formation of spiro-OMeTAD$^+$. According to Figure 3a, it was vital to achieve the optimum doping concentration of iodine in the layer of spiro-OMeTAD, because iodine has a strong oxidation effect [27] on both perovskite and hole transport material. According to experimental details, the mole concentration of spiro-OMeTAD was approximately 6.452×10^{-2} mmol/mL. Meanwhile, with the gradient of the concentration of the iodine dopant increasing, the I* radicals were consequently increased from 0 mmol/mL to 2.362×10^{-2} mmol/mL, 4.724×10^{-2} mmol/mL, 7.086×10^{-2} mmol/mL, 9.448×10^{-2} mmol/mL, and 3.543×10^{-1} mmol/mL, respectively. In comparison to the constant molar concentration of spiro-OMeTAD, iodine was fully reduced by spiro-OMeTAD initially. Once the concentration was over 9 mg/mL, polarons [27] were formed after reaction with I2 as the excessive oxidant. Herein, the absorption intensity decreased (in Figure S2) after dopant concentration was over 9 mg/mL due to the photobleaching effect.

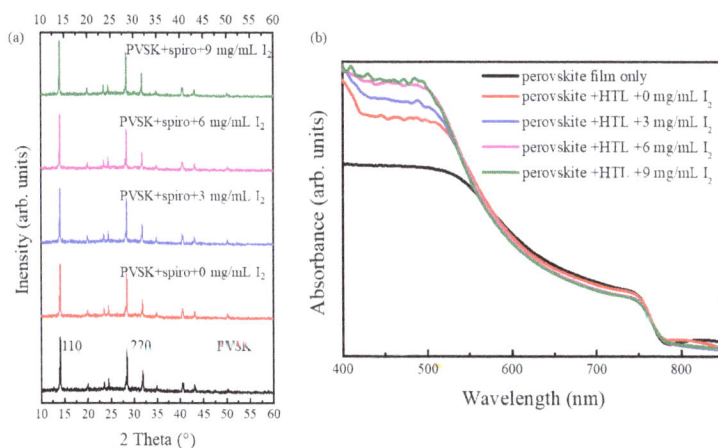

Figure 1. (**a**) XRD patterns of MAPbI3 thin film after coating with doped spiro-OMeTAD hole transport layer (HTL) layers; (**b**) UV–vis absorption of perovskite intrinsic film and with different doped spiro-OMeTAD HTL layers.

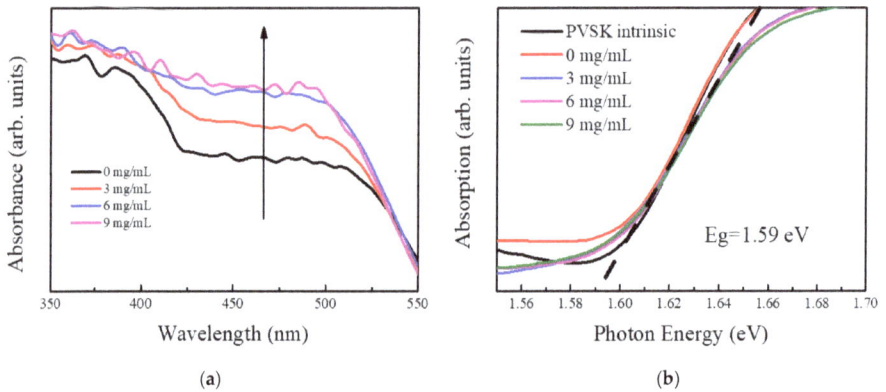

Figure 2. (**a**) Absorption spectra of films of pristine and oxidized spiro-OMeTAD layers and the reduction of the polaronic band (≈500 nm) by reaction with I⁻; (**b**) bandgap of perovskite films; the tangent line of the cutoff peak shows that the bandgap of all films is around 1.59 eV.

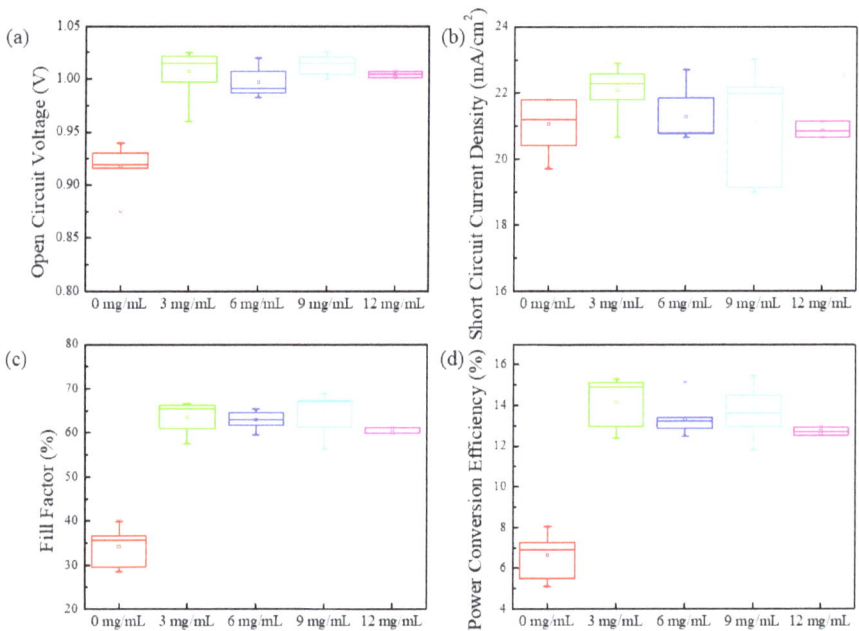

Figure 3. Photovoltaic parameters of PSCs evaporated with gold electrode.

The photovoltaic properties of the MAPbI₃ PSCs were sequentially characterized under solar simulator after counter electrode evaporation. It is worth mentioning that the silver electrode is not suitable for iodine dopant-based HTLs due to silver iodide (AgI) formation, according to our records in Figure S3. Thermal evaporation is a homogeneous deposition process; however, we have observed a gradient variation of rough surfaces after silver evaporation. Obviously, the quality of electrode morphology relied on the overweighting of iodine. To exclude the formation of AgI, we applied gold as the counter electrode material for the PSCs. The photovoltaic parameters were summarized and are shown in Figure 3. It can be seen that the V_{oc} was immediately increased from 0.92 V (average) to over 1 V due to spiro-OMeTAD$^+$ formation. According to the definition, the V_{oc} is related to both light-generated current and saturation current [28], as shown by Equation (1):

$$V_{oc} = \frac{nkT}{q} \ln\left(\frac{I_L}{I_0} + 1\right) \tag{1}$$

where n is the ideality factor, T is the absolute temperature, I_L is the photo-generated current, I_{sc} is the short-circuit current, and I_0 is the reverse saturation current. In the open-circuit condition, most charges remained in excited states and decayed to ground states through a nongeminate recombination. As we have described, the dopant has slight impacts on perovskite crystals, and the photo-generated currents were approximately close to each other. As a result, the V_{oc} depends on the saturation current. *J–V* measurements in the dark condition are summarized in Figure S4 (Supplementary Materials). It was found that the I_s was decreased from 100 μA to only 14 μA once the iodine was doped into the HTL. Meanwhile, under the forward bias condition, the depletion zone was extended because of spiro-OMeTAD$^+$ formation. As a result, the threshold voltage was increased from 1.47 V to 1.76 V. Hence, the V_{oc} was immediately increased when the dopant was employed in the HTL.

The spiro-OMeTAD$^+$ was instantly formed by the iodine dopant, leading to an improved charge extraction, leading to a FF promotion. Meanwhile, it was found that the grain boundaries were improved as well (Figure S5). The perovskite crystal grain grew larger in low dopant concentration, which contributed to the charge extraction as well. In comparison to the pristine 35%, the FF was dramatically increased to over 60%. However, though the charge extraction was improved, the charge density was relatively stable because of the limitation of photon absorption. Furthermore, apparently, the charge transport remained stable since the thicknesses of all layers were the same, as described in the Experimental section. Hence, the short-circuit current density (J_{sc}) remained stable, indicating that the charge density was rarely decreased after iodine doping. All these parameters contributed to the Power Conversion Efficiency (PCE). Although the average and highest PCE of a single channel was from the group of 3 mg/mL, we believe that the group of 6 mg/mL was better for the practical application because its PCE was smaller in the dispersion ratio compared with others. Also, we provided photovoltaic properties of PSCs with silver electrodes in Figure S6. In low dopant concentration, the FF suddenly increased to 60% because of the interfacial improvement. It is clear to see that the highest PCE was from the group of 6 mg/mL, corresponding to the molar ratio as we have mentioned. In the condition of dopant concentration of 6 mg/mL, most spiro-OMeTAD can be oxidized into spiro-OMeTAD$^+$ without the introduction of iodine impurity. However, once the dopant concentration was over 9 mg/mL, excess iodine remained in the HTL, leading to the short-circuit current collapse. These results were in accordance with Figure S6; both indicated that high concentration of iodine is unsuitable for HTLs and PSCs.

The interfacial charge extraction was characterized by the time-resolved photoluminescence based on samples of perovskite film and perovskite/HTL dual films. The results were summarized in Figure S1. The intersystem conversion rate was rapidly increased by at least one order of magnitude once the HTL was spin-coated onto the perovskite film. Simultaneously, the singlet transition was around 2 ns, independent of the HTL contact. Note that the time-resolved photoluminescence and fluorescence have invariant radiation efficiency and peak positions, indicating that the oxidant has a negligible influence on perovskite crystals.

Further transient photovoltage (TPV) and transient photocurrent (TPC) investigations were carried out to verify the interfacial interactions as drawn in Figure S8. The laser was integrated with the polarizers and a half-wave plate inside. Conventionally, the laser was perpendicularly injected into PSCs with p-polarized light, triggering the -electron photoexcitation perpendicular to the interface, while other information such as the interfacial interactions was eliminated. We modified the TPV/TPC setups with a home-built sample holder so that the PSC could rotate 45 degrees along with the tangential direction. Once the sample was deviated from its original position, the excited state of perovskite crystal would be polarized across the interface perpendicular with the Poynting vector. Hence, the interfacial interactions can be traced by comparing with different angles, as shown in Figure 4.

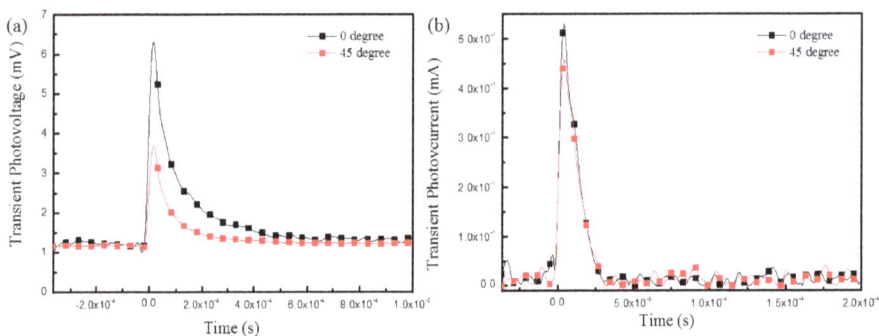

Figure 4. (**a**) shows the transient photovoltage (TPV) results via angle shift and (**b**) exhibits the transient photocurrent (TPC) changed by switching angles.

The charge dynamics confirmed that iodine doping has a positive effect on the interfacial charge transfer of perovskite crystals. The excited polarons were polarized across the interface once the electric field of light was shifted 45°, as we assumed. As plotted in Figure 4, the excited electron transfer was 5 μs faster than instantly after the sample was tilted, due to the interfacial interactions. Meanwhile, the iodine doping has a slight influence on perovskite crystals, as we have reiterated. The nongeminate recombination then plays a dominant role in the open-circuit condition. The detrapping rate constant used of 2.57×10^7 s^{-1} qualitatively reproduces the current tail after light-off, confirming that charges in perovskite crystal are rapidly detrapped. The current intensity and trapping rates were slightly changed compared with the perpendicular incident condition. Although the transient photovoltage was lower at 45° of incidence angle, the charge extraction remained stable with a constant trapping rate of 9.48×10^4 s^{-1} in the short-circuit condition.

However, the long-term stability was unsatisfactory, compared with dopant-free samples. After one-week exposure in ambient air, PSCs with doped HTLs lost power generation ability, while PSCs with pure spiro-OMeTAD HTLs maintained stability in PCE. One reason [29] is that the I$^-$ diffused to the anode, leading to a light-induced reactivity of the gold iodide; a new possible degradation mechanism in PSCs. The formation of $MA_2Au_2I_6$ blocked charge extraction, and then the PCE was dramatically decreased after long-term exposure in air. After one-week exposure in ambient air, samples were characterized by XRD measurements; the XRD patterns are shown in Figure 5. It can be seen that a split peak appeared close to 14.1° once the iodine was doped into the HTL. The result was in accordance with the formation of $MA_2Au_2I_6$, as we have mentioned [29]. Meanwhile, an additional peak emerged at 6.79° after one week in air. This peak was compared with related research [30]. A new phase of the Pb_3I_8 intermediate was formed due to the perovskite decomposition. The excess iodine played an important role in the perovskite decomposition since the dopant-free sample remained in the tetragonal phase. As a result, in order to avoid the ion diffusion, an additional buffer layer such as polymethyl methacrylate (PMMA) is essential for further studies of long-term stability based on this type of structure.

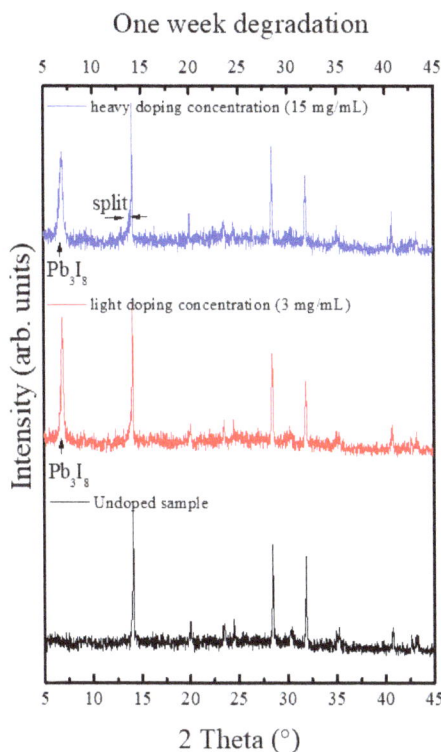

Figure 5. The XRD patterns of PSCs after one-week exposure in air.

4. Conclusions

In this article, iodine effects on the perovskite crystal and the relevant device are investigated. It was found that when it is applied in an HTL, the oxidation time for the HTL can be significantly decreased. The experimental results found that iodine has no substantial impacts on crystal quality or the perovskite optical bandgap. The PSC performance was best when the molar ratio of dopant and spiro-OMeTAD was equal; that is, in equivalent mole concentration. The improvement was mainly benefited from V_{oc} and FF enhancement, possibly due to the charge extraction enhancement. However, iodine doping devices exhibit decreased stability. These findings can be used to guide the design of other optoelectronic devices such as light-emitting diodes and lasers, which have recently involved hybrid perovskite materials.

Supplementary Materials: The following are available online at www.mdpi.com/2073-4352/8/5/185/s1. Figure S1: The fluorescence spectrum (a) and time-resolve photoluminescence spectra (b), Figure S2: absorption spectra of perovskite films with oxidized spiro-OMeTAD layers when the dopant was over 9 mg/mL, Figure S3: Rough morphology of sliver electrode with increasing iodine doping, Figure S4: Dark current of perovskite solar cells after the electrode evaporation, Figure S5: cross-section images of perovskite samples with doped HTL, the HTLs were doped with various concentration of iodine, Figure S6: photovoltaic parameters of PSCs evaporated with sliver electrode, Figure S7: Photovoltaic parameters of PSCs evaporated with gold electrodes. All samples were characterized after 48 h oxygen aging in sealed cabin, Figure S8: Schematic figure of TPV/TPC setups, Table S1: Lifetime extracted from Figure S1, where the t_1 was the lifetime of radiative de-excitation and t_2 was the time of delayed fluorescence.

Author Contributions: H.Z. and Q.C. conceived and designed the experiments; G.L. performed and analyzed the experiments; L.L. conceived the transient experiments and contributed analysis; X.N. contributed to the device fabrication and characterization. G.L. and Q.C. wrote the paper.

Acknowledgments: This work is supported by the Fundamental Research Funds for the Central Universities (Grant No. JUSRP11834). We sincerely thank Enli technology for offering the steady-state solar simulator.

Conflicts of Interest: The authors declare no conflict of interest.

References

1. Zhou, H.; Chen, Q.; Li, G.; Luo, S.; Song, T.B.; Duan, H.S.; Hong, Z.; You, J.; Liu, Y.; Yang, Y. Interface engineering of highly efficient perovskite solar cells. *Science* **2014**, *345*, 542–546. [CrossRef] [PubMed]
2. Sun, W.; Li, Y.; Xiao, Y.; Zhao, Z.; Ye, S.; Rao, H.; Ting, H.; Bian, Z.; Xiao, L.; Huang, C.; et al. An ammonia modified PEDOT: PSS for interfacial engineering in inverted planar perovskite solar cells. *Org. Electron.* **2017**, *46*, 22–27. [CrossRef]
3. Burschka, J.; Nazeeruddin, M.K.; Gratzel, M. Sequential deposition as a route to high performance perovskite-sensitized solar cells. *Nature* **2013**, *499*, 316–319. [CrossRef] [PubMed]
4. Heo, J.H.; Shin, D.H.; Jang, M.H.; Lee, M.L.; Kang, M.G.; Im, S.H. Highly flexible, high-performance perovskite solar cells with adhesion promoted $AuCl_3$-doped graphene electrodes. *J. Mater. Chem. A* **2017**, *5*, 21146–21152. [CrossRef]
5. Wang, Z.; Peng, L.; Lin, Z.; Ni, J.; Yi, P.; Lai, X.; He, X.; Lei, Z. Flexible semiconductor technologies with nanohole-provided high area coverages and their application in plasmonic-enhanced thin film photovoltaics. *Sci. Rep.* **2017**, *7*, 13155. [CrossRef] [PubMed]
6. Chen, L.; Xie, X.; Liu, Z.; Lee, E.C. A transparent poly(3,4-ethylenedioxylenethiophene): Poly (styrene sulfonate) cathode for low temperature processed, metal-oxide free perovskite solar cells. *J. Mater. Chem. A* **2017**, *5*, 6974–6980. [CrossRef]
7. Qiao, B.; Song, P.; Cao, J.; Zhao, S.; Shen, Z.; Gao, D.; Liang, Z.; Xu, Z.; Song, D.; Xu, X. Water-resistant, monodispersed and stably luminescent $CsPbBr_3/CsPb2Br_5$ core-shell-like structure lead halide perovskite nanocrystals. *Nanotechnology* **2017**, *28*, 445602. [CrossRef] [PubMed]
8. Hou, X.; Huang, S.; Ou-Yang, W.; Pan, L.; Sun, Z.; Chen, X. Constructing efficient and stable perovskite solar cells via interconnecting perovskite grains. *ACS Appl. Mater. Interfaces* **2017**, *9*, 35200–35208. [CrossRef] [PubMed]
9. Shukla, S.; Shukla, S.; Haur, L.J.; Dintakurti, S.S.; Han, G.; Priyadarshi, A.; Baikie, T.; Mhaisalkar, S.G.; Mathews, N. Effect of formamidinium/Cesium substitution and PbI_2 on the long-term stability of triple-cation perovskites. *ChemSusChem* **2017**, *10*, 3804–3809. [CrossRef] [PubMed]
10. Moumni, B.; Jaballah, A.B. Correlation between oxidant concentrations, morphological aspects and etching kinetics of silicon nanowires during sliver-assist electroless etching. *Appl. Surf. Sci.* **2017**, *425*, 1–7. [CrossRef]
11. Jin, J.; Shen, H.; Zheng, P.; Chan, K.S.; Zhang, X.; Jin, H. >20.5% diamond wire sawn multicrystalline silicon solar cells with maskless inverted pyramid like texturing. *IEEE J. Photovolt.* **2017**, *7*, 1264–1269. [CrossRef]
12. Dullweber, T.; Hannebauer, H.; Dorn, S.; Schimanke, S.; Merkle, A.; Hampe, C.; Brendel, R. Emitter saturation current densities of $22fA/cm^2$ applied to industrial PERC solar cells approaching 22% conversion efficiency. *Prog. Photovolt.* **2017**, *25*, 509–514. [CrossRef]
13. Green, M.A.; Emery, K.; Hishikawa, Y.; Warta, W. Solar cell efficiency tables (version 51). *Prog. Photovolt.* **2017**. [CrossRef]
14. Chen, Q.; Zhou, H.; Hong, Z.; Luo, S.; Duan, H.S.; Wang, H.H.; Liu, Y.; Li, G.; Yang, Y. Planar heterojunction perovskite solar cells via vapor-assisted solution process. *J. Am. Chem. Soc.* **2014**, *136*, 622–625. [CrossRef] [PubMed]
15. Liu, M.; Johnston, M.B.; Snaith, H.J. Efficient planar heterojunction perovskite solar cells by vapour deposition. *Nature* **2013**, *501*, 395–398. [CrossRef] [PubMed]
16. Conings, B.; Baeten, L.; De Dobbelaere, C.; D'Haen, J.; Manca, J.; Boyen, H.G. Perovskite-based hybrid solar cells exceeding 10% efficiency with high reproducibility using a thin film sandwich approach. *Adv. Mater.* **2014**, *26*, 2041–2046. [CrossRef] [PubMed]
17. Saliba, M.; Tan, K.W.; Sai, H.; Moore, D.T.; Scott, T.; Zhang, W.; Estroff, L.A.; Wiesner, U.; Snaith, H.J. Influence of thermal processing protocol upon the crystallization and photovoltaic performance of organic-inorganic lead trihalide perovskites. *J. Phys. Chem. C* **2014**, *118*, 17171–17177. [CrossRef]

18. Xiao, M.; Huang, F.; Huang, W.; Dkhissi, Y.; Zhu, Y.; Etheridge, J.; Gray-Weale, A.; Bach, U.; Cheng, Y.B.; Spiccia, L. A fast deposition-crystallization procedure for highly efficient lead iodide perovskite thin-film solar cells. *Angew. Chem.* **2014**, *126*, 10056–10061. [CrossRef]

19. Jeng, J.Y.; Chiang, Y.F.; Lee, M.H.; Peng, S.R.; Guo, T.F.; Chen, P.; Wen, T.C. $CH_3NH_3PbI_3$ perovskite/fullerene planar heterojunction hybrid solar cells. *Adv. Mater.* **2013**, *25*, 3727–3732. [CrossRef] [PubMed]

20. Da, Y.; Xuan, Y.M.; Li, Q. Quantifying energy losses in planar perovskite solar cells. *Sol. Energy Mater. Sol. Cells* **2017**, *174*, 206–213. [CrossRef]

21. Si, H.; Liao, Q.; Zhang, Z.; Li, Y.; Yang, X.; Zhang, G.; Kang, Z.; Zhang, Y. An innovative design of perovskite solar cells with Al_2O_3 inserting at ZnO/perovskite interface for improving the performance and stability. *Nano Energy* **2016**, *22*, 223–231. [CrossRef]

22. Zheng, K.; Žídek, K.; Abdellah, M.; Chen, J.; Chábera, P.; Zhang, W.; Al-Marri, M.J.; Pullerits, T. High excitation intensity opens a new trapping channel in organic-inorganic hybrid perovskite nanoparticles. *ACS Energy Lett.* **2016**, *1*, 1154–1161. [CrossRef]

23. Sun, X.; Ji, L.Y.; Chen, W.W.; Guo, X.; Wang, H.H.; Lei, M.; Wang, Q.; Li, Y.F. Halide anion-fullerene pi noncovalent interactions: N-doping and a halide anion migration mechanism in p-i-n perovskite solar cells. *J. Mater. Chem. A* **2017**, *5*, 20720–20728. [CrossRef]

24. Li, L.; Liu, N.; Xu, Z.; Chen, Q.; Wang, X.; Zhou, H. Precise Composition Tailoring of Mixed-Cation Hybrid Perovskites for Efficient Solar Cells by Mixture Design Methods. *ACS Nano* **2017**, *11*, 8804–8813. [CrossRef] [PubMed]

25. Yang, W.S.; Park, B.W.; Seok, S.I. Iodine management in formamidinium-ldea-halide-based perovskite layers for efficient solar cells. *Science* **2017**, *356*, 1376–1379. [CrossRef] [PubMed]

26. Carrillo, J.; Guerrero, A.; Rahimnejad, S.; Almora, O.; Zarazua, I.; Mas-Marza, E.; Bisquert, J.; Garcia-Belmonte, G. Ionic reactivity at contacts and aging of methylammonium lead triiodode perovskite solar cells. *Adv. Energy Mater.* **2016**, *6*, 1502246. [CrossRef]

27. Tsai, H.; Nie, W.; Cheruku, P.; Mack, N.H.; Xu, P.; Gupta, G.; Mohite, A.D.; Wang, H.L. Optimizing composition and morphology for large-grain perovskite solar cells, via chemical control. *Chem. Mater.* **2015**, *27*, 5570–5576. [CrossRef]

28. Park, I.J.; Seo, S.; Park, M.A.; Lee, S.; Kim, D.H.; Zhu, K.; Shin, H.; Kim, J.Y. Effect of rubidium incorporation on the structural, electrical, and photovoltaic properties of Methylammonium lead iodide-based perovskite solar cells. *ACS Appl. Mater. Interfaces* **2017**, *9*, 41898–41905. [CrossRef] [PubMed]

29. Shlenskaya, N.N.; Belich, N.A.; Grätzel, M.; Goodilin, E.A.; Tarasov, A.B. Light-induced reactivity of gold and hybrid perovskite as a new possible degradation mechanism in perovskite solar cells. *J. Mater. Chem. A* **2018**, *6*, 1780–1786. [CrossRef]

30. Cao, J.; Jing, X.; Yan, J.; Hu, C.; Chen, R.; Yin, J.; Li, J.; Zheng, N. Identifying the molecular structures of intermediates for optimizing the fabrication of high-quality perovskite films. *J. Am. Chem. Soc.* **2016**, *138*, 9919–9926. [CrossRef] [PubMed]

crystals

MDPI

Article

Enhancing Optically Pumped Organic-Inorganic Hybrid Perovskite Amplified Spontaneous Emission via Compound Surface Plasmon Resonance

Xiaoyan Wu [1], Yanglong Li [1], Wei Li [1], Lingyuan Wu [1], Bo Fu [1], Weiping Wang [1], Guodong Liu [1,*], Dayong Zhang [1], Jianheng Zhao [1] and Ping Chen [2,*]

[1]   Institute of Fluid Physics, China Academy of Engineering Physics, Mianyang 621900, China; wuxiaoyan1219@caep.cn (X.W.); jesseliyl@163.com (Y.L.); vleefoxtrot@outlook.com (W.L.); wubaly@163.com (L.W.); fubo_phoenix@163.com (B.F.); wwpwzc@yeah.net (W.W.); zdyzdy113@163.com (D.Z.); jianh_zhao@caep.cn (J.Z.)
[2]   School of Physical Science and Technology, MOE Key Laboratory on Luminescence and Real-Time Analysis, Southwest University, Chongqing 400715, China
*   Correspondence: guodliu@126.com (G.L.); chenping206@126.com (P.C.);
    Tel.: +86-0816-2487-145 (G.L.); +86-188-8378-1341 (P.C.)

Received: 31 January 2018; Accepted: 1 March 2018; Published: 7 March 2018

Abstract: Organic-inorganic hybrid perovskite has attracted intensive attention from researchers as the gain medium in lasing devices. However, achieving electrically driven lasing remains a significant challenge. Modifying the devices' structure to enhance the optically pumped amplified spontaneous emission (ASE) is the key issue. In this work, gold nanoparticles (Au NPs) are first doped into PEDOT: PSS buffer layer in a slab waveguide device structure: Quartz/PEDOT: PSS (with or w/o Au NPs)/$CH_3NH_3PbBr_3$. As a result, the facile device shows a significantly enhanced ASE intensity and a narrowed full width at half maximum. Based on experiments and theoretical simulation data, the improvement is mainly a result of the compound surface plasmon resonance, including simultaneous near- and far-field effects, both of which could increase the density of excitons excited state and accelerate the radiative decay process. This method is highly significant for the design and development and fabrication of high-performance organic-inorganic hybrid perovskite lasing diodes.

Keywords: perovskite lasing diodes; amplified spontaneous emission; gold nanoparticles; compound surface plasmon resonance

1. Introduction

Organic-inorganic hybrid perovskites have recently attracted intensive attention for solution-processed optoelectronic devices such as photovoltaics, light-emitting diodes, lasing, and photodetectors, because of their easily tunable optical bandgap, as well as their attractive absorption, emission, and charge transport properties [1–6]. Among these applications, the use of perovskites as gain medium in lasing devices has attracted a lot of attention by an increasing number of scientists [7–9]. In the last few years, although a number of optically pumped driving perovskite-lasing devices have been confirmed, they have not yet been able to achieve electrical driving stimulated emission [10,11]. Among the many methods to achieve electrical pumping lasing, enhancing amplified spontaneous emission (ASE) is a key factor [1]. Motivated by this idea, researchers not only tried material modification, but also searched for ways to optimize the device structure [12–19]. For material modification, Sum et al. demonstrated ultra-stable ASE under remarkably low thresholds with a large visible spectrum tenability (390–790 nm) with halide perovskites [12]. Zhu et al. reported single-crystal halide perovskite nanowires with low lasing thresholds (220 nJ·cm^{-2}) and high quality factors

(Q ~3600) [13]. For device structure optimization, Friend et al. reported flexible solution-processed crystalline films with prominent high-photoluminescence (PL) quantum efficiency (up to 70%). They found that the free charge carrier formation in $CH_3NH_3PbI_{3-x}Cl_x$ perovskite was within 1 ps, and the bimolecular recombination time scales of these free charge carriers were within 10 s to 100 s of ns [14]. Song et al. transferred $CH_3NH_3PbBr_3$ (MAPbBr$_3$) perovskite micro-rod onto a few-layered graphene slice; it was found that the total output intensity was significantly enhanced more than four-fold, and the threshold was reduced by around 20% [15]. From the above descriptions, device structure optimization is a highly feasible and effective route for constructing high-performance lasing devices.

Surface plasmon resonance by metal nanoparticles (NPs) has been confirmed to result in many unique optical characteristics, such as surface plasmon resonance sensing and detection, metal-enhanced fluorescence (MEF), focusing and concentrating light, surface-enhanced Raman scattering (SERS), plasmon metamaterials, and so on [20–28]. In the last few years, metal NPs have been successfully utilized in perovskite solar cells and light-emitting diodes (PeLEDs) [29–33]. However, few studies have reported on perovskites lasing devices. In theory, gold nanoparticles (Au NPs) could enhance the density of excitons excited state and accelerate the radiative decay process, which is the primary process in ASE.

In this work, we doped Au NPs into a PEDOT: PSS buffer layer in a slab waveguide device (Quartz/PEDOT: PSS (with or w/o Au NPs)/MAPbBr$_3$) and obtained ~16-fold improvement in the gain medium MAPbBr$_3$. Based on experiments and theoretical simulation data, the device's performance improvement can be attributed to compound surface plasmon resonance, including simultaneous near- and far-field effects, both of which could increase the density of the excitons' excited state and accelerate the radiative decay process.

2. Materials and Methods

Au NPs synthesis methods: 20 nm diameter Au NPs were synthesized in accordance with our previous reports [34]. In a 250 mL flask, 100 mL sample aqueous HAuCl$_4$ (0.25 mM, Sinopharm Chemical Reagent Co., Ltd., Shanghai, China) was prepared. The solution was vigorously stirred until boiling, and 1 mL 5% $Na_3C_6H_5O_7$ (Enox) was added until the solution reached a wine red color, indicating that the desired size Au NPs had been synthesized.

Perovskite synthesis methods: The MAPbBr$_3$ solution was prepared by dissolving MABr and PbBr$_2$ (MABr > 99.9%, PbBr$_2$ > 99.9%, 1.5:1 molar ratio, Xi'an Polymer Light Technology Corp., Xi'an, China) in DMF solution (reported in our previous research work [33,35]). Then chlorobenzene solution was dropped into the DMF solution until saturation, accelerating perovskite material crystallization. In the glove-box, precursor solution was spin-coated onto PEDOT: PSS layer (4000 rpm, 60 s) and annealed (80 °C, 10 min).

Characterization of devices and thin films: A fluorospectrophotometer (Hitachi F-2500) tested the PL spectrum. UV-vis spectrum were recorded by SHIMADZU UV-2600. Transmission electron microscopy (TEM) was performed using a JEM-100 CXII. A fluorescence lifetime spectrometer (Horiba Jobin Yvon FL-TCSPC) studied the time-resolved PL spectrum. Finite-different time-domain (FDTD) solutions simulated the near-field around Au NPs. A Spitfire Ace (Spectrum Physics) was selected as the optical pump source; its repetition rate and pulse width were 1 kHz and 100 fs, respectively. A Dektak 150 Surface Profiler measured the film thickness. The crystalline characterizations were determined by X-ray diffraction (XRD) patterns characterized by Shimadzu XRD-7000. The fluorescence was collected by an Pixis 100B CCD from the edge of the sample.

Metal NPs classic model of far-field effect: The enhancement originated from the retardation effect (between original light and reflected light) in the luminescence of molecules (considered as a classical linear harmonic oscillator) in front of a reflecting boundary. This indicates that the far-field effect comes from the one light-wave coupling. This phenomenon is explained by the Hertz classical equation for considering the radiation field of the dipole [36–38]. When the mirror is parallel to the dipole axis, the enhancement $Z_{theory\text{-}ASE}$ is given by

$$Z_{thoery-ASE} = \frac{q_d}{q} = \left\{ q + (1-q) \cdot \left(1 + \tfrac{3}{2}qR\left(-\frac{\sin(\gamma-\delta)}{\gamma^3} + \frac{\sin(\gamma-\delta)}{\gamma} + \frac{\cos(\gamma-\delta)}{\gamma^2} \right) \right)^{-1} \right\}^{-1} \tag{1}$$

from Equation (1) we can see that the decay time (τ) and the corresponding quantum efficiency (q) markedly depend on the phase shift (δ), the reflectivity of metal NPs (R), the distance between oscillator and the metal mirror (d), and the emitting light wavelength of the materials (λ). γ is defined as:

$$\gamma = \frac{4\pi n d}{\lambda} \tag{2}$$

where n is luminescent material refractive index.

3. Results and Discussion

In order to investigate the ASE performance, the slab waveguide device structure with Quartz/PEDOT: PSS (with or w/o Au NPs)/MAPbBr$_3$ (Scheme 1) was made. The experimental optical path for the ASE test is also shown in Scheme 1. Light scattering data for Au NPs size distribution showed a mean diameter of 20 nm in Figure 1a, which is well matched to the TEM images (Figure 1b). The XRD pattern shown in Figure 2a showed MAPbBr$_3$ crystal film diffraction peaks at 14.90° (100), 21.08° (110), 29.98° (200), 33.62° (210), 36.92° (211), 42.94° (220) and 45.68° (300). The XRD data indicated pure perovskite phase, which suggested a lamellar structure PbBr$_2$ 2D-layer parallel to the substrate [39]. In Figure 2a, the inset images display the MAPbBr$_3$ film photographs (spin-coated on quartz/PEDOT: PSS) before and after ~365 nm UV lamp irradiation, respectively, and show a smooth surfaces. Top-view and cross-section SEM images of MAPbBr$_3$ can be seen in Figure 2b,c, respectively. The MAPbBr$_3$ film coverage shows an average 300–500 nm grain size with highly uniform and dense stacking. Such high-quality film is critical for generating ASE phenomena [33].

Figure 3 shows the normalized absorption and PL spectrum of MAPbBr$_3$ film and extinction spectrum of the Au NPs solution. The absorption spectrum of MAPbBr$_3$ shows 350 nm to 650 nm, which matches the 520 nm extinction peak of the Au NPs. The MAPbBr$_3$ PL spectrum has its central wavelength peak at around 534 nm, with a full width at half maximum (FWHM) at ~24 nm, which also overlaps with the Au NP extinction peak. The overlap of the spectra indicates that surface plasmon resonance would induce an effective absorption or emission enhancement [40,41].

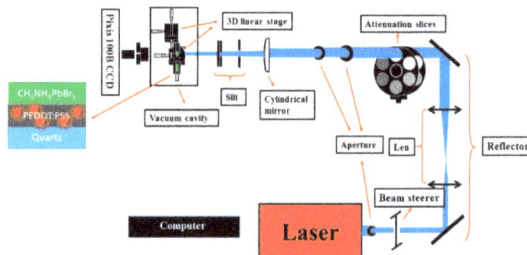

Scheme 1. The ASE test experimental optical path and the device structure.

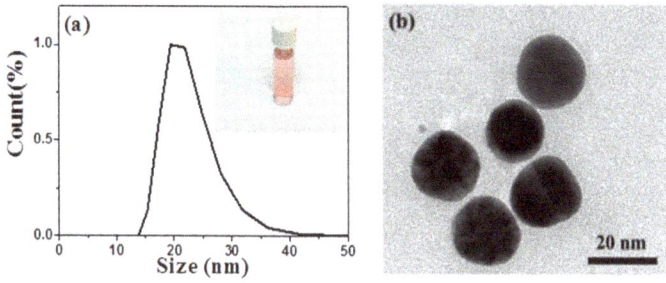

Figure 1. (**a**) Light scattering data for Au NPs size distribution, inset: Au NPs solution. (**b**) TEM image of Au NPs.

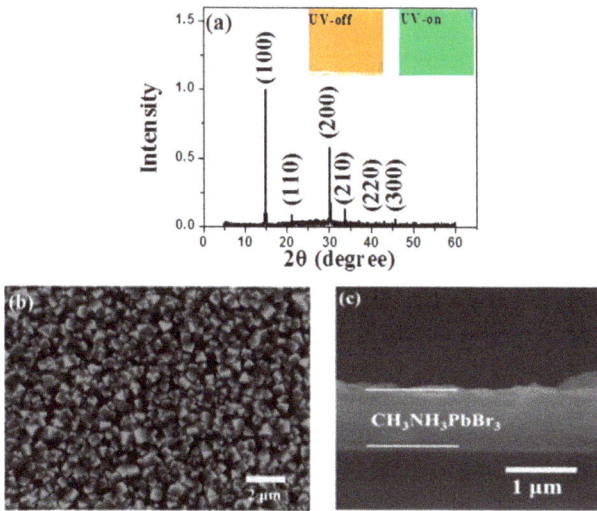

Figure 2. (**a**) XRD spectrum of MAPbBr$_3$ crystal film, inset: image of MAPbBr$_3$ film when UV-365 nm off and on, (**b,c**) Top view and cross-section SEM images of MAPbBr$_3$ film.

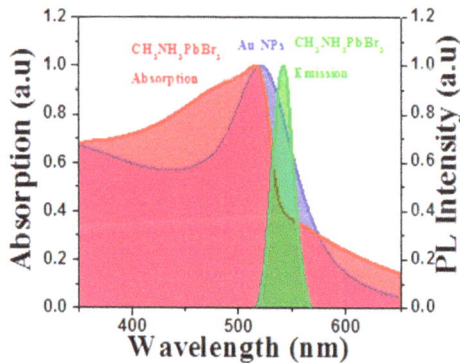

Figure 3. Absorption spectra of MAPbBr$_3$ film and Au NPs aqueous solution, PL spectrum of MAPbBr$_3$ film.

The ASE emission spectra are taken as a function of optical pump energy in the device structure Quartz/PEDOT: PSS (with Au NPs)/MAPbBr$_3$, where Quartz/PEDOT: PSS (w/o Au NPs)/MAPbBr$_3$ acts as control device. Figure 4a shows the typical ASE behavior of the control device [12], showing a significant dependence of the edge emission spectrum with increased pump energy intensity. When the device is optically pumped under low energy, it exhibits a broad ASE emission spectrum, with a FWHM of ~30 nm in the control device. Once the excitation energy is high enough, the ASE emission spectrum becomes much narrower, with FWHM ~7 nm. The emission intensity and FWHM with and w/o Au NP doping are compared, as shown in Figure 4a–d. Figure 4c,d shows the ASE intensity and FWHM, which are taken as a function of the optical pump energy curve. The ASE intensity of the device with Au NPs shows a 16-fold enhancement compared to the control one (Figure 4c). The FWHM narrows from 7.0 to 6.4 nm, which also shows an obvious narrowing after doping with Au NPs (Figure 4d).

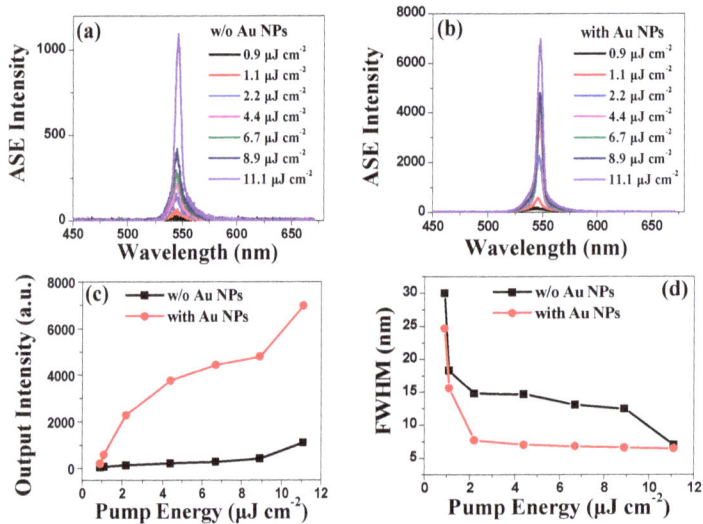

Figure 4. The emission spectrum Device Quartz/PEDOT: PSS/MAPbBr$_3$ (**a**) w/o Au NPs, (**b**) with Au NPs. The ASE peak intensity and FWHM of the emission spectra as a function of the pump energy (**c**) w/o Au NPs, (**d**) with Au NPs.

Figure 5a,b shows the absorbance and time-resolved PL spectra, in order to demonstrate the mechanism of ASE intensity enhancement by Au NPs. On one hand, due to the improved absorbance probability, the absorbance intensity of MAPbBr$_3$ improved by 38% after Au NP doping (Figure 5a) [42]. Thus, the density of the excitons' excited state is increased. On the other hand, the MAPbBr$_3$ time-resolved PL spectrum in Figure 5b shows that the exciton lifetime of 8.23 ns in the control devices surprisingly decreases to 0.28 ns with Au NPs, which means that the radiative decay process is also accelerated.

The enhancement mechanism could be contributed to the Au NP compound's near- and far-field effects. The near- and far-field distributions are simulated as shown in Figure 6a,b. For near-field, Figure 6a shows that the working distance is within the range of 1–10 nm, as simulated by FDTD around the neighborhood of Au NPs. The intensity degraded with distance very quickly, and nearly vanished beyond 10 nm. The thickness of PEDOT: PSS is the key factor for determining the improvements of ASE intensity. The thin PEDOT: PSS thickness would induce Au NP contact with MAPbBr$_3$, which could directly quench the fluorescence by Förster energy transfer [43,44]. The thick PEDOT: PSS would hinder the near-field effect. In our research work, the thickness of PEDOT: PSS is optimized at 23 nm, which would make full use of near-field. For far-field, the detailed mechanism was described in the Experimental Section. Based on the simulated far-field classic model, Figure 6b displays the main improved multi-enhancement peaks. The main improved peaks are 95 nm, 245 nm and 385 nm, which are within the scope of the MAPbBr$_3$ thickness of 500 nm. Considering the device structure and the thickness of MAPbBr$_3$ and PEDOT: PSS, the near- and far-field could both overlap with the gain medium, as shown in Figure 6c. There are two advantages to using compound near- and far-field surface plasmon resonance, in comparison to a single optical effect: (1) the enhancement ratio of ASE intensity using compound surface plasmon resonance would be greater than with a single-optical effect; and (2) considering the device structure (shown in Scheme 1), it's easy to utilize compound surface plasmon resonance in lasing devices based on perovskites.

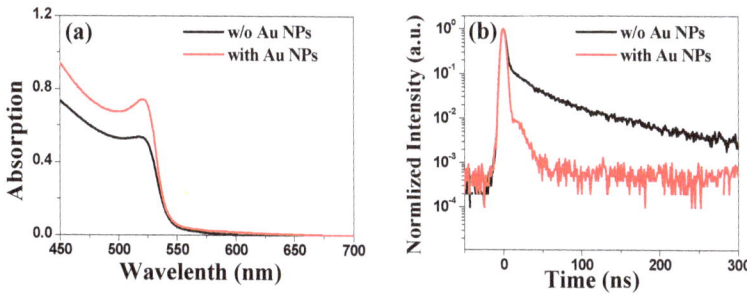

Figure 5. (a) UV-vis spectra, (b) time-resolved PL spectra in the devices' structure: Quartz/PEDOT: PSS/MAPbBr$_3$ (with and w/o Au NPs).

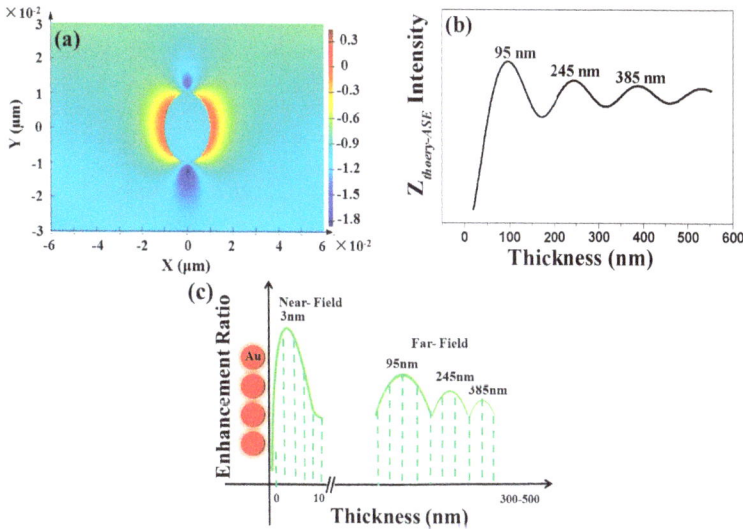

Figure 6. Theoretical simulation of (a) near-electric field profile, (b) far-field, (c) distribution in devices.

4. Conclusions

In summary, we demonstrated an obvious performance improvement of lasing devices by doping Au NPs into the optically pumped slab waveguide device. The ASE intensity and FWHM were both effectively promoted. The results indicate that the compound near- and far-field surface plasmon resonance is the mechanism of enhancement, both of which could increase the density of the excitons' excited state and accelerate the radiative decay process. Our results provide a flexible and effective route for obtaining high-performance lasing devices based on perovskites.

Acknowledgments: This work was supported by the National Natural Science Foundation of China (Grant No. 51473052, 11504300, 11602243), NSFA of China (Grant No. U1630125), Natural Science Foundation Project of CQ CSTC (Grant No. cstc2015jcyjA50002).

Author Contributions: Xiaoyan Wu designed the experiments; Xiaoyan Wu, Yanglong Li, Wei Li, Lingyuan Wu, Bo Fu performed the experiments; Xiaoyan Wu, Weiping Wang, Dayong Zhang and Ping Chen analyzed the data; Guodong Liu and Jianheng Zhao supervised the project.

Conflicts of Interest: The authors declare no conflict of interest.

References

1. Veldhuis, S.A.; Boix, P.P.; Yantara, N.; Li, M.J.; Sum, T.C.; Mathews, N.; Mhaisalkar, S.G. Perovskite Materials for Light-Emitting Diodes and Lasers. *Adv. Mater.* **2016**, *28*, 6804–6834. [CrossRef] [PubMed]
2. Schmidt, L.C.; Pertegás, A.; González-Carrero, S.; Malinkiewicz, O.; Agouram, S.; Mínguez Espallargas, G.; Bolink, H.J.; Galian, R.E.; Pérez-Prieto, J. Nontemplate. Synthesis of $CH_3NH_3PbBr_3$ Perovskite Nanoparticles. *J. Am. Chem. Soc.* **2014**, *136*, 850–852. [CrossRef] [PubMed]
3. Tan, Z.K.; Moghaddam, R.S.; Lai, M.L.; Docampo, P.; Higler, R.; Deschler, F.; Price, M.; Sadhanala, A.; Pazos, L.M.; Credgington, D.; et al. Bright Light-Emitting Diodes Based on Organometal Halide Perovskite. *Nat. Nanotechnol.* **2014**, *9*, 687–692. [CrossRef] [PubMed]
4. Yadav, P.; Prochowicz, D.; Saliba, M.; Boix, P.P.; Zakeeruddin, S.M.; Grätzel, M. Interfacial Kinetics of Efficient Perovskite Solar Cells. *Crystals* **2017**, *7*, 252. [CrossRef]
5. Kim, Y.; Cho, H.; Heo, J.H.; Kim, T.; Myoung, N.; Lee, C.; Im, S.H.; Lee, T. Multicolored Organic/Inorganic Hybrid Perovskite Light-Emitting Diodes. *Adv. Mater.* **2015**, *27*, 1248–1254. [CrossRef] [PubMed]
6. Feng, J.G.; Yan, X.X.; Zhang, Y.F.; Wang, X.D.; Wu, Y.C.; Su, B.; Fu, H.B.; Jiang, L. "Liquid Knife" to Fabricate Patterning Single-Crystalline Perovskite Microplates toward High-Performance Laser Arrays. *Adv. Mater.* **2016**, *28*, 3732–3741. [CrossRef] [PubMed]
7. Kuehne, A.J.; Gather, M.C. Organic Lasers: Recent Developments on Materials, Device Geometries, and Fabrication Techniques. *Chem. Rev.* **2016**, *116*, 12823–12864. [CrossRef] [PubMed]
8. Yang, D.C.; Xie, C.; Sun, J.H.; Zhu, H.; Xu, X.H.; You, P.; Lau, S.P.; Yan, F.; Yu, S.F. Amplified Spontaneous Emission from Organic-Inorganic Hybrid Lead Iodide Perovskite Single Crystals under Direct Multiphoton Excitation. *Adv. Opt. Mater.* **2016**, *4*, 1053–1059. [CrossRef]
9. Yu, Z.Y.; Wu, Y.S.; Liao, Q.; Zhang, H.H.; Bai, S.M.; Li, H.; Xu, Z.Z.; Sun, C.L.; Wang, X.D.; Yao, J.N.; Fu, H.B. Self-Assembled Microdisk Lasers of Perylenediimides. *J. Am. Chem. Soc.* **2015**, *137*, 15105–15111. [CrossRef] [PubMed]
10. Stehr, J.; Crewett, J.; Schindler, F.; Sperling, R.; Holleitner, A.W. A Low Threshold Polymer Laser Based on Metallic Nanoparticles Gratings. *Adv. Mater.* **2003**, *15*, 1726–1729. [CrossRef]
11. Tessler, N.; Harrison, N.T.; Friend, R.H. High Peak Brightness Polymer Light-Emitting Diodes. *Adv. Mater.* **1998**, *10*, 64–68. [CrossRef]
12. Xing, G.C.; Mathews, N.; Lim, S.S.; Natalia, Y.; Liu, X.F.; Sabba, D.; Grätzel, M.; Mhaisalkar, S.; Sum, T.C. Low-temperature solution-processed wavelength-tunable perovskites for lasing. *Nat. Mater.* **2014**, *13*, 476–480. [CrossRef] [PubMed]
13. Zhu, H.M.; Fu, Y.P.; Meng, F.; Wu, X.X.; Gong, Z.Z.; Ding, Q.; Gustaffson, M.V.; Trinh, T.M.; Jin, S.; Zhu, X.Y. Lead halide perovskite nanowire lasers with low lasing thresholds and high quality factors. *Nat. Mater.* **2015**, *14*, 636–642. [CrossRef] [PubMed]
14. Deschler, F.; Price, M.; Pathak, S.; Klintberg, L.E.; Jarausch, D.D.; Higler, R.; Hüttner, S.; Leijtens, T.; Stranks, S.D.; Snaith, H.J.; et al. High Photoluminescence Efficiency and Optically Pumped Lasing in Solution-Processed Mixed Halide Perovskite Semiconductors. *J. Phys. Chem. Lett.* **2014**, *5*, 1421–1426. [CrossRef] [PubMed]
15. Zhang, C.; Wang, K.Y.; Yi, N.B.; Gao, Y.S.; Zhu, M.X.; Sun, W.Z.; Liu, S.; Xu, K.; Xiao, S.M.; Song, Q.H. Improving the Performance of a $CH_3NH_3PbBr_3$ Perovskite Microrod Laser through Hybridization with Few-Layered Graphene. *Adv. Opt. Mater.* **2016**, *4*, 2057–2062. [CrossRef]

16. Wang, Y.; Li, X.M.; Zhao, X.; Xiao, L.; Zeng, H.B.; Sun, H.D. Nonlinear Absorption and Low-Threshold Multiphoton Pumped Stimulated Emission from All-Inorganic Perovskite Nanocrystals. *Nano. Lett.* **2016**, *16*, 448–453. [CrossRef] [PubMed]

17. Li, X.L.; Guo, Y.; Luo, B.B. Improved Stability and Photoluminescence Yield of Mn^{2+}-Doped $CH_3NH_3PbCl_3$ Perovskite Nanocrystals. *Crystals* **2018**, *8*, 4. [CrossRef]

18. Li, Y.H.; Zhang, T.Y.; Xu, F.; Wang, Y.; Li, G.; Yang, Y.; Zhao, Y.X. CH_3NH_3Cl Assisted Solvent Engineering for Highly Crystallized and Large Grain Size Mixed-Composition $(FAPbI_3)_{0.85}(MAPbBr_3)_{0.15}$ Perovskites. *Crystals* **2017**, *7*, 272. [CrossRef]

19. Saliba, M.; Wood, S.M.; Patel, J.B.; Nayak, P.K.; Huang, J.; Alexander-Webber, J.A.; Wenger, B.; Stranks, S.D.; Hörantner, M.T.; Wang, J.T.W.; et al. Structured Organic-Inorganic Perovskite toward a Distributed Feedback Laser. *Adv. Mater.* **2016**, *28*, 923–929. [CrossRef] [PubMed]

20. Liu, Y.L.; Fang, C.Y.; Yu, C.C.; Yang, T.C.; Chen, H.L. Controllable Localized Surface Plasmonic Resonance Phenomena in Reduced Gold Oxide Films. *Chem. Mater.* **2014**, *26*, 1799–1806. [CrossRef]

21. Li, Y.F.; Feng, J.; Dong, F.X.; Ding, R.; Zhang, Z.Y.; Zhang, X.L.; Chen, Y.; Bi, Y.G.; Sun, H.B. Surface plasmon-enhanced amplified spontaneous emission from organic single crystals by integrating graphene/copper nanoparticle hybrid nanostructures. *Nanoscale* **2017**, *9*, 19353–19359. [CrossRef] [PubMed]

22. Li, Y.L.; Ye, Y.; Fan, Y.D.; Zhou, J.; Jia, L.; Tang, B.; Wang, X.G. Silver Nanoprism-Loaded Eggshell Membrane: A Facile Platform for In Situ SERS Monitoring of Catalytic Reactions. *Crystals* **2017**, *7*, 45. [CrossRef]

23. Su, Y.H.; Ke, Y.F.; Cai, S.L.; Yao, Q.Y. Surface plasmon resonance of layer-by-layer gold nanoparticles induced photoelectric current in environmentally-friendly plasmon-sensitized solar cell. *Light Sci. Appl.* **2012**, *1*, e14. [CrossRef]

24. Gu, M.; Li, X.P.; Cao, Y.Y. Optical storage arrays: A perspective for future big data storage. *Light Sci. Appl.* **2014**, *3*, e177. [CrossRef]

25. Wu, X.Y.; Liu, L.L.; Choy, W.C.H.; Yu, T.C.; Cai, P.; Gu, Y.J.; Xie, Z.Q.; Zhang, Y.N.; Du, L.Y.; Mo, Y.Q.; et al. Substantial performance improvement in inverted polymer light-emitting diodes via surface plasmon resonance induced electrode quenching control. *ACS Appl. Mater. Interfaces* **2014**, *6*, 11001–11006. [CrossRef] [PubMed]

26. Wu, X.Y.; Liu, L.L.; Yu, T.C.; Yu, L.; Xie, Z.Q.; Mo, Y.Q.; Xu, S.P.; Ma, Y.G. Gold nanoparticles modified ITO anode for enhanced PLEDs brightness and efficiency. *J. Mater. Chem. C* **2013**, *1*, 7020–7025. [CrossRef]

27. Wu, X.Y.; Liu, L.L.; Deng, Z.C.; Nian, L.; Zhang, W.Z.; Hu, D.H.; Xie, Z.Q.; Mo, Y.Q.; Ma, Y.G. Efficiency improvement in polymer light-emitting diodes by "far-field" effect of gold nanoparticles. *Part. Part. Syst. Charact.* **2015**, *32*, 686–692. [CrossRef]

28. Wu, X.Y.; Zhuang, Y.Q.; Feng, Z.T.; Zhou, X.H.; Yang, Y.Z.; Liu, L.L.; Xie, Z.Q.; Chen, X.D.; Ma, Y.G. Simultaneous red-green-blue electroluminescent enhancement directed by surface plasmonic "far-field" of facile gold nanospheres. *Nano Res.* **2018**, *11*, 151–162. [CrossRef]

29. Wang, J.Y.; Hsu, F.C.; Huang, J.Y.; Wang, L.; Chen, Y.F. Bifunctional Polymer Nanocomposites as Hole-Transport Layers for Efficient Light Harvesting: Application to Perovskite Solar Cells. *ACS Appl. Mater. Interfaces* **2015**, *7*, 27676–27684. [CrossRef] [PubMed]

30. Lee, D.S.; Kim, W.; Cha, B.G.; Kwon, J.; Kim, S.J.; Kim, M.; Kim, J.; Wang, D.H.; Park, J.H. Self-Position of Au NPs in Perovskite Solar Cells: Optical and Electrical Contribution. *ACS Appl. Mater. Interfaces* **2016**, *8*, 449–454. [CrossRef] [PubMed]

31. Balakrishnan, S.K.; Kamat, P.V. $Au-CsPbBr_3$ Hybrid Architecture: Anchoring Gold Nanoparticles on Cubic Perovskite Nanocrystals. *ACS Energy Lett.* **2017**, *2*, 88–93. [CrossRef]

32. Zhang, X.; Xu, B.; Wang, W.; Liu, S.; Zheng, Y.; Chen, S.; Wang, K.; Sun, X.W. Plasmonic Perovskite Light-Emitting Diodes Based on the $Ag-CsPbBr_3$ System. *ACS Appl. Mater. Interfaces* **2017**, *9*, 4926–4931. [CrossRef] [PubMed]

33. Chen, P.; Xiong, Z.Y.; Wu, X.Y.; Shao, M.; Meng, Y.; Xiong, Z.H.; Gao, C.H. Nearly 100% Efficiency Enhancement of $CH_3NH_3PbBr_3$ Perovskite Light-Emitting Diodes by Utilizing Plasmonic Au Nanoparticles. *J. Phys. Chem. Lett.* **2017**, *8*, 3961–3969. [CrossRef] [PubMed]

34. Frens, G. Controlled Nucleation for the Regulation of the Particle Size in Monodisperse Gold Suspensions. *Nat. Phys. Sci.* **1973**, *241*, 20–22. [CrossRef]

35. Wu, X.Y.; Li, Y.L.; Wu, L.Y.; Fu, B.; Liu, G.D.; Zhang, D.Y.; Zhao, J.H.; Chen, P.; Liu, L.L. Enhancing perovskite film fluorescence by simultaneous near-and far-field effects of gold nanoparticles. *RSC Adv.* **2017**, *7*, 35752–35756. [CrossRef]

36. Kummerlen, J.; Leitner, A.; Brunner, H.; Aussenegg, F.R.; Wokaun, A. Enhanced dye fluorescence over silver island films: Analysis of the distance dependence. *Mol. Phys.* **1993**, *80*, 1031–1046. [CrossRef]

37. Kuhn, H. Classical Aspects of Energy Transfer in Molecular Systems. *J. Chem. Phys.* **1970**, *53*, 101–108. [CrossRef]

38. Drexhage, K.H. Influence of a dielectric interface on fluorescence decay time. *J. Lumin.* **1970**, *1*, 693–701. [CrossRef]

39. Yang, B.; Mao, X.; Yang, S.Q.; Li, Y.J.; Wang, Y.Q.; Wang, M.S.; Deng, W.Q.; Han, K.L. Low Threshold Two-Photon-Pumped Amplified Spontaneous Emission in $CH_3NH_3PbBr_3$ Microdisks. *ACS Appl. Mater. Interfaces* **2016**, *8*, 19587–19592. [CrossRef] [PubMed]

40. Wu, X.Y.; Liu, L.L.; Xie, Z.Q.; Ma, Y.G. Advance in Metal-based Nanoparticles for the Enhanced Performance of Organic Optoelectronics Devices. *Chem. J. Chin. Univ.* **2016**, *37*, 409–425.

41. Choi, H.; Ko, S.J.; Choi, Y.; Joo, P.; Kim, T.; Lee, B.R.; Jung, J.W.; Choi, H.J.; Cha, M.; Jeong, J.R.; et al. Versatile surface plasmon resonance of carbon-dot-supported silver nanoparticles in polymer optoelectronic devices. *Nat. Photonics* **2013**, *7*, 732–738. [CrossRef]

42. Stratakis, E.; Kymakis, E. Nanoparticle-based plasmonic organic photovoltaic devices. *Mater. Today* **2013**, *16*, 133–146. [CrossRef]

43. Xiao, Y.; Yang, J.P.; Cheng, P.P.; Zhu, J.J.; Xu, Z.Q.; Deng, Y.H.; Lee, S.T.; Li, Y.Q.; Tang, J.X. Surface plasmon-enhanced electroluminescence in organic light-emitting diodes incorporating Au nanoparticles. *Appl. Phys. Lett.* **2012**, *100*, 013308–013311. [CrossRef]

44. Kim, T.; Kang, H.; Jeong, S.; Kang, D.J.; Lee, C.; Lee, C.H.; Seo, M.K.; Lee, J.Y.; Kim, B. Au@Polymer Core-Shell Nanoparticles for Simultaneously Enhancing Efficiency and Ambient Stability of Organic Optoelectronic Devices. *ACS Appl. Mater. Interfaces* **2014**, *6*, 16956–16965. [CrossRef] [PubMed]

crystals

MDPI

Article

A Feasible and Effective Post-Treatment Method for High-Quality CH$_3$NH$_3$PbI$_3$ Films and High-Efficiency Perovskite Solar Cells

Yaxiao Jiang, Limin Tu, Haitao Li, Shaohua Li, Shi-E Yang and Yongsheng Chen *

Key Lab of Material Physics, Department of Physics and Engineering, Zhengzhou University, Zhengzhou 450052, China; JIANGYAX1992@163.com (Y.J.); tulimin12345@163.com (L.T.); 18239940667@163.com (H.L.); 17638164964@163.com (S.L.); yangshie@zzu.edu.cn (S.-E.Y.)
* Correspondence: chysh2003@zzu.edu.cn

Received: 14 December 2017; Accepted: 13 January 2018; Published: 18 January 2018

Abstract: The morphology control of CH$_3$NH$_3$PbI$_3$ (MAPbI$_3$) thin-film is crucial for the high-efficiency perovskite solar cells, especially for their planar structure devices. Here, a feasible and effective post-treatment method is presented to improve the quality of MAPbI$_3$ films by using methylamine (CH$_3$NH$_2$) vapor. This post-treatment process is studied thoroughly, and the perovskite films with smooth surface, high preferential growth orientation and large crystals are obtained after 10 s treatment in MA atmosphere. It enhances the light absorption, and increases the recombination lifetime. Ultimately, the power conversion efficiency (PCE) of 15.3% for the FTO/TiO$_2$/MAPbI$_3$/spiro-OMeTAD/Ag planar architecture solar cells is achieved in combination with this post-treatment method. It represents a 40% improvement in PCE compared to the best control cell. Moreover, the whole post-treatment process is simple and cheap, which only requires some CH$_3$NH$_2$ solution in absolute ethanol. It is beneficial to control the reaction rate by changing the volume of the solution. Therefore, we are convinced that the post-treatment method is a valid and essential approach for the fabrication of high-efficiency perovskite solar cells.

Keywords: perovskite; solar cells; methylamine vapor; post-treatment; low-cost fabrication

1. Introduction

In recent years, hybrid organic-inorganic perovskite solar cells (PSCs) have developed rapidly. The power conversion efficiencies (PCEs) of these PSCs span from 3.8% [1] to a verified PCE of 22.7% [2], stems from a number of remarkable properties, including high absorption coefficient (~10^4 cm^{-1}) [3,4], low exciton binding energy (<100 MeV) [5–7], long charge carrier diffusion length (10^2–10^5 nm) [8–11] and high charge carrier mobility (10–10^2 cm^2 V^{-1} s^{-1}) [12–15]. In addition, organometallic halide perovskite is suitable for solution and low temperature fabrication, because the precursor material in several organic solvents (dimethyl sulfoxide (DMSO), N,N-Methylformamide (DMF), γ-butyrolactone (γ-GBL), etc.) has a perfect solubility [16–19]. Multifarious methods are successfully used to improve the efficiencies of PSCs, including interface engineering [20,21], compositional engineering [22], solvent engineering [23], new materials [24] and device structure modification [25], etc. Among them, which one is more effective is indistinguishable.

To date, the PCE of PSCs has been enhanced by minimizing charge recombination in bulk perovskite film and/or at the interfaces between perovskite and charge transport layers. The reduction of charge recombination in bulk film requires high quality perovskite film with full surface coverage [26–28], large crystal size [15,29,30] and low defect density [31–33]. The post-treatment using a strong Lewis base is a very effective method to improve the quality of lead halide perovskite films [34–42]. In 2014, Zhao et al. [34] first found a room-temperature phase transformation of CH$_3$NH$_3$PbI$_3$ induced by ammonia. After one year,

Zhou et al. [35] reported CH$_3$NH$_2$ (MA) induced defect-healing (MIDH) of MAPbI$_3$ perovskite thin films based on the formation of MAPbI$_3$·xMA intermediate, resulting in an ultra-smooth and dense film with a higher degree of crystallinity and texture. Chih et al. [36] also certified that the post-treatment by MA gas promotes the regrowth of MAPbBr$_3$ crystallites following the preferred orientations and significantly enhances the film quality and photoluminescence properties. Zhao et al. [38] thoroughly studied the post-treatment process, and provided design rules for the broad, rational extension of this process to new systems and scales. Jiang et al. [39] introduced MA gas to eliminate grain boundaries and gaps, and Jacobs et al. [40] employed a thermally induced recrystallization of MAPbI$_3$ during the liquidation process to enhance grain size. However, MA gas is supplied mainly through a solid reaction between MACl and KOH powders at room temperature, which is more complex and expensive. In this paper, commercially cheap methylamine solution in absolute ethanol is used directly to supply MA vapor, and the quality of MAPbI$_3$ perovskite films can be optimized feasibly by changing the post-treatment times in MA atmosphere. Only after 10 s treatment can the perovskite films with surface smoothness, compact structure and preferential orientation in (110) planes be obtained, which obviously improves the photoluminescence properties and carrier extraction efficiency of perovskite. Introducing this post-treatment method in the PSCs fabrication process, the PCE of FTO/TiO$_2$/MAPbI$_3$/spiro-OMeTAD/Ag planar heterojunction device is increased by 40% compared to the control devices without treatment.

2. Results and Discussion

Figure 1a presents the X-ray diffraction (XRD) spectrum of MAPbI$_3$ films with different treatment times in MA gas. The XRD patterns of as-deposited MAPbI$_3$ film without MA gas treating (0 s) exhibits a main peak at 14.2, which is the characteristic of (110) plane in the tetragonal crystal structure, with other peaks corresponding to the (112), (211), (202), (220), (310), (312), (224), and (314) planes in the plot. The positions of these peaks are accorded with the results published previously [29]. The XRD intensity of (110) peak increases seven-fold with 5 s treating and nine-fold with 15 s treating approximately compared to that without MA treating, then decreases, further extending the MA treatment time. However, as shown in Figure 1a, the intensities of (220) and (310) peaks decrease firstly and then increase with prolonging the time. They ultimately reach the minimum value at 5 s treatment. The XRD intensity ratio of (110) and (310) peaks ($I_{(110)}/I_{(310)}$) with MA treatment time is shown in Figure 1b, and a maximum ratio is obtained with 10 s treatment. This suggests that the regrowth of MAPbI$_3$ crystallites by MA gas treating prefers the formation of (110) planes on the substrate.

The average crystal size of (110) plane deduced from Scherrer Formula is presented in Figure 1b. The crystal size showed a trend of first increasing and then decreasing with MA treatment time, which is similar to the change of $I_{(110)}/I_{(310)}$ ratios. A maximum size of 47 nm is achieved with 10 s treatment, which is about twofold compared to the as-deposited film, indicating that the optimum MA treatment time is 10 s. Figure 1c shows the UV-Visible light absorbance spectra of the corresponding samples. There is a steep cutoff at the wavelength about 785 nm in accordance with a band bap of about 1.58 eV. The light absorbance at the range of 400–550 nm is more sensitive to the MA treatment time, and the largest absorbance is obtained for the sample with 10 s MA treating. The implementation of unreasonable MA treatment time results in a lower absorbance than that of the as-deposited MAPbI$_3$ film.

Figure 1d illustrates steady-state photoluminescence (PL) measurement of MAPbI$_3$ films deposited on FTO/TiO$_2$ substrates. The position of the stable PL emission peaks center at ~768 nm varies with <10 MeV, which is consistent with the optical gap deduced from the absorption spectra. The PL intensity of as-deposited MAPbI$_3$ film is very high as shown in Figure 1c, suggesting the inefficient carrier transfer at the TiO$_2$/MAPbI$_3$ interface. The PL intensity decreases with MA treatment time from 0 s to 10 s, and then increases with the further treating. In contrast, the PL measurement of MAPbI$_3$ film treated with ethanol vapor, as shown in Figure A1, remains unchanged regardless of whether the post-processing time of ethanol vapor is 10 s or 30 s. This indicates that the post-treatment of ethanol vapor is not helpful for the properties of MAPbI$_3$ film or the carrier transfer at TiO$_2$/MAPbI$_3$

interface. MAPbI$_3$ film treated with 10 s MA exhibits a minimum PL intensity, which is reduced by 50% compared with the film without MA treating (0 s), indicating a more efficient carrier transfer at interface. This is directly related to the (110) preferential orientation of the perovskite film after 10 s post treatment. Wang et al. [43] reported that the preferential orientation of MAPbI$_3$ perovskite crystal with (110) crystal plane was better for perovskite solar cells, based on synchrotron radiation two-dimensional X-ray grazing incidence diffraction. The crystallographic orientation of the (110) crystal grains is favorable for the propagation of electrons and holes, and the migration rate of the carriers, which facilitates the rapid propagation of electrons and holes along the out-of-plane direction to the electron and hole transport layer. The carriers convert into an effective photocurrent by the electrode collection, thereby enhancing the device's photoelectric conversion performance.

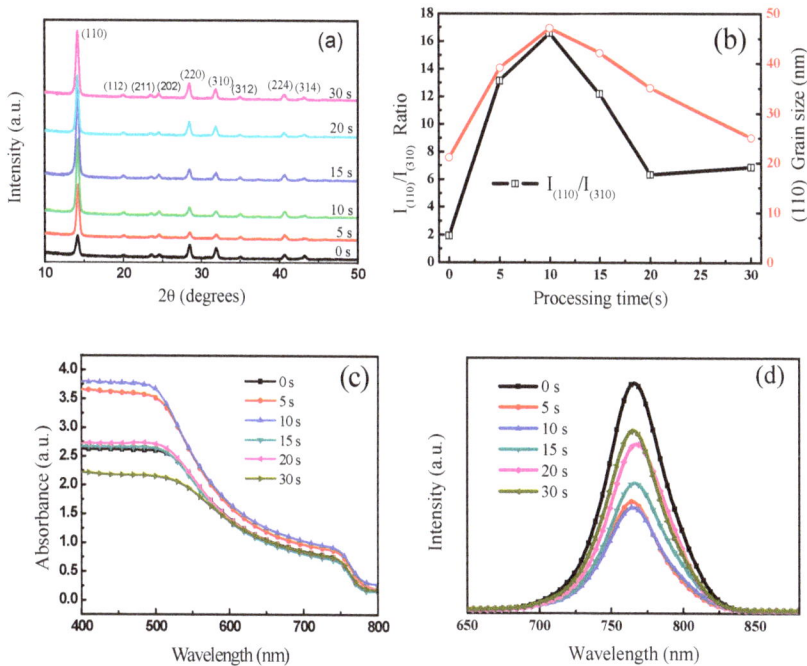

Figure 1. (**a**) X-ray diffraction (XRD) spectrum and the corresponding; (**b**) I(110)/I(310) Ratio and the average crystal size of (110) plane of the CH$_3$NH$_3$PbI$_3$ (MAPbI$_3$) films at different CH$_3$NH$_2$ (MA) treatment times; (**c**) UV–Vis absorption spectra of MAPbI$_3$ films; and (**d**) photoluminescence (PL) spectrum of FTO/TiO$_2$/MAPbI$_3$ with different times of MA treatment.

The scanning electron microscopy (SEM) images of MAPbI$_3$ film with different MA gas treatment times are shown in Figure 2a–f (see more data in Figure A2). The as-deposited MAPbI$_3$ film is composed of densely packed grains with the domain size distributing in a range of 200–300 nm, and the crystal boundaries are visible on the surface of the polycrystalline film, in agreement with the previous report of MAPbI$_3$ film prepared by the one-step anti-solvent method [44]. After exposing the MAPbI$_3$ film to MA gas for 5 s, and 10 s, respectively, as shown in Figure 2b–c, the boundary interfaces become unclear, which infers that the gas-solid reaction benefits the regrowth of MAPbI$_3$ crystallites through the grain boundaries. The corresponding tapping-mode atomic force microscope (AFM) height images in 2.0 μm × 2.0 μm area are presented in Figure 2g–i, and the root-mean-square roughness (RMS) decreases from 7.624 nm for the as-deposited sample to 4.585 nm for the perovskite with 10 s MA treating (three-dimensional images are shown Figure A3). As the post-treatment time increases to 15 s

(Figure 2d), some pinholes appear on the MAPbI$_3$ film surface. Further increasing the post-processing time to 20 s and 30 s (Figure 2e–f), the MAPbI$_3$ film is not uniformly covered on the substrate with many voids between grains. Through the low magnification (Figure A2), the film completely covers the substrate when post-processing time is below 15 s. However, if the post-processing time is over 20 s, the substrate is exposed. Therefore, the quality of MAPbI$_3$ film treated with MA gas is affected by post-processing times.

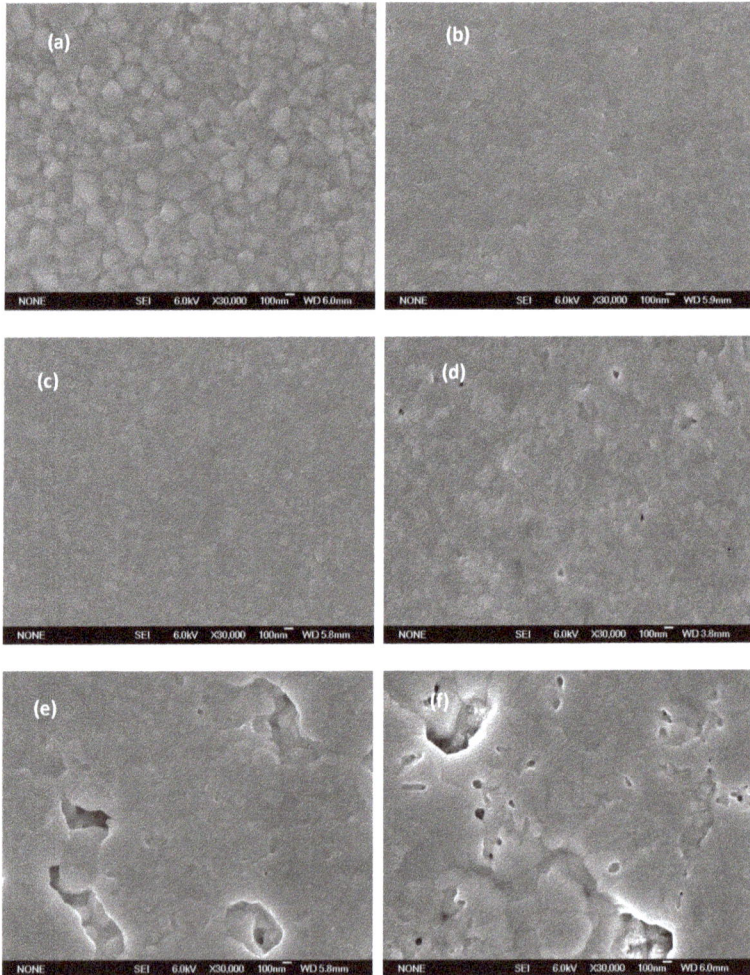

Figure 2. *Cont.*

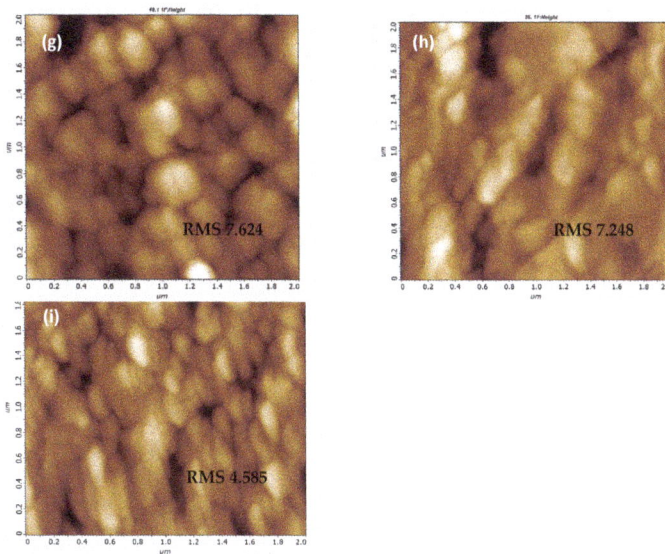

Figure 2. SEM images of MAPbI$_3$ films at different MA treatment times: (**a**) 0 s, (**b**) 5 s, (**c**) 10 s, (**d**) 15 s, (**e**) 20 s, (**f**) 30 s. Tapping-mode AFM height images of perovskite films treated with MA gas at different times: (**g**) 0 s, (**h**) 5 s, (**i**) 10 s.

Figure 3a depicts the cross-sectional view of the PSC with glass/FTO/TiO$_2$/MAPbI$_3$/spiro-OMeTAD/Ag planar structure, in which the thicknesses of TiO$_2$, MAPbI$_3$, spiro-OMeTAD, and Ag are about 80 nm, 360 nm, 250 nm, and 100 nm, respectively. Figure 3b shows the J-V characteristic curves of PSCs and the key parameters are listed in Table 1. The PSCs that without MA treating exhibit a J$_{SC}$ of 17.29 ± 0.83 mA/cm^2, an open-circuit voltage (V$_{OC}$) of 0.99 ± 0.02 V, and a fill factor (FF) of 59.5 ± 1.7%, leading to a corresponding average PCE of 10.2%. The best device shows a PCE of 10.7% with J$_{SC}$ is 18.1 mA/cm^2, V$_{OC}$ is 0.98 V, and FF is 60.7%. It is obvious that the low J$_{SC}$ is the main factor that limits the PCE, which is consistent with the reported perovskite solar cell with a similar device structure [45]. For the device with 5 s and 10 s MA treating, J$_{SC}$ and V$_{OC}$ are strikingly enhanced, resulting in an increase in PCE from 12.5% to 14.4%. The best device made of MAPbI$_3$ layer with 10 s MA treating yields a J$_{SC}$ of 22.1 mA/cm^2, a V$_{OC}$ of 1.1 V, a FF of 61.5% and a PCE of 15.3%. The crystallinity of perovskite is improved step by step with the increasing of post-treatment times, and the (110) preferred growth is more pronounced (Figure 1a). However, the prolonged MA treatment, 15 s, decreases the performance of devices to an average PCE of 8.9%.

Figure 3c shows the incident photon-to-current conversion efficiency (IPCE) and the integrated product of the EQE curves with the AM1.5G photon flux for the corresponding devices. The J$_{SC}$ data calculated from the IPCE spectra for the devices with 0 s, 5 s, 10 s, and 15 s MA treating is 16.6, 18.4, 20.9, and 15.1 (mA/cm^2) respectively, in excellent agreement with the J-V measurements. To check the reproducibility of the device performance, we collected the power conversion efficiency of one batch of devices (25 devices, Figure A4). When the post-treatment time is 10 s, the statistics of the PCEs distribution demonstrates the reliability and repeatability. The results followed a Gaussian distribution with small relative standard deviations.

Furthermore, electrochemical impedance spectroscopy (EIS) is employed to investigate the charge transport processes and recombination dynamics in PSCs. The EIS measurement of PSCs is conducted under 0.6 V applied bias and dark conditions, as shown in Figure 3d. The traditional front high frequency arc has disappeared, confirming that the interface contact capacitance (perovskite/TiO$_2$ or spiro-OMeTAD/Ag) is nearly zero. To further understand the impedance spectra, the inset equivalent circuit is used to fit the test results to get detailed information, and the results are listed in Table 1. The difference of R_S is minute, leading to a minor variation of V_{OC} [46]. R_{rec} (recombination resistance) increases with the increase of MA treatment times from 0 s to 10 s, suggesting the significant improvement in electron injection and the effective suppression in charge recombination, resulting in the enhanced device performance. However, once the treatment exceeds 10 s, R_{rec} rapidly decreases, revealing deteriorative electron extraction at the interfaces that directly related to the formation of pinholes in the MAPbI$_3$ films [47]. These results suggest that the defects at the grain boundaries and the coverage of film are the two crucial parameters that directly affect the device's performance. It is confirmed that the MA treatment method is a very effective process to improve the quality of MAPbI$_3$ films again. The optimum MA treatment time is 10 s longer than the reported 2–3 s or <1 s [34,37], which is directly involved with the MA gas pressure and the film coverage. So, further optimization of the MA treatment, such as temperature, pressure, etc., is very necessary to obtain high quality thin films and high-performance devices.

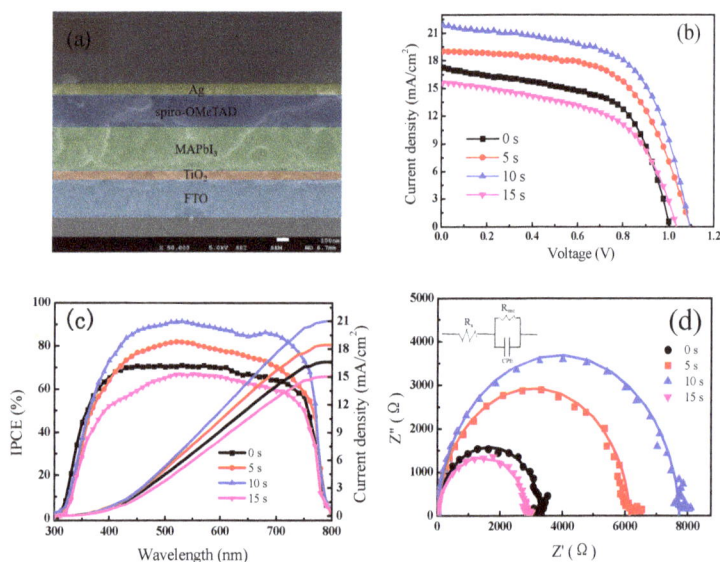

Figure 3. (a) The cross-sectional view of glass/FTO/TiO$_2$/MAPbI$_3$/spiro-OMeTAD/Ag planar heterojunction perovskite solar cell; (b) J-V curves and (c) incident photon-to-electron conversion efficiency (IPCE) spectra and integrated current density of the perovskite solar cells (PSCs) at different MA treatment times; (d) The impedance spectra of different MA treatment times of PSCs under 0.6 V applied bias in dark, the inset picture is the equivalent circuit.

Table 1. Parameters of Current Density-Voltage Measurement and Electrochemical Impendence of the PSCs at different MA treatment times.

Time (s)	V_{OC} (V)	J_{SC} (mA/cm^2)	FF (%)	PCE$_{ave}$ (Best) (%)	R_s (Ω cm^2)	R_{rec} (Ω cm^2)
0	0.99 ± 0.02	17.29 ± 0.83	59.5 ± 1.7	10.2 (10.7)	5.8	3265
5	1.09 ± 0.01	19.02 ± 0.63	61.3 ± 1.3	12.5 (13.6)	9.5	6076
10	1.09 ± 0.01	21.85 ± 0.52	61.5 ± 1.2	14.4 (15.3)	8.4	7719
15	1.02 ± 0.01	15.64 ± 0.76	54.5 ± 1.6	8.9 (9.8)	4.1	2876

3. Materials and Methods

3.1. Materials

Dimethylsulfoxide (DMSO), chlorobenzene and *N,N*-Methylformamide (DMF) obtained from Acros Organics. Titanium (IV) Chloride (TiCl$_4$, 99.0%) purchased from Kermel (Tianjin, China). Spiro-OMeTAD (99.5%), Li-TFSI (99.95%), 4-tert-butyl pyridine (TBP, 96%), Lead (II) Iodide (PbI$_2$), Methylammonium iodide (MAI) was purchased from Xi'an Polymer Light Technology Corp (Xi'an, China). Methylamine solution in absolute ethanol (38 wt %) was purchased from Sigma Aldrich (Shanghai, China). All the chemicals were directly used without further purification.

3.2. Fabrication of Perovskite Solar Cells

The FTO was wiped with a cotton swab dipped in glass cleaner and then cleaned ultrasonically for 20 min in deionized water, acetone, alcohol and isopropanol orderly. Finally, the washed FTO substrates were treated with UV for 10 min.

Preparation of TiO$_2$ blocking layer by chemical bath deposition method: the cooled TiCl$_4$ (99.9%) solution of 2200 μL was added dropwise to ice-water of 100 mL (Divided by 22 times, each time added dropwise 100 μL). Then the solution stirred for another 30 min to prepare a homogeneous solution at a concentration of 200 mM. The FTO substrates were immersed into the precursor solution and kept in an oven at 70 °C for 50 min. After 50 min, the FTO substrates were washed with water and ethanol and then dried at 100 °C in air for an hour.

Lead iodide (PbI$_2$) and methyl ammonium iodide (MAI) were dissolved in organic solvent (DMSO:DMF = 3:7, *v/v*) with molar ratio of 1:1, stirred in an N$_2$ filled glove box at 65 °C for 12 h. A yellow precursor solution of methylamine lead iodine at a concentration of 1.25 M was obtained. The spin-coating procedure was performed in the glovebox: first, 700 rpm for 15 s; second, 3000 rpm for 25 s, then anti-solvent chlorobenzene was dropped onto the spinning substrates 18 s after the start of the 3000 rpm spin stage. Finally, the film was annealed at 60 °C for 3 min, 80 °C for 5 min, and 100 °C for 10 min to obtain MAPbI$_3$ film (Figure 4).

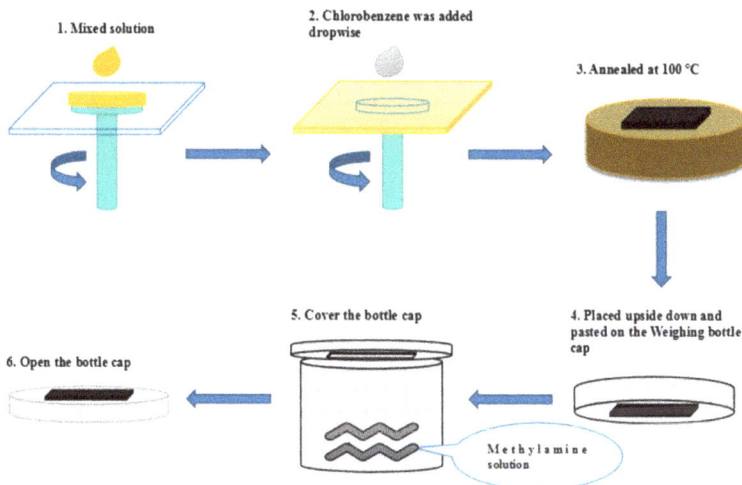

Figure 4. The schematic drawing illustrates the process to prepare MAPbI$_3$ polycrystalline film and the setup to vapor treatment process.

A portion of the prepared MAPbI$_3$ films were taken out randomly and subjected to post-treatment in MA vapor. Then we added 80 µL of Methylamine solution in absolute ethanol into the middle of the weighing bottle (50.60 mL). MAPbI$_3$ film was placed upside down and pasted on the weighing bottle cap. When the cap was covered, the black MAPbI$_3$ film immediately became colorless. Then we started timing. The post-processing times were 5 s, 10 s, 15 s, 20 s, 30 s. We opened the bottle cap to stop the process and the colorless film immediately turned black (Figure 4). As a comparison, 80 uL of ethanol was dropped into the middle of the weighing bottle, but black MAPbI$_3$ film did not become colorless in 5 s, 10 s, 15 s, 20 s or 30 s (Figure A5).

The hole transporting layer was deposited by spin coating at 3000 rpm for 30 s. The spin-coating formulation was prepared by dissolving 72.5 mg of (2,2,7,7-tetrakis (*N,N*-dip-methoxyphenylamine)-9,9-spirobifluorene) (spiro-OMeTAD) powder in Chlorobenzene solution (1 mL), and then 17.5 µL of a lithium salt solution (the concentration was 520 mg/mL, the solvent was acetonitrile) was added, followed by the addition of 28.5 µL of 4-tert-butylpyridine (TBP). Finally, silver electrodes were deposited on the hole transporting layer.

3.3. Instruments

The phase structure was characterized on a Rigaku (D/MAX-2400) X-ray diffractometer (Cu Ka radiation, λ = 1.5425 Å). Scanning electron microscopy (SEM) images were obtained via a field emission scanning electron microscope (JSM-6700F) (JEOL, Tokyo, Japan). Ultraviolet–visible absorption spectra were recorded on a Shimadzu UV-3150 spectrophotometer in the 400–800 nm wavelength range at room temperature. The surface roughness of the films was obtained by atomic force microscopy (AFM, SII Nano Technology Ltd., Nanonavi, Shanghai, China) in noncontact mode. Steady-state photoluminescence (PL) spectra were measured on a FlouroMax-4 Spectrophotometer, using a pulsed diode laser as an excitation source. The current density-voltage (J-V) characteristics were recorded with a Keithley 2400 source meter and 300 W collimated Xenon lamp (Newport) calibrated with the light intensity to 100 mW cm^{-2} under AM1.5G solar light conditions by the certified silicon solar cell. Incident photon-to-electron conversion efficiency (IPCE) was obtained on a computer-controlled IPCE system (Newport) containing a Xenon lamp, a monochromator and Keithley multimeter. Electrochemical impedance spectroscopy was measured using an IM6e Electrochemical Workstation (ZAHENR, Kronach, Germany), in the dark and at 0.6 V, with the frequency ranging between 100 KHz and 0.1 Hz. The Z-View software (Scribner Associates Inc., Southern Pines, NC, USA) was used to fit the impedance spectra.

4. Conclusions

In conclusion, we demonstrate a feasible and effective post-treatment method using MA vapor for improving the quality of MAPbI$_3$ films. MA treatment can repair the micro-gaps between grains, and eliminate the negative effect of pinhole and grain boundary on the device performance. This is benefit for improving the surface morphology, microstructure, and photo-electrical properties of MAPbI$_3$ films varied along with the treatment times in MA vapor. XRD and SEM results show that the MAPbI$_3$ films with surface smoothness, compact structure and preferential orientation in (110) planes can be obtained after 10 s treatment. Under this processing condition, the migration rate of the carriers is enhanced and the absorbance of the MAPbI$_3$ film is increased through the optical performance test. As a result, the best efficiency of 15.3% is achieved with J$_{SC}$ of 22.1 (mA/cm^2), V$_{OC}$ of 1.1 V, FF of 61.5%. Considering that MA vapor is introduced by commercially cheap methylamine ethanol solution, it is expected that this feasible and effective post-treatment method will benefit the easy-control and low-cost fabrication of highly efficient perovskite solar cells.

Acknowledgments: This work was supported by the National Natural Science Foundation of China (61574129), and Basic and Advanced Technology Research Project of Henan Province of China (152300410035).

Author Contributions: Yongsheng Chen and Shi-E Yang conceived and designed the experiments; Yaxiao Jiang performed the experiments; Limin Tu and Haitao Li analyzed the data; Shaohua Li contributed reagents/materials/ analysis tools; Yaxiao Jiang wrote the paper.

Conflicts of Interest: The authors declare no conflict of interest.

Appendix A

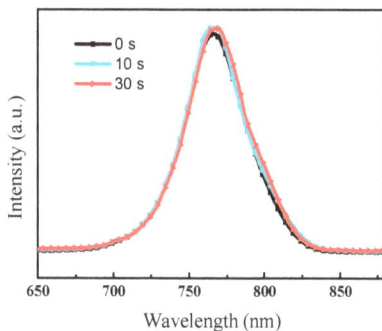

Figure A1. PL spectrum of FTO/TiO$_2$/MAPbI$_3$ with different ethanol vapor treatment times.

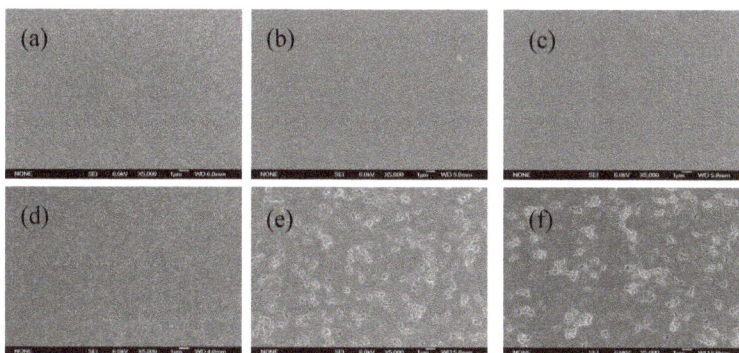

Figure A2. SEM image of MAPbI$_3$ films at low magnification with different MA treatment times. Processing time: (**a**) 0 s, (**b**) 5 s, (**c**) 10 s, (**d**) 15 s, (**e**) 20 s, (**f**) 30 s.

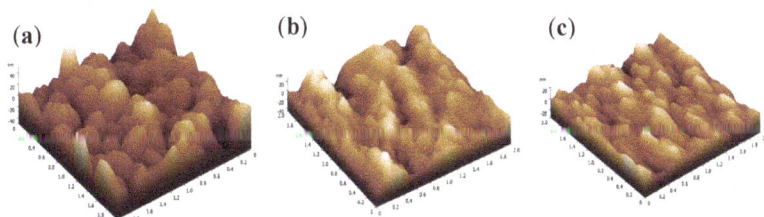

Figure A3. AFM three-dimensional images of perovskite films with different MA treatment times: (**a**) 0 s (**b**) 5 s (**c**) 10 s.

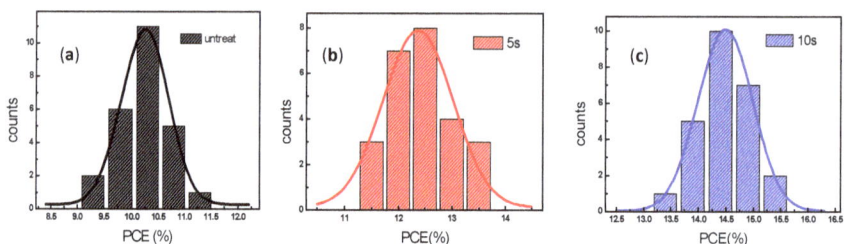

Figure A4. Histograms of PCEs measured for 25 cells that MA gas treating with different times. Processing time: (**a**) 0 s, (**b**) 5 s, (**c**) 10 s.

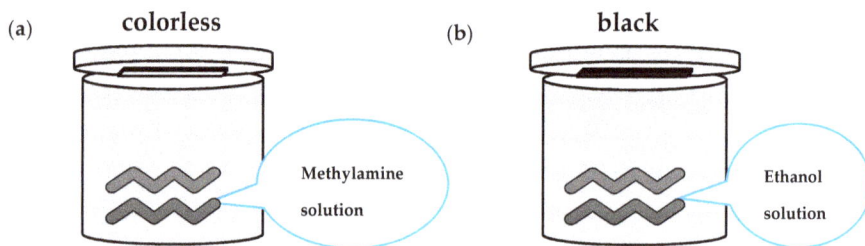

Figure A5. The schematic drawing illustrates the difference to the MA vapor treatment (**a**) and the ethanol vapor treatment (**b**).

References

1. Kojima, A.; Teshima, K.; Shirai, Y.; Miyasaka, T. Organometal halide perovskites as visible-light sensitizers for photovoltaic cells. *J. Am. Chem. Soc.* **2009**, *131*, 6050–6051. [CrossRef] [PubMed]
2. National Renewable Energy Laboratory. Best Research-Cell Efficiencies. Available online: https://www.nrel.gov/pv/assets/images/efficiency-chart.png (accessed on 30 October 2017).
3. Park, N.-G. Perovskite solar cells: An emerging photovoltaic technology. *Mater. Today* **2015**, *18*, 65–72. [CrossRef]
4. Xie, Z.; Liu, S.; Qin, L.; Pang, S.; Wang, W.; Yan, Y.; Yao, L.; Chen, Z.; Wang, S.; Du, H.; et al. Refractive index and extinction coefficient of CH$_3$NH$_3$PbI$_3$ studied by spectroscopic ellipsometry. *Opt. Mater. Express* **2015**, *5*, 29–43.
5. Ishihara, T. Optical properties of PbI$_2$-based perovskite structures. *J. Lumin.* **1994**, *60*, 269–274. [CrossRef]
6. Lin, Q.; Armin, A.; Nagiri, R.C.R.; Burn, P.L.; Meredith, P. Electro-optics of perovskite solar cells. *Nat. Photonics* **2015**, *9*, 106–112. [CrossRef]
7. Miyata, A.; Mitioglu, A.; Plochocka, P.; Portugall, O.; Wang, J.T.-W.; Stranks, S.D.; Snaith, H.J.; Nicholas, R.J. Direct measurement of the exciton binding energy and effective masses for charge carriers in organic-inorganic tri-halide perovskites. *Nat. Phys.* **2015**, *11*, 582–587. [CrossRef]
8. Xing, G.; Mathews, N.; Sun, S.; Lim, S.S.; Lam, Y.M.; Grätzel, M.; Mhaisalkar, S.; Sum, T.C. Long-range balanced electron- and hole-transport lengths in organic-inorganic CH$_3$NH$_3$PbI$_3$. *Science* **2013**, *342*, 344–347. [CrossRef] [PubMed]
9. Stranks, S.D.; Eperon, G.E.; Grancini, G.; Menelaou, C.; Alcocer, M.J.P.; Leijtens, T.; Herz, L.M.; Petrozza, A.; Snaith, H.J. Electron-hole diffusion lengths exceeding 1 μm in an organometal trihalide perovskite absorber. *Science* **2013**, *342*, 341–344. [CrossRef] [PubMed]
10. Edri, E.; Kirmayer, S.; Henning, A.; Mukhopadhyay, S.; Gartsman, K.; Rosenwaks, Y.; Hodes, G.; Cahen, D. Why lead methylammonium tri-iodide perovskite-based solar cells require a mesoporous electron transporting scaffold (but not necessarily a hole conductor). *Nano Lett.* **2014**, *14*, 1000–1004. [CrossRef] [PubMed]

11. Dong, Q.; Fang, Y.; Shao, Y.; Mulligan, P.; Qiu, J.; Cao, L.; Huang, J. Electron-hole diffusion lengths >175 μm in solution-grown CH₃NH₃PbI₃ single crystals. *Science* **2015**, *347*, 967–970. [CrossRef] [PubMed]
12. Ponseca, C.S., Jr.; Savenije, T.J.; Abdellah, M.; Zheng, K.; Yartsev, A.; Pascher, T.; Harlang, T.; Chabera, P.; Pullerits, T.; Stepanov, A.; et al. Organometal halide perovskite solar cell materials rationalized: Ultrafast charge generation, high and microsecond-long balanced mobilities, and slow recombination. *J. Am. Chem. Soc.* **2014**, *136*, 5189–5192. [CrossRef] [PubMed]
13. Stoumpos, C.C.; Malliakas, C.D.; Kanatzidis, M.G. Semiconducting tin and lead iodide perovskites with organic cations: Phase transitions, high mobilities, and near-infrared photoluminescent properties. *Inorg. Chem.* **2013**, *52*, 9019–9038. [CrossRef] [PubMed]
14. Leijtens, T.; Stranks, S.D.; Eperon, G.E.; Lindblad, R.; Johansson, E.M.J.; McPherson, I.J.; Rensmo, H.; Ball, J.M.; Lee, M.M.; Snaith, H.J. Electronic properties of meso-superstructured and planar organometal halide perovskite films: Charge trapping, photodoping, and carrier mobility. *ACS Nano* **2014**, *8*, 7147–7155. [CrossRef] [PubMed]
15. Wehrenfennig, C.; Eperon, G.E.; Johnston, M.B.; Snaith, H.J.; Herz, L.M. High charge carrier mobilities and lifetimes in organolead trihalide perovskites. *Adv. Mater.* **2014**, *26*, 1584–1589. [CrossRef] [PubMed]
16. Jeng, J.-Y.; Chiang, Y.-F.; Lee, M.-H.; Peng, S.-R.; Guo, T.-F.; Chen, P.; Wen, T.-C. CH₃NH₃PbI₃ perovskite/fullerene planar-heterojunction hybrid solar cells. *Adv. Mater.* **2013**, *25*, 3727–3732. [CrossRef] [PubMed]
17. Xiao, Z.; Dong, Q.; Bi, C.; Shao, Y.; Yuan, Y.; Huang, J. Solvent annealing of perovskite-induced crystal growth for photovoltaic-device efficiency enhancement. *Adv. Mater.* **2014**, *26*, 6503–6509. [CrossRef] [PubMed]
18. Etgar, L.; Gao, P.; Qin, P.; Graetzel, M.; Nazeeruddin, M.K. A hybrid lead iodide perovskite and lead sulfide QD heterojunction solar cell to obtain a panchromatic response. *J. Mater. Chem. A* **2014**, *2*, 11586–11590. [CrossRef]
19. Jeon, N.J.; Noh, J.H.; Kim, Y.C.; Yang, W.S.; Ryu, S.; Seok, S.I. Solvent engineering for high-performance inorganic–organic hybrid perovskite solar cells. *Nat. Mater.* **2014**, *13*, 897–903. [CrossRef] [PubMed]
20. Zhou, H.; Chen, Q.; Li, G.; Luo, S.; Song, T.-B.; Duan, H.-S.; Hong, Z.; You, J.; Liu, Y.; Yang, Y. Interface engineering of highly efficient perovskite solar cells. *Science* **2014**, *345*, 542–546. [CrossRef] [PubMed]
21. Subbiah, J.; Mitchell, V.D.; Hui, N.K.C.; Jones, D.J.; Wong, W.W.H. A Green Route to Conjugated Polyelectrolyte Interlayers for High-Performance Solar Cells. *Angew. Chem. Int. Ed.* **2017**, *56*, 8431–8434. [CrossRef] [PubMed]
22. Jeon, N.; Noh, J.; Yang, W.S.; Kim, Y.C.; Ryu, S.; Seo, J.; Seok, S.I. Highly efficient planar perovskite solar cells through band alignment engineering. *Nature* **2015**, *517*, 476–480. [CrossRef] [PubMed]
23. Jeon, N.J.; Noh, J.H.; Kim, Y.C.; Yang, W.S.; Ryu, S.; Seok, S.I. Highly efficient and stable planar perovskite solar cells with reduced graphene oxide nanosheets as electrode interlayer. *Nat. Mater.* **2014**, *13*, 897–903. [CrossRef] [PubMed]
24. Yang, W.S.; Noh, J.H.; Jeon, N.J.; Kim, Y.C.; Ryu, S.; Seok, S.I. High-performance photovoltaic perovskite layers fabricated through intramolecular exchange. *Science* **2015**, *348*, 1234–1237. [CrossRef] [PubMed]
25. Hwang, K.; Jung, Y.-S.; Heo, Y.-J.; Scholes, F.H.; Watkins, S.E.; Subbiah, J.; Jones, D.J.; Kim, D.-Y.; Kim, D.V. Toward Large Scale Roll-to-Roll Production of Fully Printed Perovskite Solar Cells. *Adv. Mater.* **2015**, *27*, 1241–1247. [CrossRef] [PubMed]
26. Eperon, G.E.; Burlakov, V.M.; Docampo, P.; Goriely, A.; Snaith, H.J. Morphological control for high performance, solution-processed planar heterojunction perovskite solar cells. *Adv. Funct. Mater.* **2014**, *24*, 151–157. [CrossRef]
27. Kim, H.-B.; Choi, H.; Jeong, J.; Kim, S.; Walker, B.; Song, S.; Kim, J.Y. Mixed solvents for the optimization of morphology in solution-processed, inverted-type perovskite/fullerene hybrid solar cells. *Nanoscale* **2014**, *6*, 6679–6683. [CrossRef] [PubMed]
28. Shen, D.; Yu, X.; Cai, X.; Peng, M.; Ma, Y.; Su, X.; Xiao, L.; Zou, D. Understanding the solvent-assisted crystallization mechanism inherent in efficient organic–inorganic halide perovskite solar cells. *J. Mater. Chem. A* **2014**, *2*, 20454–20461. [CrossRef]
29. Xiao, M.; Huang, F.; Huang, W.; Dkhissi, Y.; Zhu, Y.; Etheridge, J.; Gray-Weale, A.; Bach, U.; Cheng, Y.-B.; Spiccia, L. A fast deposition-crystallization procedure for highly efficient lead iodide perovskite thin-film solar cells. *Angew. Chem. Int. Ed.* **2014**, *53*, 9898–9903. [CrossRef] [PubMed]

30. Nie, W.; Tsai, H.; Asadpour, R.; Blancon, J.-C.; Neukirch, A.J.; Gupta, G.; Crochet, J.J.; Chhowalla, M.; Tretiak, S.; Alam, M.A.; et al. High-efficiency solution-processed perovskite solar cells with millimeter-scale grains. *Science* **2015**, *347*, 522–525. [CrossRef] [PubMed]

31. Yang, Y.; Yang, M.; Li, Z.; Crisp, R.; Zhu, K.; Beard, M.C. Comparison of recombination dynamics in CH₃NH₃PbBr₃ and CH₃NH₃PbI₃ perovskite films: Influence of exciton binding energy. *J. Phys. Chem. Lett.* **2015**, *6*, 4688–4692. [CrossRef] [PubMed]

32. Fu, K.; Zhou, Q.; Chen, Y.; Lu, J.; Yang, S.-E. The simulation of physical mechanism for HTM-free perovskite organic lead iodide planar heterojunction solar cells. *J. Opt.* **2015**, *17*, 105904. [CrossRef]

33. Shao, Y.; Xiao, Z.; Bi, C.; Yuan, Y.; Huang, J. Origin and elimination of photocurrent hysteresis by fullerene passivation in CH₃NH₃PbI₃ planar heterojunction solar cells. *Nat. Commun.* **2014**, *5*, 5784. [CrossRef] [PubMed]

34. Zhao, Y.; Zhu, K. Optical bleaching of perovskite (CH₃NH₃)PbI₃ through room-temperature phase transformation induced by ammonia. *Chem. Commun.* **2014**, *50*, 1605–1607. [CrossRef] [PubMed]

35. Zhou, Z.; Wang, Z.; Zhou, Y.; Pang, S.; Wang, D.; Xu, H.; Liu, Z.; Padture, N.P.; Cui, G. Methylamine gas induced defect-healing behavior of CH₃NH₃PbI₃ thin films for perovskite solar cells. *Angew. Chem. Int. Ed.* **2015**, *54*, 9705–9709. [CrossRef] [PubMed]

36. Chih, Y.-K.; Wang, J.-C.; Yang, R.-T.; Liu, C.-C.; Chang, Y.-C.; Fu, Y.-S.; Lai, W.-C.; Chen, P.; Wen, T.-C.; Huang, Y.-C.; et al. NiOₓ Electrode interlayer and CH₃NH₂/CH₃NH₃PbBr₃ interface treatment to markedly advance hybrid perovskite-based light-emitting diodes. *Adv. Mater.* **2016**, *28*, 8687–8694. [CrossRef] [PubMed]

37. Jain, S.M.; Philippe, B.; Johansson, E.M.J.; Park, B.-W.; Rensmo, H.; Edvinsson, T.; Boschloo, G. Frustrated lewis pair-mediated recrystallization of CH₃NH₃PbI₃ for improved optoelectronic quality and high voltage planar perovskite solar cells. *Energy Environ. Sci.* **2016**, *9*, 3770–3782. [CrossRef]

38. Zhao, T.; Williams, S.T.; Chueh, C.-C.; deQuilettes, D.W.; Liang, P.-W.; Ginger, D.S.; Jen, A.K.-Y. Design rules for the broad application of fast (<1 s) methylamine vapor based, hybrid perovskite post deposition treatments. *RSC Adv.* **2016**, *6*, 27475–27484.

39. Jiang, Y.; Juarez-Perez, E.J.; Ge, Q.; Wang, S.; Leyden, M.R.; Ono, L.K.; Raga, S.R.; Hu, J.; Qi, Y. Post-annealing of MAPbI₃ perovskite films with methylamine for efficient perovskite solar cells. *Mater. Horiz.* **2016**, *3*, 548–555. [CrossRef]

40. Jacobs, D.L.; Zang, L. Thermally induced recrystallization of MAPbI₃ perovskite under methylamine atmosphere: An approach to fabricating large uniform crystalline grains. *Chem. Commun.* **2016**, *52*, 10743–10746. [CrossRef] [PubMed]

41. Zhang, T.; Guo, N.; Li, G.; Qian, X.; Li, L.; Zhao, Y. A controllable fabrication of grain boundary PbI₂ nanoplates passivated lead halide perovskites for high performance solar cells. *Nano Energy* **2016**, *26*, 50–56. [CrossRef]

42. Li, C.; Pang, S.; Xu, H.; Cui, G. Methylamine gas based synthesis and healing process toward upscaling of perovskite solar cells: Progress and perspective. *Sol. RRL* **2017**, *1*, 1700076. [CrossRef]

43. Wang, Z.-K.; Li, M.; Yang, Y.-G.; Hu, Y.; Ma, H.; Gao, X.-Y.; Liao, L.-S. High efficiency Pb-In binary metal perovskite solar cells. *Adv. Mater.* **2016**, *28*, 6695–6703. [CrossRef] [PubMed]

44. Cai, Q.; Li, H.; Jiang, Y.; Tu, L.; Ma, L.; Wu, X.; Yang, S.-E.; Shi, Z.; Zang, J.; Chen, Y. High-efficiency perovskite solar cells based on MAI(PbI₂)₁₋ₓ(FeCl₂)ₓ absorber layers. *Sol. Energy* **2018**, *159*, 786–793. [CrossRef]

45. Liang, C.; Wu, Z.; Li, P.; Fan, J.; Zhang, Y.; Shao, G. Chemical bath deposited rutile TiO₂ compact layer toward efficient planar heterojunction perovskite solar cells. *Appl. Surf. Sci.* **2016**, *391*, 337–344. [CrossRef]

46. Bisquert, J.; Bertoluzzi, L.; Mora-Sero, I.; Garcia-Belmonte, G. Theory of impedance and capacitance spectroscopy of solar cells with dielectric relaxation, drift-diffusion transport, and recombination. *J. Phys. Chem. C* **2014**, *118*, 18983–18991. [CrossRef]

47. Suarez, B.; Gonzalez-Pedro, V.; Ripolles, T.S.; Sanchez, R.S.; Otero, L.; Mora-Sero, I. Recombination study of combined halides (Cl, Br, I) perovskite solar cells. *J. Phys. Chem. Lett.* **2014**, *5*, 1628–1635. [CrossRef] [PubMed]

crystals

MDPI

Article

Improved Stability and Photoluminescence Yield of Mn^{2+}-Doped CH$_3$NH$_3$PbCl$_3$ Perovskite Nanocrystals

Xianli Li [1,2,†], Yan Guo [1,2,†] and Binbin Luo [1,2,*]

1 Department of Chemistry, Shantou University, Guangdong 515063, China; lixianli@stu.edu.cn (X.L.);
   17yguo1@stu.edu.cn (Y.G.)
2 Department of Chemistry and Key Laboratory for Preparation and Application of Ordered Structural
   Materials of Guangdong Province, Shantou University, Guangdong 515063, China
* Correspondence: bbluo@stu.edu.cn
† These authors contributed equally to this work.

Received: 21 November 2017; Accepted: 20 December 2017; Published: 23 December 2017

Abstract: Organic–inorganic CH$_3$NH$_3$PbCl$_3$ perovskite nanocrystals (PNCs) doped with Mn^{2+}, CH$_3$NH$_3$Pb$_x$Mn$_{1-x}$Cl$_3$, have been successfully prepared using a reprecipitation method at room temperature. Structural and morphological characterizations reveal that the CH$_3$NH$_3$Pb$_x$Mn$_{1-x}$Cl$_3$ PNCs with cubic phase transforms from particles to cubes and increases in size from 16.2 ± 4.4 nm in average diameter to 25.3 ± 7.2 nm in cubic length after the addition of Mn^{2+} precursor. The CH$_3$NH$_3$Pb$_x$Mn$_{1-x}$Cl$_3$ PNCs exhibit a weak exciton emission at ~405 nm with a low absolute quantum yield (QY) of around 0.4%, but a strong Mn^{2+} dopant emission at ~610 nm with a high QY of around 15.2%, resulting from efficient energy transfer from the PNC host to the Mn^{2+} dopant via the $^4T_1 \rightarrow ^6A_1$ transition. In addition, the thermal and air stability of CH$_3$NH$_3$Pb$_x$Mn$_{1-x}$Cl$_3$ PNCs are improved due to the passivation with (3-aminopropyl) triethoxysilane (APTES), which is important for applications such as light emitting diodes (LEDs).

Keywords: perovskite; Mn^{2+} doping; CH$_3$H$_3$PbCl$_3$; stability

---

1. Introduction

As a promising candidate for photovoltaics and lighting application, lead halide perovskite (LHP) afford many intriguing advantages over traditional semiconductor materials such as simple fabrication [1–4], abundant precursors source, defect tolerance [5,6], high photoluminescence (PL) quantum yields (QY) [5,7–9], halogen dependent emission over the entire visible spectral range with narrow full width at half maximum (FWHM) [10,11], as well as long carrier lifetime and diffusion length [12]. Compared with the intensive investigations of lead bromide and iodide perovskite [13–17], lead chloride perovskites have received less attention due to its low PL QY and large bandgap (~2.97 eV), which is less desirable for light emitting diodes (LEDs) and solar cells applications [18].

Doping metal ions into nanocrystals is an effective approach to tuning their optical, electronic and magnetic properties, as has been widely applied to II-VI metal chalcogenide nanocrystals [19,20]. Recently, several heterovalent or isovalent metal ions such as Bi^{3+} [21], Al^{3+} [22], Sn^{2+} [23], Cd^{2+} [23], Zn^{2+} [23], and Mn^{2+} [24–30] have been introduced into LHP as dopants to modulate their optical properties. Among these metal ions, Mn^{2+}-doped all inorganic CsPbCl$_3$ perovskite nanocrystals (PNCs) have attracted increasing attention owing to the appropriate band alignment of CsPbCl$_3$ and d-d transition of Mn^{2+} (2.15 eV, $^6A_1 \rightarrow ^4T_1$), resulting in efficient intraparticle energy transfer between the host exciton and dopant ions [25,31]. Moreover, the incorporation of Mn^{2+} may provide the possibility of imparting paramagnetism to the host materials [31].

In general, hot injection approach has been applied in the preparation of Mn^{2+}-doped lead halide PNCs because of the homogeneous products and relatively simple purification process [32,33].

However, the highly strict conditions of this method and poor stability of the resulting PNCs greatly hinder their applications. In contrast, organic-inorganic hybrid $CH_3NH_3PbCl_3$ PNCs can be easily prepared via solution-based reprecipitation approach at room temperature when (3-aminopropyl) triethoxysilane (APTES) is used as a capping ligand [34,35]. Additionally, the SiO_2 shell resulting from the hydrolysis of APTES capping ligand can greatly improve the product yield and stability of $CH_3NH_3PbCl_3$ PNCs towards water and air [34].

In this work, we demonstrate that Mn^{2+} can be doped into $CH_3NH_3PbCl_3$ PNCs via a reprecipitation method. By adjusting the doping concentration of Mn^{2+}, $CH_3NH_3Pb_xMn_{1-x}Cl_3$ PNCs with a strong dopant emission located at 610 nm were obtained. Besides the greatly improved overall QY of $CH_3NH_3Pb_xMn_{1-x}Cl_3$ PNCs (from ~0.4% to ~15.6%) upon Mn^{2+} doping, the thermal and air stability have also been enhanced by the SiO_2 shell formed from hydrolysis of APTES. The combined stability and high PL QY make these PNCs potentially useful for various photonics applications including LEDs.

2. Experimental Section

2.1. Materials

All the chemicals were used as received without further purification, including $PbCl_2$ (99.99%, Aladdin Chemical Co., Ltd., Shanghai, China), CH_3NH_2 (40%, TCI Chemical Co., Ltd., Tokyo, Japan), $MnCl_2$ (99%, Aladdin Chemical Co., Ltd., Shanghai, China), APTES (99%, Aladdin Chemical Co., Ltd., Shanghai, China), HCl (37.5%, Xilong Scientific Co., Ltd., Shantou, China), toluene (AR, Xilong Scientific Co., Ltd., Shantou, China), N,N-dimethylformamide (DMF, spectroscopic grade, 99.9%, Aladdin Chemical Co., Ltd., Shanghai, China), oleic acid (85%, TCI Chemical Co., Ltd., Tokyo, Japan), octylamine (99%, Fisher Scientific Co., Ltd., Pittsburgh, PA, US), ethyl acetate (99%, Aladdin Chemical Co., Ltd., Shanghai, China).

Mn^{2+}-doped $CH_3NH_3PbCl_3$ PNCs: CH_3NH_3Cl was prepared according to a reported procedure [18]. The Mn^{2+}-doped $CH_3NH_3PbCl_3$ PNCs were synthesized following a method reported previously with some small changes [1]. 1.5 mL precursor solution of PNCs was prepared by dissolving 0.05 mmol CH_3NH_3Cl, 0.05 mmol ($PbCl_2$ + $MnCl_2$), 0.05 mmol APTES or octylamine and 40 μL oleic acid in 1.5 mL DMF solvent and ultrasonicated until the solution became transparent. Then, 10 μL, 25 μL, 50 μL and 100 μL of the precursor solution were injected slowly into 5 mL toluene to form $CH_3NH_3PbCl_3$ PNCs solution. Mn^{2+}-doped $CH_3NH_3PbCl_3$ PNCs with different doping concentration of Mn^{2+} were prepared by varying the ratio between $PbCl_2$ and $MnCl_2$. The solid samples were obtained by adding ethyl acetate in toluene to precipitate $CH_3NH_3PbCl_3$ PNCs and centrifuged at 4000 rpm subsequently, then re-dispersed in toluene. These steps were repeated three times. At last, the dry samples were collected by placing the samples at 50 °C oven overnight.

Mn^{2+}-doped $CH_3NH_3PbCl_3$ PNCs films: The films were prepared by dropping purified solution on microslides and dried at room temperature.

2.2. Characterization

Absorption and photoluminescence properties of the samples were collected with circular dichroism spectra (MOS450, Biologic Science Instruments, Seyssinet-Pariset, France) and fluorescence spectra (PTI QM-TM, Photon Technology International, Ontario, Canada), respectively. Absolute QY was recorded on HAMAMATSU C11347 spectrometer (Hamamatsu Photonics Instruments, Hamamatsu City, Japan) by testing samples in toluene. Luminescence lifetime measurements were collected on EDINBURGH INSTRUMENTS (FLS920, Edinburgh Instruments Ltd., Livingston, UK). X-Ray diffraction (XRD, MiniFlex 600, Rigaku Corporation, Tokyo, Japan) analysis was used to obtain the crystalline phase. The scanning angle range was 10–50° (2θ) with a rate of 3°/min. Temperature-dependent Powder XRD (Ultima IV, Rigaku Corporation, Tokyo, Japan) were studied to indicate the thermal stability of samples. Transmission electron microscopy (TEM, Tecnai G2 F20,

FEI Technologies, Hillsboro, OR, USA) were carried out to obtain the morphology and interlayer spacing of samples at an accelerating voltage of 200 kV. The elemental analysis was conducted on inductively coupled plasma atomic emission spectroscopy (ICP-AES, Agilent 720, Agilent Technologies, Santa Clara, CA, USA).

3. Results and Discussion

3.1. Structural and Morphological Characterizations

The structure and morphology of the prepared samples were determined by XRD and TEM measurements, respectively. For the nominal doping concentration of Mn^{2+} in 0~75 at.% range, the XRD pattern (Figure 1) shows a series of diffraction peaks that can all be attributed to cubic phase $CH_3NH_3PbCl_3$ perovskite structure (space group: Pm$\bar{3}$m), demonstrating that the incorporation of Mn^{2+} has little effect on the crystal structure of $CH_3NH_3PbCl_3$ [18]. With the addition of 90 at. % $MnCl_2$, some diffraction peaks attributed to CH_3NH_3Cl and $MnCl_2$ precursors could be clearly observed due to the excess amount of precursors. Note that the broad band ranging from 15° to 38° is characteristic of amorphous silica resulting from the hydrolysis of APTES, which has been clearly elucidated in our previous work [34]. To further determine the concentration of Mn^{2+} dopant, ICP-AES was used, with results summarized in Table S1. In the synthesis of Mn^{2+}-doped $CH_3NH_3PbCl_3$ PNCs, it was found that APTES would react with $MnCl_2$ to form a puce complex at a high concentration of Mn^{2+} precursor, as shown in Figure S1. The color of $CH_3NH_3Pb_xMn_{1-x}Cl_3$ PNCs capped with APTES under room light is deeper than that of $CH_3NH_3Pb_xMn_{1-x}Cl_3$ PNCs passivated with octylamine owing to the adsorption of Mn^{2+}-APTES complex, indicating that the detected concentration in Table S1 are higher than the actual concentration of Mn^{2+} in the crystal lattice.

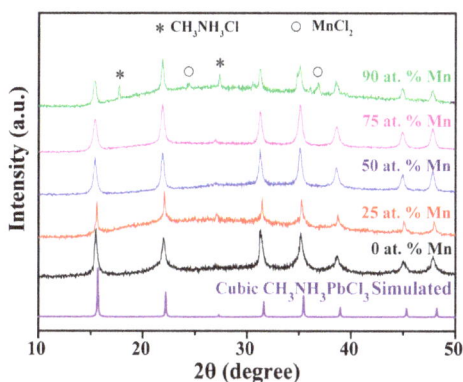

Figure 1. XRD pattern of Mn^{2+}-doped $CH_3NH_3PbCl_3$ PNCs with different concentration of Mn^{2+} (nominal concentration). The asterisks and circles denote the characteristic XRD peaks of CH_3NH_3Cl and $MnCl_2$, respectively.

To determine the morphology evolution and size distribution, TEM images were recorded, as shown in Figure 2. For undoped $CH_3NH_3PbCl_3$ PNCs (Figure 2a), the PNCs are mostly spherical with an average diameter of 16.2 ± 4.4 nm. The PNCs are embedded in amorphous silica matrix (Figure S2), which results from the hydrolysis of APTES. When $MnCl_2$ precursor was added, the resulting Mn^{2+}-doped PNCs are mostly cubic with an average dimension increasing from 10.4 ± 5.1 nm (Figure 2b) to 25.3 ± 7.2 nm (Figure 2e) as the Mn^{2+} concentration was increased from 25 to 90 at. %. Due to the complex reaction between APTES and the Mn^{2+} precursor, the concentration of APTES ligands decreased after adding $MnCl_2$ precursor. Higher concentration of APTES can facilitate the formation of small spherical PNCs due to the large steric hindrance of branched APTES [34]. Likewise, with the increase of Mn^{2+} precursor concentration, the lower concentration of APTES will result in large cubic PNCs, which is consistent with the TEM results of this work. High-resolution TEM (HRTEM, Figure 2f) reveals that the lattice space of undoped PNCs is around 0.28 nm, corresponding to the (200) crystal plane of cubic $CH_3NH_3PbCl_3$.

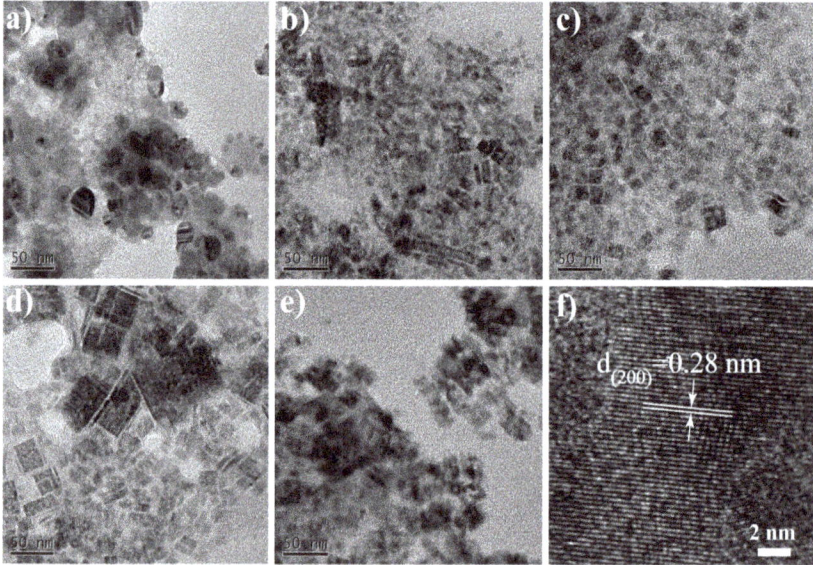

Figure 2. TEM images of Mn^{2+}-doped $CH_3NH_3PbCl_3$ PNCs doped with different Mn^{2+} concentrations. (**a**) 0 at. % Mn^{2+}, (**b**) 25 at. % Mn^{2+}, (**c**) 50 at. % Mn^{2+}, (**d**) 75 at. % Mn^{2+}, (**e**) 90 at. % Mn^{2+} and (**f**) HRTEM image of 0 at. % Mn^{2+}.

3.2. Optical Properties

Similar to Mn^{2+}-doped all inorganic $CsPbCl_3$ PNCs, new emission band may be introduced in Mn^{2+}-doped organic-inorganic $CH_3NH_3PbCl_3$ PNCs, as shown in the PL spectra in Figure 3, where the electronic absorption spectra are also shown. For undoped $CH_3NH_3PbCl_3$ PNCs, the absorption spectrum exhibits a sharp rise starting around 400 nm and an excitonic peak at ~373 nm. In comparison, the excitonic peak positions of all the Mn^{2+}-doped $CH_3NH_3PbCl_3$ PNCs shift 4–10 nm towards long wavelength owing to a significant increase in average particle size. For the PL spectra, Mn^{2+}-doped $CH_3NH_3PbCl_3$ PNCs feature a broad emission band peaked at 610 nm, attributed to the spin-forbidden $^4T_1 \rightarrow ^6A_1$ transition of Mn^{2+} [31], along with a narrow luminescence band located at ~400 nm, which is assigned to intrinsic excitonic emission of the $CH_3NH_3PbCl_3$ host. Compared to the Mn^{2+} dopant emission (~590 nm) in $CsPbCl_3$ PNCs [28,31], Mn^{2+} dopant emission in the $CH_3NH_3PbCl_3$ system significantly red shifts by ~20 nm, possibly caused by the crystal lattice difference between cubic $CH_3NH_3PbCl_3$ and cubic/tetragonal $CsPbCl_3$. The lattice parameters of cubic $CH_3NH_3PbCl_3$ (space group: $Pm\bar{3}m$, a = b = c = 5.685 Å) are a little larger than that of cubic $CsPbCl_3$ (a = b = c = 5.605 Å) and tetragonal $CsPbCl_3$ (a = b = 5.590 Å, c = 5.630 Å) [18,36], resulting in the smaller energy difference (Δ_O) of Mn^{2+} d-orbitals.

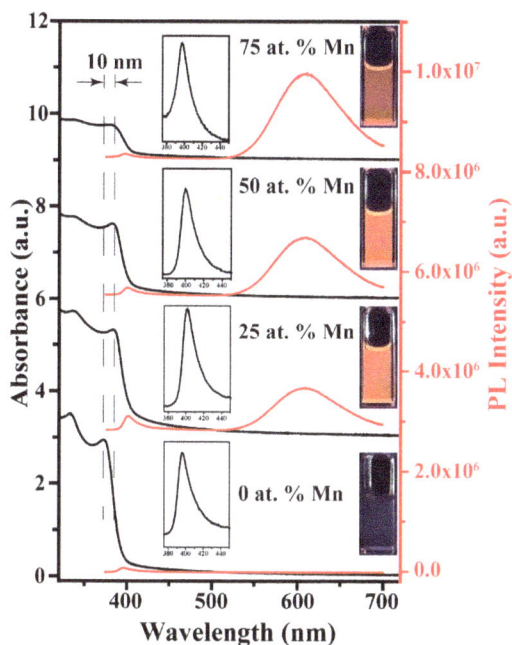

Figure 3. Absorption and PL spectra of Mn^{2+}-doped $CH_3NH_3PbCl_3$ PNCs solution with different nominal Mn^{2+} doping concentration. Inset: Enlarged views of PL spectra located in 375–450 nm range and digital pictures of Mn^{2+}-doped $CH_3NH_3PbCl_3$ PNCs under 365 nm UV light.

In addition, all excitonic emissions of Mn^{2+}-doped $CH_3NH_3PbCl_3$ PNCs show a slight red-shift (~4 nm) with respect to that of undoped $CH_3NH_3PbCl_3$ PNCs, consistent with red-shifted absorption spectra, attributed to the increase in size of PNCs. Intriguingly, with the increased concentration of Mn^{2+} dopant, the PL intensity of host increases first and then decreases, which is highly unusual. On one hand, the Mn^{2+} dopant ions may remove the preexisting structural defects and enhance the excitonic emission of hosts [25,37]. At the low doping concentration, both the intensity of excitionic emission and dopant emission are increased after 25 at. % Mn^{2+} doping, attributed to the dominant defect passivation of Mn^{2+} and exciton-to-Mn^{2+} energy transfer. On the other hand, the excitonic emission will be quenched upon the incorporation of Mn^{2+}. With the addition of a higher Mn^{2+} concentration, the transfer of excitonic energy from the host to Mn^{2+} dopants becomes more dominant, giving rise to the decrease of excitonic emission intensity and enhancement of dopant emission intensity. Importantly, the overall PL QY was highly improved from 0.4% (undoped) to 15.6% (75 at. % Mn^{2+}-doped), indicating efficient excitons-to-Mn^{2+} energy transfer, which is desired for light emitting applications.

To demonstrate that the Mn^{2+} dopant emission does not originate from the Mn^{2+}-APTES complex, contrast test was conducted as shown in Figure S3. Without the formation of $CH_3NH_3PbCl_3$ perovskite, no absorption and emission band can be observed. In the range of 380–700 nm, no emission can be seen for $MnCl_2$ precursor, indicating that the Mn^{2+} dopant emission results from the host absorption. In addition, PLE spectra are collected to further illustrate the emission mechanism of Mn^{2+}-doped $CH_3NH_3PbCl_3$ PNCs. As shown in Figure S4, a sharp onset located at ~400 nm can be clearly observed for all Mn^{2+}-doped $CH_3NH_3PbCl_3$ PNCs, consistent with the absorption of undoped $CH_3NH_3PbCl_3$ PNCs. These results display that the emission band centered at 610 nm originates from the absorption of $CH_3NH_3PbCl_3$ PNCs, manifesting the energy transfer from host to dopants.

The volume ratio between precursor solution and anti-solvent (toluene) is an important factor in the synthesis of $CH_3NH_3PbCl_3$ PNCs when adopting the reprecipitation strategy. Figure S5 displays the PL spectrum of Mn^{2+}-doped $CH_3NH_3PbCl_3$ PNCs solution prepared with different volume ratios. The Mn^{2+} dopant emission greatly increases in the range of 50–200 volume ratio after normalizing to excitonic emission, implying that more Mn^{2+} ions are incorporated into the crystal lattice of $CH_3NH_3PbCl_3$. However, the intensity of dopant emission is diminished when reaching 500 toluene/precursor volume ratio, owing to the poor crystallinity of the host at high Mn^{2+} dopant concentration, thereby lower the transfer efficiency of photo-induced excitons between the host and dopants due to the competition of charge transfer to trap states [28].

To investigate the PL decay of $CH_3NH_3Pb_xMn_{1-x}Cl_3$ PNCs, time-resolved PL spectra are obtained. All the PL decay kinetics were fitted with a single exponential curve, as shown in Figure 4. An expected decrease of excitonic lifetime from 2.40 ns to 2.04 ns can be observed with increasing the Mn^{2+} concentration from 0% to 75% (Figure 4a), due to the energy transfer from the host to Mn^{2+} dopants. Interestingly, the small change of exciton emission lifetime indicates the insignificant effect of dopant states on the overall exciton decay in the $CH_3NH_3PbCl_3$ PNCs. Since the introduction of Mn^{2+} dopants provides a new and efficient decay pathway of the excitons, which leads to the radiative decay of Mn^{2+} via the $^4T_1 \rightarrow {}^6A_1$ transition, the overall PL QY of the doped PNCs was improved [28].

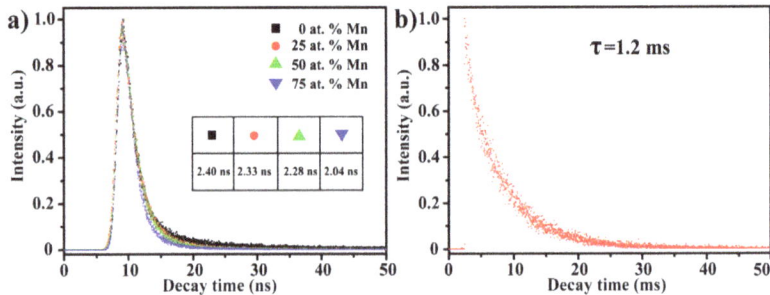

Figure 4. PL decay spectrum of Mn^{2+}-doped $CH_3NH_3PbCl_3$ PNCs films prepared with different $MnCl_2$ concentration. (**a**) Monitored at exciton emission (~410 nm) and (**b**) Mn^{2+} dopant emission of 50 at.% Mn (~610 nm).

PL decay curve of Mn^{2+} dopant emission located at 610 nm was also studied and presented in Figure 4b. A very long lifetime of 1.2 ms was observed resulting from the spin-forbidden nature of Mn^{2+} $^4T_1 \rightarrow {}^6A_1$ transition, suggesting the successful incorporation of Mn^{2+} [28,38].

3.3. Stability Investigation

Although moisture stability of $CH_3NH_3PbCl_3$ has been greatly improved, their practical applications are still restricted by their low decomposition temperature, which is determined by their low formation energies [39]. A recent study found that the formation energy of Mn^{2+}-doped $CsPbBr_3$ is calculated to be ~5% larger than that of pure $CsPbBr_3$ using DFT simulation, suggesting the better thermal stability of $CsPbBr_3$ after Mn^{2+} doping with a specific concentration [39]. In this work, the thermal stability of pure $CH_3NH_3PbCl_3$ and $CH_3NH_3Pb_xMn_{1-x}Cl_3$ PNCs has also been studied using temperature-dependent XRD, as presented in Figure 5. Without the addition of $MnCl_2$ precursor, $CH_3NH_3PbCl_3$ PNCs capped with octylamine (Figure 5a) show poor thermal stability. Some diffraction peaks due to impurities could be observed when the temperature was increased up to 170 °C. $CH_3NH_3Pb_xMn_{1-x}Cl_3$ PNCs passivated with octylamine were relatively stable under 170 °C, but still to be decomposed at 180 °C (Figure 5b). In contrast, both $CH_3NH_3PbCl_3$ PNCs and $CH_3NH_3Pb_xMn_{1-x}Cl_3$ PNCs passivated with APTES show better thermal stability, as shown in

Figure 5c,d. The decomposition temperature was significantly improved by 20–30 °C compared to PNCs passivated with octylamine. The improved thermal stability of $CH_3NH_3PbCl_3$ PNCs passivated with APTES may be due to the SiO_2 coating layer, which maintains the morphology and crystal structure of $CH_3NH_3PbCl_3$ as a result of the higher thermal stability of SiO_2.

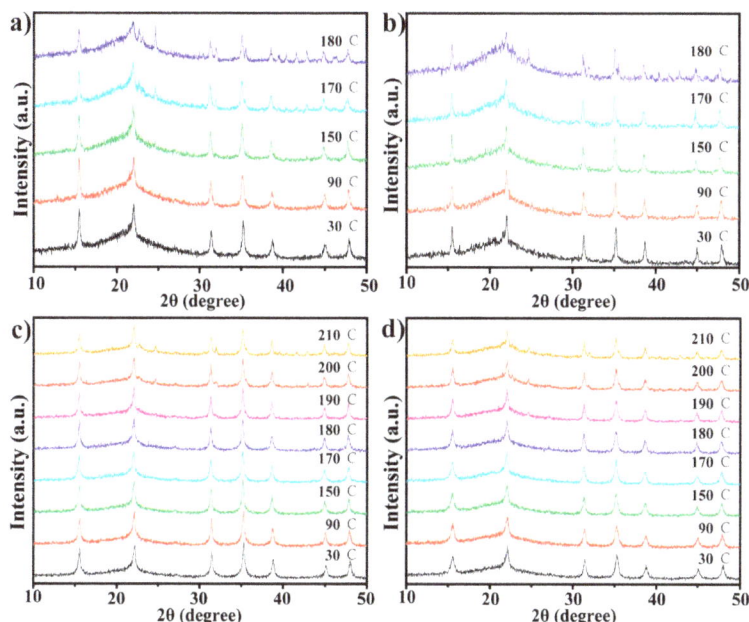

Figure 5. Temperature-dependent Powder XRD test of (**a**) $CH_3NH_3PbCl_3$ PNCs passivated with octylamine, (**b**) 50 at.% Mn^{2+}-doped $CH_3NH_3PbCl_3$ PNCs passivated with octylamine, (**c**) $CH_3NH_3PbCl_3$ PNCs passivated with APTES and (**d**) 50 at.% Mn^{2+}-doped $CH_3NH_3PbCl_3$ PNCs passivated with APTES.

The air stability of $CH_3NH_3Pb_xMn_{1-x}Cl_3$ PNCs films was also studied, as shown in Figure S6. The PL intensity $CH_3NH_3Pb_xMn_{1-x}Cl_3$ PNCs passivated with APTES and octylamine drop ~9% and ~73% after one week, indicating the better air stability of $CH_3NH_3Pb_xMn_{1-x}Cl_3$ PNCs capped with APTES.

4. Conclusions

Doping of Mn^{2+} into organic-inorganic $CH_3NH_3PbCl_3$ PNCs has been demonstrated using a versatile reprecipitation approach. The PL QY of the obtained samples are greatly improved after Mn^{2+} doping as a result of efficient energy transfer from the host to dopants. Furthermore, both thermal and air stability are enhanced due to the presence of SiO_2 on the PNC surface. Importantly, compared to the Mn^{2+} dopant emission (~590 nm) in all inorganic $CsPbCl_3$ system, the larger crystal lattice of $CH_3NH_3PbCl_3$ results in a lower energy splitting of Mn^{2+} d-orbitals, thereby emitting at a longer wavelength (~610 nm). In this manner, Mn^{2+} dopant emission may be tuned by modulating the crystal lattice through substitution of different components in perovskites.

Supplementary Materials: Supplementary materials can be found at www.mdpi.com/2073-4352/8/1/4/s1, Figure S1: Photographs of 50 at.% Mn^{2+}-doped $CH_3NH_3PbCl_3$ PNCs passivated with octylamine (left) and APTES (right) under room light (A) and 365 nm UV light (B); Figure S2: TEM image of $CH_3NH_3PbCl_3$ PNCs passivated with APTES; Figure S3: (a) Absorption and (b) PL spectra of solutions prepared using $MnCl_2$ and

APTES precursors (black line), and using $MnCl_2$, APTES, $PbCl_2$ and CH_3NH_3Cl precursors (red line). (c) PL spectra of $MnCl_2$ (λ_{ex} = 360 nm); Figure S4: PLE spectra of Mn doped $CH_3NH_3PbCl_3$ PNCs with different doping concentration (λ_{em} = 610 nm); Figure S5: PL spectra of 50 at.% Mn^{2+}-doped $CH_3NH_3PbCl_3$ PNCs solutions with different toluene/precursor volume ratios; Figure S6: Air stability test of Mn^{2+}-doped $CH_3NH_3PbCl_3$ PNCs passivated with octylamine and APTES; Table S1. Elemental analysis of $CH_3NH_3Pb_xMn_{1-x}Cl_3$ PNCs passivated with APTES under different doping concentration using ICP-AES.

Acknowledgments: This project was supported by the support of STU Scientific Research Foundation for Talents (NTF17001) and Science and technology planning project of Guangdong Province (No. 2014A020216045).

Author Contributions: X.L. and B.L. conceived and designed the experiments; X.L. and Y.G. performed the experiments and analyzed the data; B.L. wrote the paper.

Conflicts of Interest: The authors declare no conflicts of interest.

References

1. Zhang, F.; Zhong, H.; Chen, C.; Wu, X.; Hu, X.; Huang, H.; Han, J.; Zou, B.; Dong, Y. Brightly luminescent and color-tunable colloidal $CH_3NH_3PbX_3$ (X = Br, I, Cl) quantum dots: Potential alternatives for display technology. *ACS Nano* **2015**, *9*, 4533–4542. [CrossRef] [PubMed]
2. Luo, B.; Naghadeh, S.B.; Allen, A.L.; Li, X.; Zhang, J.Z. Peptide-passivated lead halide perovskite nanocrystals based on synergistic effect between amino and carboxylic functional groups. *Adv. Funct. Mater.* **2017**, *27*, 1604018. [CrossRef]
3. Xuan, T.; Yang, X.; Lou, S.; Huang, J.; Liu, Y.; Yu, J.; Li, H.; Wong, K.; Wang, C.; Wang, J. High stable $CsPbBr_3$ quantum dots coated with alkyl phosphate for white light-emitting diodes. *Nanoscale* **2017**, *9*, 15286–15290. [CrossRef] [PubMed]
4. Konstantakou, M.; Perganti, D.; Falaras, P.; Stergiopoulos, T. Anti-solvent crystallization strategies for highly efficient perovskite solar cells. *Crystals* **2017**, *7*, 291. [CrossRef]
5. Dirin, D.N.; Protesescu, L.; Trummer, D.; Kochetygov, I.V.; Yakunin, S.; Krumeich, F.; Stadie, N.P.; Kovalenko, M.V. Harnessing defect-tolerance at the nanoscale: Highly luminescent lead halide perovskite nanocrystals in mesoporous silica matrixes. *Nano Lett.* **2016**, *16*, 5866–5874. [CrossRef] [PubMed]
6. Huang, H.; Bodnarchuk, M.I.; Kershaw, S.V.; Kovalenko, M.V.; Rogach, A.L. Lead Halide perovskite nanocrystals in the research spotlight: Stability and defect tolerance. *ACS Energy Lett.* **2017**, *2*, 2071–2083. [CrossRef] [PubMed]
7. Koscher, B.A.; Swabeck, J.K.; Bronstein, N.D.; Alivisatos, A.P. Essentially trap-free $CsPbBr_3$ colloidal nanocrystals by postsynthetic thiocyanate surface treatment. *J. Am. Chem. Soc.* **2017**, *139*, 6566–6569. [CrossRef] [PubMed]
8. Protesescu, L.; Yakunin, S.; Bodnarchuk, M.I.; Krieg, F.; Caputo, R.; Hendon, C.H.; Yang, R.X.; Walsh, A.; Kovalenko, M.V. Nanocrystals of cesium lead halide perovskites ($CsPbX_3$, X = Cl, Br, and I): Novel optoelectronic materials showing bright emission with wide color gamut. *Nano Lett.* **2015**, *15*, 3692–3696. [CrossRef] [PubMed]
9. Nedelcu, G.; Protesescu, L.; Yakunin, S.; Bodnarchuk, M.I.; Grotevent, M.J.; Kovalenko, M.V. Fast anion-exchange in highly luminescent nanocrystals of cesium lead halide perovskites ($CsPbX_3$, X = Cl, Br, I). *Nano Lett.* **2015**, *15*, 5635–5640. [CrossRef] [PubMed]
10. Zhang, D.; Yang, Y.; Bekenstein, Y.; Yu, Y.; Gibson, N.A.; Wong, A.B.; Eaton, S.W.; Kornienko, N.; Kong, Q.; Lai, M.; et al. Synthesis of composition tunable and highly luminescent cesium lead halide nanowires through anion-exchange reactions. *J. Am. Chem. Soc.* **2016**, *138*, 7236–7239. [CrossRef] [PubMed]
11. Vybornyi, O.; Yakunin, S.; Kovalenko, M.V. Polar-solvent-free colloidal synthesis of highly luminescent alkylammonium lead halide perovskite nanocrystals. *Nanoscale* **2016**, *8*, 6278–6283. [CrossRef] [PubMed]
12. Huang, H.; Polavarapu, L.; Sichert, J.A.; Susha, A.S.; Urban, A.S.; Rogach, A.L. Colloidal lead halide perovskite nanocrystals: Synthesis, optical properties and applications. *NPG Asia Mater.* **2016**, *8*, e328. [CrossRef]
13. Luo, B.; Pu, Y.-C.; Yang, Y.; Lindley, S.A.; Abdelmageed, G.; Ashry, H.; Li, Y.; Li, X.; Zhang, J.Z. Synthesis, optical properties, and exciton dynamics of organolead bromide perovskite nanocrystals. *J. Phys. Chem. C* **2015**, *119*, 26672–26682. [CrossRef]
14. Tyagi, P.; Arveson, S.M.; Tisdale, W.A. Colloidal organohalide perovskite nanoplatelets exhibiting quantum confinement. *J. Phys. Chem. Lett.* **2015**, *6*, 1911–1916. [CrossRef] [PubMed]

15. Akkerman, Q.A.; Park, S.; Radicchi, E.; Nunzi, F.; Mosconi, E.; De Angelis, F.; Brescia, R.; Rastogi, P.; Prato, M.; Manna, L. Nearly monodisperse insulator Cs_4PbX_6 (X = Cl, Br, I) nanocrystals, their mixed halide compositions, and their transformation into $CsPbX_3$ nanocrystals. *Nano Lett.* **2017**, *17*, 1924–1930. [CrossRef] [PubMed]

16. Noh, J.H.; Im, S.H.; Heo, J.H.; Mandal, T.N.; Seok, S.I. Chemical management for colorful, efficient, and stable inorganic-organic hybrid nanostructured solar cells. *Nano Lett.* **2013**, *13*, 1764–1769. [CrossRef] [PubMed]

17. Johnston, M.B.; Herz, L.M. Hybrid perovskites for photovoltaics: charge-carrier recombination, diffusion, and radiative efficiencies. *Acc. Chem. Res.* **2016**, *49*, 146–154. [CrossRef] [PubMed]

18. Liu, Y.; Yang, Z.; Cui, D.; Ren, X.; Sun, J.; Liu, X.; Zhang, J.; Wei, Q.; Fan, H.; Yu, F.; et al. Two-inch-sized perovskite $CH_3NH_3PbX_3$ (X = Cl, Br, I) crystals: Growth and characterization. *Adv. Mater.* **2015**, *27*, 5176–5183. [CrossRef] [PubMed]

19. Fitzmorris, B.C.; Pu, Y.C.; Cooper, J.K.; Lin, Y.F.; Hsu, Y.J.; Li, Y.; Zhang, J.Z. Optical properties and exciton dynamics of alloyed core/shell/shell $Cd_{(1-X)}Zn_{(X)}Se/ZnSe/ZnS$ quantum dots. *ACS Appl. Mater. Interfaces* **2013**, *5*, 2893–2900. [CrossRef] [PubMed]

20. Zeng, R.; Zhang, T.; Dai, G.; Zou, B. Highly emissive, color-tunable, phosphine-free Mn:ZnSe/ZnS core/shell and Mn:ZnSeS shell-alloyed doped nanocrystals. *J. Phys. Chem. C* **2011**, *115*, 3005–3010. [CrossRef]

21. Begum, R.; Parida, M.R.; Abdelhady, A.L.; Murali, B.; Alyami, N.M.; Ahmed, G.H.; Hedhili, M.N.; Bakr, O.M.; Mohammed, O.F. Engineering interfacial charge transfer in $CsPbBr_3$ perovskite nanocrystals by heterovalent doping. *J. Am. Chem. Soc.* **2017**, *139*, 731–737. [CrossRef] [PubMed]

22. Liu, M.; Zhong, G.; Yin, Y.; Miao, J.; Li, K.; Wang, C.; Xu, X.; Shen, C.; Meng, H. Aluminum-doped cesium lead bromide perovskite nanocrystals with stable blue photoluminescence used for display backlight. *Adv. Sci.* **2017**, 1700335. [CrossRef] [PubMed]

23. Van der Stam, W.; Geuchies, J.J.; Altantzis, T.; van den Bos, K.H.; Meeldijk, J.D.; Van Aert, S.; Bals, S.; Vanmaekelbergh, D.; de Mello Donega, C. Highly emissive divalent-ion-doped colloidal $CsPb_{1-x}M_xBr_3$ perovskite nanocrystals through cation exchange. *J. Am. Chem. Soc.* **2017**, *139*, 4087–4097. [CrossRef] [PubMed]

24. Lin, C.C.; Xu, K.Y.; Wang, D.; Meijerink, A. Luminescent manganese-doped $CsPbCl_3$ perovskite quantum dots. *Sci. Rep.* **2017**, *7*, 45906. [CrossRef] [PubMed]

25. Parobek, D.; Roman, B.J.; Dong, Y.; Jin, H.; Lee, E.; Sheldon, M.; Son, D.H. Exciton-to-dopant energy transfer in mn-doped cesium lead halide perovskite nanocrystals. *Nano Lett.* **2016**, *16*, 7376–7380. [CrossRef] [PubMed]

26. Das Adhikari, S.; Dutta, S.K.; Dutta, A.; Guria, A.K.; Pradhan, N. Chemically tailoring the dopant emission in Manganese-doped $CsPbCl_3$ perovskite nanocrystals. *Angew. Chem.* **2017**, *56*, 8746–8750. [CrossRef] [PubMed]

27. Huang, G.; Wang, C.; Xu, S.; Zong, S.; Lu, J.; Wang, Z.; Lu, C.; Cui, Y. Postsynthetic doping of $MnCl_2$ molecules into preformed $CsPbBr_3$ perovskite nanocrystals via a halide exchange-driven cation exchange. *Adv. Mater.* **2017**, *29*. [CrossRef] [PubMed]

28. Liu, H.; Wu, Z.; Shao, J.; Yao, D.; Gao, H.; Liu, Y.; Yu, W.; Zhang, H.; Yang, B. $CsPb_xMn_{1-x}Cl_3$ perovskite quantum dots with high mn substitution ratio. *ACS Nano* **2017**, *11*, 2239–2247. [CrossRef] [PubMed]

29. Wang, Q.; Zhang, X.; Jin, Z.; Zhang, J.; Gao, Z.; Li, Y.; Liu, S.F. Energy-Down-Shift $CsPbCl_3$: Mn quantum dots for boosting the efficiency and stability of perovskite solar cells. *ACS Energy Lett.* **2017**, *2*, 1479–1486. [CrossRef]

30. Mir, W.J.; Jagadeeswararao, M.; Das, S.; Nag, A. Colloidal mn-doped cesium lead halide perovskite nanoplatelets. *ACS Energy Lett.* **2017**, 537–543. [CrossRef]

31. Liu, W.; Lin, Q.; Li, H.; Wu, K.; Robel, I.; Pietryga, J.M.; Klimov, V.I. Mn^{2+}-doped lead halide perovskite nanocrystals with dual-color emission controlled by halide content. *J. Am. Chem. Soc.* **2016**, *138*, 14954–14961. [CrossRef] [PubMed]

32. Shamsi, J.; Dang, Z.; Bianchini, P.; Canale, C.; Stasio, F.D.; Brescia, R.; Prato, M.; Manna, L. colloidal synthesis of quantum confined single crystal $CsPbBr_3$ nanosheets with lateral size control up to the micrometer range. *J. Am. Chem. Soc.* **2016**, *138*, 7240–7243. [CrossRef] [PubMed]

33. Bekenstein, Y.; Koscher, B.A.; Eaton, S.W.; Yang, P.; Alivisatos, A.P. Highly luminescent colloidal nanoplates of perovskite cesium lead halide and their oriented assemblies. *J. Am. Chem. Soc.* **2015**, *137*, 16008–16011. [CrossRef] [PubMed]

34. Luo, B.; Pu, Y.C.; Lindley, S.A.; Yang, Y.; Lu, L.; Li, Y.; Li, X.; Zhang, J.Z. Organolead halide perovskite nanocrystals: Branched capping ligands control crystal size and stability. *Angew. Chem.* **2016**, *55*, 8864–8868. [CrossRef] [PubMed]

35. Sun, C.; Zhang, Y.; Ruan, C.; Yin, C.; Wang, X.; Wang, Y.; Yu, W.W. Efficient and stable white leds with silica-coated inorganic perovskite quantum dots. *Adv. Mater.* **2016**, *28*, 10088–10094. [CrossRef] [PubMed]

36. Moller, C.K. A phase transition in cesium plumbochloride. *Nature* **1957**, *180*, 981–982. [CrossRef]

37. Rossi, D.; Parobek, D.; Dong, Y.; Son, D.H. Dynamics of exciton-mn energy transfer in mn-doped CsPbCl$_3$ perovskite nanocrystals. *J. Phys. Chem. C* **2017**, *121*, 17143–17149. [CrossRef]

38. Guria, A.K.; Dutta, S.K.; Adhikari, S.D.; Pradhan, N. Doping Mn^{2+} in lead halide perovskite nanocrystals: Successes and challenges. *ACS Energy Lett.* **2017**, *2*, 1014–1021. [CrossRef]

39. Zou, S.; Liu, Y.; Li, J.; Liu, C.; Feng, R.; Jiang, F.; Li, Y.; Song, J.; Zeng, H.; Hong, M.; et al. Stabilizing cesium lead halide perovskite lattice through Mn(II) substitution for air-stable light-emitting diodes. *J. Am. Chem. Soc.* **2017**, *139*, 11443–11450. [CrossRef] [PubMed]

crystals

MDPI

Article

Atomic Characterization of Byproduct Nanoparticles on Cesium Lead Halide Nanocrystals Using High-Resolution Scanning Transmission Electron Microscopy

Mian Zhang, Hongbo Li, Qiang Jing, Zhenda Lu and Peng Wang *

National Laboratory of Solid State Microstructures, College of Engineering and Applied Sciences and Collaborative Innovation Center of Advanced Microstructures, Nanjing University, Nanjing 210093, China; mg1534034@smail.nju.edu.cn (M.Z.); lihongbo@nju.edu.cn (H.L.); jq05103113@163.com (Q.J.); luzhenda@nju.edu.cn (Z.L)
* Correspondence: wangpeng@nju.edu.cn

Received: 23 November 2017; Accepted: 20 December 2017; Published: 22 December 2017

Abstract: Recent microstructural studies on lead halide perovskite nanocrystals have consistently reported the coexistence of byproduct nanoparticles (NPs). However, the nature of these NPs and their formation mechanism are still a matter of debate. Herein, we have investigated the structure and compositions of the NPs located on colloidal cesium lead bromide nanocrystals ($CsPbBr_3$ NCs), mainly through aberration-corrected transmission electron microscopy and spectroscopy. Our results show that these NPs can be assigned to $PbBr_2$ and $CsPb_2Br_5$. The new $CsPb_2Br_5$ species are formed by reacting $CsPbBr_3$ NCs with the remaining $PbBr_2$ during the drying process. In addition, observation of the metallic Pb NPs are ascribed to the electron damage effect on $CsPbBr_3$ NCs during transmission electron microscopy imaging.

Keywords: lead halide perovskite materials; colloidal cesium lead halide nanocrystals; scanning transmission electron microscopy and spectroscopy

1. Introduction

Lead halide perovskite (LHP) materials have attracted significant attention in recent years, due to their promising applications in solar cells [1–5], light emitting diodes (LEDs) [6–9], lasers [10–13], and photo-detectors [14–17]. In the family of LHP materials, colloidal all-inorganic cesium lead halide ($CsPbX_3$, X = Cl, Br, and I) perovskite nanocrystals combine both high stability and superior high luminescent efficiency compared to their organic-inorganic counterparts [18,19]. With their colloidal properties, many works have concentrated on applying them to solution-processed devices such as display screens [20], lasers [12] and LEDs [8].

Recent studies have consistently reported the observation of nanoparticles (NPs) with high contrast to LHP nanocrystals ($CH_3NH_3PbBr_3$ [21] and $CsPbBr_3$ [22–26]) in transmission electron microscopy (TEM) images (i.e., the white dots in Figure 1). However, the structure, chemistry, and forming mechanism of these NPs are still unclear and under debate in these reports. Some reports indicate that these NPs are Pb metal, originating from either the precursor material $PbBr_2$ [26], or the electron damaged $CsPbBr_3$ lattices [23–25]. One report suggests that these NPs are $PbBr_2$ [21] and another report claims that these NPs are $CsPb_2Br_5$ [22].

Here, we demonstrate that these NPs exist as a mix of $PbBr_2$ and $CsPb_2Pb_5$ on the $CsPbBr_3$ nanocrystals (NCs). We propose a formation process of these NPs: (i) Excess reactant of $PbBr_2$ during the synthesis and, subsequently, incomplete purification of the $CsPbBr_3$ products, synergetically lead to excess Pb^{2+} and Br^- dissolved in the $CsPbBr_3$ dispersion of nonpolar solvent. After the solvent evaporate from the carbon film, the $PbBr_2$ NPs quickly precipitate on the $CsPbBr_3$ NCs; (ii) The $PbBr_2$ NPs react with part of the $CsPbBr_3$ lattices to form the $CsPb_2Br_5$ NPs, still on the $CsPbBr_3$ NCs. Additional observation of the metallic Pb NPs are ascribed to the electron damage effect on $CsPbBr_3$ NCs during TEM imaging.

2. Results and Discussion

2.1. Morphology, Size, and Position

An overview of the NPs on the $CsPbBr_3$ NCs was first investigated using scanning TEM high-angle annular dark field (STEM-HAADF) images as shown in Figure 1. $CsPbBr_3$ NCs of about 15 nm with cubic and rectangular shapes are demonstrated. The NPs on the $CsPbBr_3$ NCs are of an average size of 2.72 ± 0.52 nm (inset of Figure 1a). These NPs have a tendency to locate near the edge of each $CsPbBr_3$ NC and a NP bridging itself between two $CsPbBr_3$ NCs has been even found (indicated with an arrow in Figure 1b), which suggests that these NPs do not locate inside the $CsPbBr_3$ lattice, but are attached to the surface of $CsPbBr_3$ NCs.

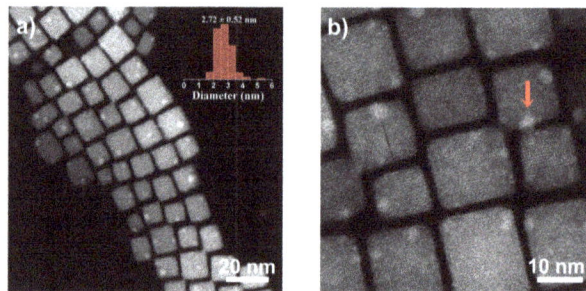

Figure 1. Overview STEM-HAADF images of the NPs on the $CsPbBr_3$ NCs at different magnifications. The inset of (**a**) shows the size distribution of over 300 NPs. The red arrow in (**b**) indicates a NP bridging itself between two $CsPbBr_3$ NCs.

2.2. Structure and Compositions

The structure of these NPs on the $CsPbBr_3$ NCs was first investigated by selected-area electron diffraction (SAED). As shown in Figure 2a, excluding the bright diffractive rings belonging to the $CsPbBr_3$ (100), (110), (200), (210), and (220) planes, additional spots indicated with arrows can be indexed as a $PbBr_2$ (121) plane and a $CsPb_2Br_5$ (114) plane. More SAED results are illustrated in Figure S1 to further verify the existence of both $PbBr_2$ NPs and $CsPb_2Br_5$ NPs. The identification of $PbBr_2$ and $CsPb_2Br_5$ is difficult to directly perform from their weak SAED patterns in the background of those from $CsPbBr_3$ NCs, as most low-index diffraction rings from $PbBr_2$ planes and $CsPb_2Br_5$ planes are overlapped with the much brighter ones from $CsPbBr_3$ planes (Table S1). Therefore, while much fewer diffraction spots corresponding to $PbBr_2$ and $CsPb_2Br_5$ are found in the SAED patterns compared with the large number of NPs observed in STEM-HAADF images, this is most likely due to the small size and poor crystallinity of the NPs expressing much weaker diffractive signals, largely covered by the signals from carbon films and $CsPbBr_3$ NCs. Later, the existence of the $PbBr_2$ and $CsPb_2Br_5$ NPs were further confirmed using high-resolution STEM-HAADF. Figure 2b shows a NP (about 3 nm in diameter) with an interplanar spacing of 2.55 Å, matching well with the $CsPb_2Br_5$ (312) plane (a PbBr textsubscript2 (230) plane is also possible). Note that other possible compounds, including Pb and CsBr, have all been taken into consideration and do not match with the above measurement results. Therefore, the NPs are preliminarily considered as a mix of $PbBr_2$ and $CsPb_2Br_5$.

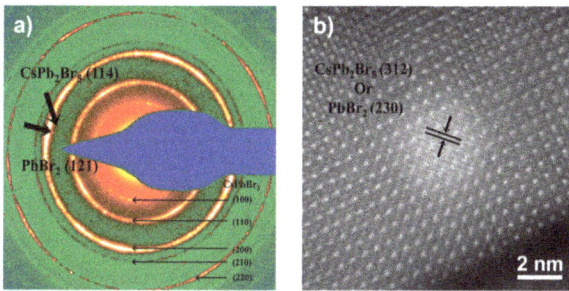

Figure 2. (a) SAED pattern of the NPs on the $CsPbBr_3$ NCs. Arrowed spots can be identified as $PbBr_2$ NPs and $CsPb_2Br_5$ NPs, excluding the main diffraction rings belonging to various $CsPbBr_3$ planes. (b) High-resolution STEM-HAADF image of a $CsPb_2Br_5$ NP (a $PbBr_2$ NP is also possible) formed on the $CsPbBr_3$ NC ($CsPbBr_3$, cubic phase, PDF#75-0412; $PbBr_2$, orthorhombic phase, PDF#31-0679; $CsPb_2Br_5$, tetragonal phase, PDF#25-0211).

In addition, metallic Pb NPs were also found on $CsPbBr_3$ NCs using high-resolution STEM-HAADF. Figure 3a,b shows the Pb (200) and Pb (020) atomic planes with a lattice spacing of 2.46 Å and an interfacial angle of 90°, oriented along a [001] zone axis. Since previous SAED results (Figure 2a) have implied no existence of metallic Pb, which contradicts the reported argument that metallic Pb NPs exist originally in the $CsPbBr_3$ dispersions [26], the Pb NPs found here at a high magnification (electron dose 1.9×10^5 e$^-$/Å^2) may, to a large extent, originate from the as-reported electron radiation effect on $CsPbBr_3$ lattice [23–25]. From Figure 3b, it is clear to see that the $CsPbBr_3$ lattices under the as-observed Pb NP were modified and damaged. Therefore, we propose that the metallic Pb NPs observed here may not be the extraneous NPs, like the $PbBr_2$ and $CsPb_2Br_5$ NPs, but rather a degradation product of $CsPbBr_3$ NCs from the focused electron beam during TEM imaging.

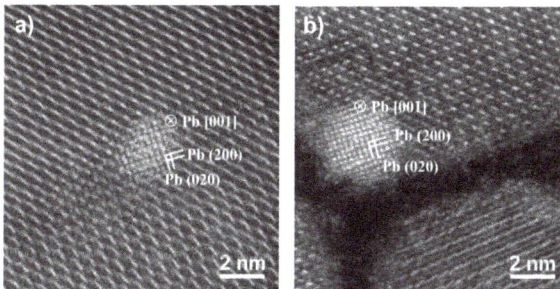

Figure 3. High-resolution STEM-HAADF images of the Pb NPs on the $CsPbBr_3$ NCs with [001] orientation (Pb, cubic phase, PDF#87-0663).

Furthermore, 2D energy dispersive spectroscopy (EDS) elemental mappings of several $CsPbBr_3$ NCs covered with the NPs (Figure 4a) was carried out to study the compositional information of the NPs. Due to the small size of the NPs and corresponding insufficient X-ray counts, the elemental maps (Figure 4a) have not shown strong variations of Cs, Pb, and Br in the position of the NPs. However, when we statistically calculated the atomic percentages of Cs (red), Pb (green), and Br (blue) from regions without the NPs (pure NC, i.e., indicated with a yellow circle in (a)) in comparison to that from regions with the NPs (NC+NP, i.e., indicated with a blue circle in (a)), as plotted in Figure 4b, it is evident to see that the NC+NP regions contain a lower atomic percentage of Br, more of Pb, and a nearly constant percentage of Cs, compared to the pure NC regions. The increment of the Pb percentage in NC+NP regions was most likely caused by the existence of $PbBr_2$ and $CsPb_2Br_5$ NPs. However, a question may arise that, if the NPs consisted of $PbBr_2$ and $CsPb_2Br_5$, the Br percentage should, in theory, increase, which is unexpectedly inconsistent with the experimental data (blue curve) in Figure 4b.

In order to understand this discrepancy, we further directly compare the integrated EDS spectra of Cs, Pb, and Br taken from both NC+NP (a) and pure NC (b) regions, as shown in Figure 5. The two spectra were normalized with the peak of Cs as the only reactant of $PbBr_2$ was in excess during the synthesis. Regarding the signal of Br, it unexpectedly decreases for the NC+NP region compared to the pure NC region. In theory, the signal of Br should increase for the NC+NP region, if the additional NPs consisted of $PbBr_2$ and $CsPb_2Br_5$. However, by considering the inevitable electron-induced Br loss during EDS measurements [25], there is a high possibility that the existence of these NPs may lead to the structural instability of $CsPbBr_3$. As a result, the NC+NP region might have undergone an even more severe electron damage effect compared to the pure NC region, so that the electron-induced Br loss may surpass the theoretical Br increase coming from the extra existence of $PbBr_2$ and $CsPb_2Br_5$ NPs, causing the total Br intensity to decrease, as shown in Figure 5b and the Br percentage to unexpectedly decrease in Figure 4b. It can be concluded that the increase of the Pb percentage found in Figure 4b is not only due to the excess of $PbBr_2$ and $CsPb_2Br_5$ NPs, but also the Br loss induced by the strong electron damage effect in the NC+NP regions. It can also be understood that the atomic Br/Pb ratio (Figure 4c) being lower for the NC+NP regions, may be due to the interplay of the existence of $PbBr_2$ and $CsPb_2Br_5$ NPs whose Br/Pb ratio are lower than the stoichiometric one (3:1), and the Br loss induced by the strong electron damage effect in the NC+NP regions during EDS measurements. In addition, the increase of Pb intensity in the NC+NP region in Figure 5b further confirms that an excess Pb content exists in the NC+NP region, most likely in the forms of $PbBr_2$ and $CsPb_2Br_5$ NPs in addition to $CsPbBr_3$.

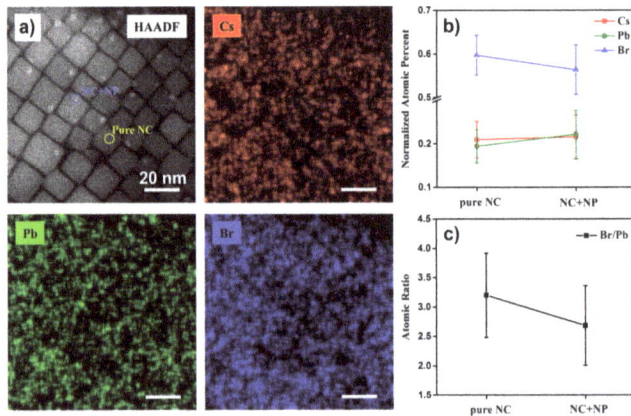

Figure 4. (a) STEM-EDS raw counts mappings of $CsPbBr_3$ NCs with the NPs. No strong variations of Cs, Pb, or Br are found in the position of the NPs. None of the auto-filter processing is applied here. (b) A plot of the calculated atomic percentages for Cs, Pb, and Br versus different regions (pure NC regions and NC+NP regions). We took an average of 10 quantified elemental atomic percentage results subtracted from 10 pure NC regions (i.e., indicated by a yellow circle in (a)). The error bar is the standard deviation. Similar calculations were applied to the NC+NP regions (i.e., indicated by a blue circle in (a)). (c) A plot of calculated atomic ratio of Br/Pb versus different regions (pure NC regions and NC+NP regions).

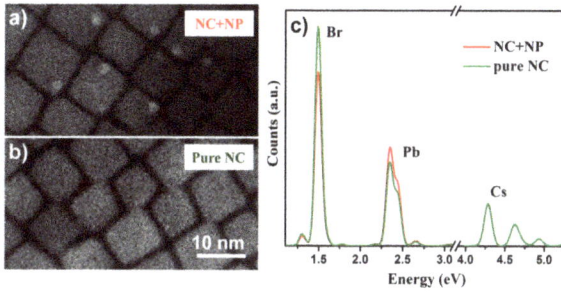

Figure 5. HAADF images of the NC+NP (**a**) and pure NC (**b**) regions. (**c**) Integrated EDS spectra acquired from NC+NP (—) and pure NC (—) regions as shown in (**a**) and (**b**), respectively.

2.3. Formation Mechanism

Here we suppose that the presence of PbBr$_2$ NPs can be ascribed to the excess PbBr$_2$ in the CsPbBr$_3$ dispersions. In our synthesis, molar ratio of reactant Cs:Pb:Br is controlled at about 1:3.3:6.6, which is similar to the most reported hot-injection procedures. Obviously, according to the stoichiometry ratio of CsPbBr$_3$ NCs, Cs species work as a limiting agent, and PbBr$_2$ is highly excessive in this reaction. In addition, incomplete purification of the CsPbBr$_3$ products during the following centrifugation step enables excess PbBr$_2$ to be left in the final CsPbBr$_3$ dispersions [27], existing as ion-complexes with oleic acid or oleylamine. When the drop of CsPbBr$_3$ dispersions on the ultrathin carbon film evaporates, the CsPbBr$_3$ NCs are spread out on the carbon film while the PbBr$_2$ NPs would precipitate on the carbon film or the upper surface of the NCs (Figure 6). This is consistent with the previous observation that the NPs are preferentially located at the corners and edges of the surface of CsPbBr$_3$ NCs. The chemical equilibrium reaction involved is as follows:

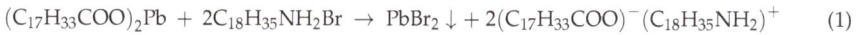

$$(C_{17}H_{33}COO)_2Pb + 2C_{18}H_{35}NH_2Br \rightarrow PbBr_2 \downarrow + 2(C_{17}H_{33}COO)^-(C_{18}H_{35}NH_2)^+ \qquad (1)$$

as the reaction shows, PbBr$_2$ would precipitate out, while oleate and oleylammonium form ion pairs. We note that if the above reaction was not complete, lead oleate and oleylammonium bromide may also precipitate out as separate species in the NPs. However, as it is difficult to directly characterize the structures of these organic non-crystalline materials through TEM imaging, the existence of lead oleate and oleylammonium bromide are not discussed here and are subject to future investigation.

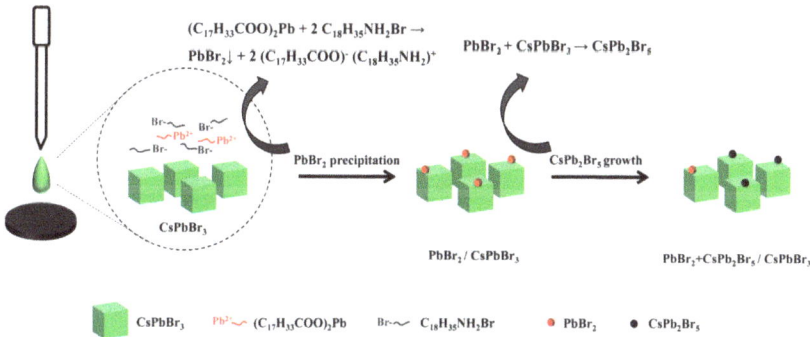

Figure 6. Schematic of the formation mechanism of the PbBr$_2$ NPs and the CsPb$_2$Br$_5$ NPs on CsPbBr$_3$ NCs.

Few reports have referred to the source of $CsPb_2Br_5$. One report demonstrated the existence of 10% $CsPb_2Br_5$ in the $CsPbBr_3$ solutions at room temperature [28]. Another report suggested that the excess $PbBr_2$ facilitates a structural transformation of small $CsPbBr_3$ nanocrystals in the precursor into $CsPb_2Br_5$ at a low temperature [22]. A recent report verified $CsPbBr_3$ nanocrystals can transfer to $CsPb_2Br_5$ nanosheets as the reaction time prolongs, clearly indicating this conversion (from $CsPbBr_3$ to $CsPb_2Br_5$) is thermodynamically favorable [29]. Therefore, we propose that the generation of $CsPb_2Br_5$ can be ascribed to the following process:

$$PbBr_2 + CsPbBr_3 \rightarrow CsPb_2Br_5 \tag{2}$$

The precipitated $PbBr_2$ NPs touch the $CsPbBr_3$ NC surface and tend to facilitate the generation of $CsPb_2Br_5$ NPs, as illustrated in Figure 6.

To confirm the contribution of excess $PbBr_2$ to the existence of the NPs, the remaining amount of $PbBr_2$ in the $CsPbBr_3$ dispersions were changed. Since directly changing the reactant amount may destroy the synthesis, we changed the purification times of the as-synthesized $CsPbBr_3$ products instead. Therefore, a comparison was made through TEM characterization between the following two samples: $CsPbBr_3$ NCs washed one time and $CsPbBr_3$ NCs washed three times. For the one-time wash sample, it is clear to see in Figure 7 that the NPs are evenly distributed both on the $CsPbBr_3$ NCs (a) and the carbon film (b), whereas for the sample washed three times with acetone, none of the NPs were found, neither on $CsPbBr_3$ NCs (c) or the carbon film (d). Moreover, the microstructure of the NPs on the carbon film was further confirmed by high-resolution TEM imaging and EDS elemental mapping. Figure 8a,b shows the high-resolution TEM images of two NPs on the carbon film. As shown in Figure 8a, the interplanar spacings of the NPs match well with the $PbBr_2$ (231) and $(42\bar{1})$ planes with an interfacial angle of 56°, oriented along a $[5\bar{6}8]$ axis. Additionally, in Figure 8b, the interplanar spacing of the NPs match well with the $PbBr_2$ (231) and $(\bar{2}12)$ planes with an interfacial angle of 74°, oriented along a $[5\bar{6}8]$ axis. Additionally, EDS measurements reveal that these NPs on the carbon film contain both Br and Pb, but lack Cs (See Figure S2 and Table S2). As a result, the unwashed NPs on the carbon film are verified to be $PbBr_2$. Therefore, it is confirmed that the excess reactant of $PbBr_2$ and incomplete purification contribute to the existence of the $PbBr_2$ NPs on both $CsPbBr_3$ NCs and carbon films.

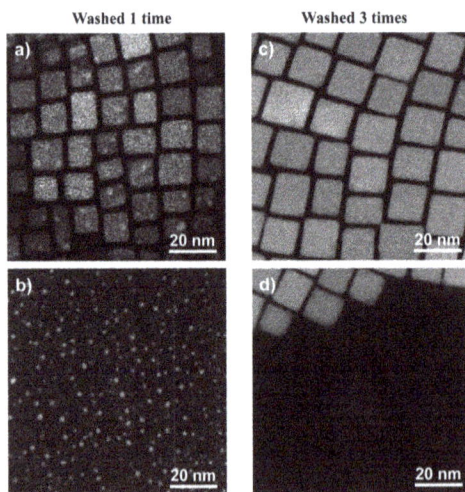

Figure 7. HAADF-STEM images of the distribution of the NPs on both $CsPbBr_3$ NCs and carbon films for the sample washed one time (**a,b**) and the sample washed three times (**c,d**), respectively.

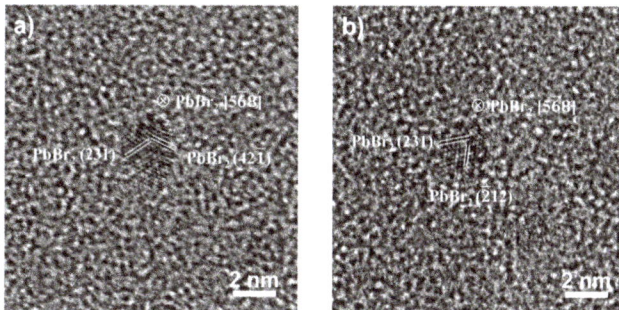

Figure 8. High-resolution TEM images of the PbBr$_2$ NPs found on carbon films along a [5̄68] axis.

As for the metallic Pb NPs, it was reported that the Pb NPs were formed in the PbBr$_2$ solution prior to injection of Cs salt [23], while another report investigated the effects of high energy electron radiation contributing to Br atoms desorption and Pb atoms crystallization [25]. Since previous SAEDs have indicated no Pb metals to exist, we assume that the original CsPbBr$_3$ dispersions have contained no Pb metals. Therefore, the observed Pb NPs by high-resolution HAADF-STEM might result from the electron damage effect on CsPbBr$_3$ NCs. Electron radiation (300 keV) induces desorption of Br atoms from the surface of the NCs and reduces Pb^{2+} cations to metallic Pb^0 atoms. Pb^0 atoms diffuse and aggregate into Pb NPs, oriented epitaxially on the CsPbBr$_3$ lattice.

3. Materials and Methods

3.1. Materials

Oleic acid (OA 90%), octadecene (ODE 90%), oleylamine (OAM 80–90%), cyclohexane, acetone, cesium carbonate (Cs$_2$CO$_3$ 99.9%), and lead bromide (PbBr$_2$ 99.999%) were purchased from Aladdin (Shanghai, China). All chemicals were used as received without any further purification.

3.2. Synthesis of CsPbBr$_3$ Perovskite NCs

The CsPbBr$_3$ perovskite NCs adopted here were fabricated following a hot injection method [30]. For the synthesis of the CsPbBr$_3$, 0.1 g of PbBr$_2$ was added into a mixture of 7.5 mL ODE, 1.5 mL OAM, and 0.75 mL OA, and then heated to 120 °C for 30 min under vacuum. The temperature was then raised to 150 °C, followed by the rapid injection of 0.7 mL of Cs-oleate solution (0.407 g Cs$_2$CO$_3$ dissolved within 20 mL ODE and 1.5 mL OA). After several seconds, the solution was rapidly cooled by the water bath. The products were then washed via centrifugation with cyclohexane/acetone for one or three times. Finally the products were dispersed in cyclohexane forming long-term colloidally-stable dispersions.

3.3. Characterization of the NPs on the CsPbBr$_3$ NCs

TEM images, selected-area electron diffraction (SAED) patterns and scanning TEM high angle annular dark field (STEM-HAADF) images were acquired using a FEI Tecnai G2 F20 S-TWIN microscope, operated at an accelerating voltage of 200 kV. High-resolution TEM images, high-resolution STEM-HAADF images [31], and energy-dispersive spectroscopy (EDS) elemental mappings were acquired using a FEI Titan^3 G2 60-300 microscope, operated at an accelerating voltage of 300 kV, equipped with double aberration correctors [32] and Super-X EDS detectors [33]. The image pixel size in the high-resolution TEM and high-resolution STEM-HAADF images was calibrated using the ratio of the CsPbBr$_3$ lattice constant listed in the PDF card (CsPbBr$_3$, cubic phase, PDF#75-0412) to the corresponding interplanar spacing pixels of the CsPbBr$_3$ NC measured in the image at the same magnification.

4. Conclusions

The TEM experiments presented here demonstrate that the byproduct NPs on CsPbBr$_3$ NCs is a mix of PbBr$_2$ and CsPb$_2$Br$_5$. We propose here that the excess reactant of PbBr$_2$ during the synthesis and, subsequently, the incomplete purification of the CsPbBr$_3$ products, synergetically led to the final excess precipitation of the PbBr$_2$ NPs on CsPbBr$_3$ NCs after solvent evaporation. The subsequent reaction between the PbBr$_2$ NPs and the CsPbBr$_3$ surface induce the formation of the CsPb$_2$Br$_5$ NPs. Moreover, the existence of these NPs may add to the structural instability of CsPbBr$_3$, so methods to remove these NPs are of great importance for maintaining their superior optical properties. Here, we propose that these byproduct NPs can be fully removed via washing CsPbBr$_3$ NCs three times with acetone. In addition, the metallic Pb NPs, which were not found in the original CsPbBr$_3$ dispersions, could be attributed to the electron damage of CsPbBr$_3$ NCs during TEM imaging. The above formation mechanism of the NPs on CsPbBr$_3$ NCs may also apply to other CsPbX$_3$ and hybrid lead halide perovskite nanocrystals.

Supplementary Materials: The following are available online at www.mdpi.com/2073-4352/8/1/2/s1; Figure S1: SAED patterns from two 0.48 μm^2 areas of the NPs on the CsPbBr$_3$ NCs; Figure S2: An integrated EDS spectrum acquired from a region with the NPs on a carbon film; Table S1: Interplanar spacings of CsPbBr$_3$, PbBr$_2$, and CsPb$_2$Br$_5$, respectively; Table S2: Atomic percentages of Cs, Pb, and Br for a region with the NPs on a carbon film.

Acknowledgments: This work was supported by the National Basic Research Program of China (grant no. 2015CB654901); the National Natural Science Foundation of China (11474147); the Natural Science Foundation of Jiangsu Province (grant no. BK20151383); the International Science and Technology Cooperation Program of China (2014DFE00200); and the Fundamental Research Funds for the Central Universities (no. 021314380077).

Author Contributions: Mian Zhang designed the experiments. Hongbo Li and Qiang Jing contributed sample materials. Mian Zhang performed the experiments and analyzed the dataset. Zhenda Lu and Peng Wang designed, supervised the project, and interpreted data.

Conflicts of Interest: The authors declare no conflict of interest.

References

1. Kojima, A.; Teshima, K.; Shirai, Y.; Miyasaka, T. Organometal halide perovskites as visible-light sensitizers for photovoltaic cells. *J. Am. Chem. Soc.* **2009**, *131*, 6050–6051. [CrossRef] [PubMed]
2. Lee, M.M.; Teuscher, J.; Miyasaka, T.; Murakami, T.N.; Snaith, H.J. Efficient hybrid solar cells based on meso-superstructured organometal halide perovskites. *Science* **2012**, *338*, 643–647. [CrossRef] [PubMed]
3. Liu, D.; Kelly, T.L. Perovskite solar cells with a planar heterojunction structure prepared using room-temperature solution processing techniques. *Nat. Photonics* **2014**, *8*, 133–138. [CrossRef]
4. Liu, M.; Johnston, M.B.; Snaith, H.J. Efficient planar heterojunction perovskite solar cells by vapour deposition. *Nature* **2013**, *501*, 395–398. [CrossRef] [PubMed]
5. Yang, W.S.; Noh, J.H.; Jeon, N.J.; Kim, Y.C.; Ryu, S.; Seo, J.; Seok, S.I. High-performance photovoltaic perovskite layers fabricated through intramolecular exchange. *Science* **2015**, *348*, 1234–1237. [CrossRef] [PubMed]
6. Cho, H.; Jeong, S.-H.; Park, M.-H.; Kim, Y.-H.; Wolf, C.; Lee, C.-L.; Heo, J.H.; Sadhanala, A.; Myoung, N.; Yoo, S. Overcoming the electroluminescence efficiency limitations of perovskite light-emitting diodes. *Science* **2015**, *350*, 1222–1225. [CrossRef] [PubMed]
7. Kim, Y.H.; Cho, H.; Heo, J.H.; Kim, T.S.; Myoung, N.; Lee, C.L.; Im, S.H.; Lee, T.W. Multicolored Organic/Inorganic Hybrid Perovskite Light-Emitting Diodes. *Adv. Mater.* **2015**, *27*, 1248–1254. [CrossRef] [PubMed]
8. Song, J.; Li, J.; Li, X.; Xu, L.; Dong, Y.; Zeng, H. Quantum Dot Light-Emitting Diodes Based on Inorganic Perovskite Cesium Lead Halides (CsPbX$_3$). *Adv. Mater.* **2015**, *27*, 7162–7167. [CrossRef] [PubMed]

9. Tan, Z.-K.; Moghaddam, R.S.; Lai, M.L.; Docampo, P.; Higler, R.; Deschler, F.; Price, M.; Sadhanala, A.; Pazos, L.M.; Credgington, D. Bright light-emitting diodes based on organometal halide perovskite. *Nat. Nanotechnol.* **2014**, *9*, 687–692. [CrossRef] [PubMed]

10. D'Innocenzo, V.; Srimath Kandada, A.R.; De Bastiani, M.; Gandini, M.; Petrozza, A. Tuning the light emission properties by band gap engineering in hybrid lead halide perovskite. *J. Am. Chem. Soc.* **2014**, *136*, 17730–17733. [CrossRef] [PubMed]

11. Deschler, F.; Price, M.; Pathak, S.; Klintberg, L.E.; Jarausch, D.-D.; Higler, R.; Huttner, S.; Leijtens, T.; Stranks, S.D.; Snaith, H.J. High photoluminescence efficiency and optically pumped lasing in solution-processed mixed halide perovskite semiconductors. *J. Phys. Chem. Lett.* **2014**, *5*, 1421–1426. [CrossRef] [PubMed]

12. Yakunin, S.; Protesescu, L.; Krieg, F.; Bodnarchuk, M.I.; Nedelcu, G.; Humer, M.; De Luca, G.; Fiebig, M.; Heiss, W.; Kovalenko, M.V. Low-threshold amplified spontaneous emission and lasing from colloidal nanocrystals of caesium lead halide perovskites. *Nat. Commun.* **2015**, *6*, 8056. [CrossRef] [PubMed]

13. Zhu, H.; Fu, Y.; Meng, F.; Wu, X.; Gong, Z.; Ding, Q.; Gustafsson, M.V.; Trinh, M.T.; Jin, S.; Zhu, X. Lead halide perovskite nanowire lasers with low lasing thresholds and high quality factors. *Nat. Mater.* **2015**, *14*, 636–642. [CrossRef] [PubMed]

14. Fang, Y.; Dong, Q.; Shao, Y.; Yuan, Y.; Huang, J. Highly narrowband perovskite single-crystal photodetectors enabled by surface-charge recombination. *Nat. Photonics* **2015**, *9*, 675–686. [CrossRef]

15. Hu, X.; Zhang, X.; Liang, L.; Bao, J.; Li, S.; Yang, W.; Xie, Y. High-performance flexible broadband photodetector based on organolead halide perovskite. *Adv. Funct. Mater.* **2014**, *24*, 7373–7380. [CrossRef]

16. Lee, Y.; Kwon, J.; Hwang, E.; Ra, C.H.; Yoo, W.J.; Ahn, J.H.; Park, J.H.; Cho, J.H. High-performance perovskite–graphene hybrid photodetector. *Adv. Mater.* **2015**, *27*, 41–46. [CrossRef] [PubMed]

17. Yakunin, S.; Sytnyk, M.; Kriegner, D.; Shrestha, S.; Richter, M.; Matt, G.J.; Azimi, H.; Brabec, C.J.; Stangl, J.; Kovalenko, M.V. Detection of X-ray photons by solution-processed lead halide perovskites. *Nat. Photonics* **2015**, *9*, 444–449. [CrossRef] [PubMed]

18. Protesescu, L.; Yakunin, S.; Bodnarchuk, M.I.; Krieg, F.; Caputo, R.; Hendon, C.H.; Yang, R.X.; Walsh, A.; Kovalenko, M.V. Nanocrystals of cesium lead halide perovskites ($CsPbX_3$, X = Cl, Br, and I): Novel optoelectronic materials showing bright emission with wide color gamut. *Nano Lett.* **2015**, *15*, 3692–3696. [CrossRef] [PubMed]

19. Wang, Y.; Li, X.; Song, J.; Xiao, L.; Zeng, H.; Sun, H. All-inorganic colloidal perovskite quantum dots: A new class of lasing materials with favorable characteristics. *Adv. Mater.* **2015**, *27*, 7101–7108. [CrossRef] [PubMed]

20. Yoon, H.C.; Kang, H.; Lee, S.; Oh, J.H.; Yang, H.; Do, Y.R. Study of perovskite QD down-converted LEDs and six-color white LEDs for future displays with excellent color performance. *ACS Appl. Mater. Interfaces* **2016**, *8*, 18189–18200. [CrossRef] [PubMed]

21. Sichert, J.A.; Tong, Y.; Mutz, N.; Vollmer, M.; Fischer, S.; Milowska, K.Z.; García Cortadella, R.; Nickel, B.; Cardenas-Daw, C.; Stolarczyk, J.K. Quantum size effect in organometal halide perovskite nanoplatelets. *Nano Lett.* **2015**, *15*, 6521–6527. [CrossRef] [PubMed]

22. Zhang, X.; Xu, B.; Zhang, J.; Gao, Y.; Zheng, Y.; Wang, K.; Sun, X.W. All-inorganic perovskite nanocrystals for high-efficiency light emitting diodes. Dual-phase cspbbr3-cspb2br5 composites. *Adv. Funct. Mater.* **2016**, *26*, 4595–4600. [CrossRef]

23. Van der, S.W.; Geuchies, J.J.; Altantzis, T.; Kh, V.D.B.; Meeldijk, J.D.; Van, A.S.; Bals, S.; Vanmaekelbergh, D.; De, M.D.C. Highly emissive divalent-ion-doped colloidal cspb1-xmxbr3 perovskite nanocrystals through cation exchange. *J. Am. Chem. Soc.* **2017**, *139*, 4087–4097. [CrossRef] [PubMed]

24. Tong, Y.; Bladt, E.; Aygüler, M.F.; Manzi, A.; Milowska, K.Z.; Hintermayr, V.A.; Docampo, P.; Bals, S.; Urban, A.S.; Polavarapu, L. Highly luminescent cesium lead halide perovskite nanocrystals with tunable composition and thickness by ultrasonication. *Angew. Chem.* **2016**, *55*, 13887–13892. [CrossRef] [PubMed]

25. Dang, Z.; Shamsi, J.; Palazon, F.; Imran, M.; Akkerman, Q.A.; Park, S.; Bertoni, G.; Prato, M.; Brescia, R.; Manna, L. In Situ Transmission Electron Microscopy Study of electron beam-induced transformations in colloidal cesiumlead halide perovskite nanocrystals. *Acs Nano* **2017**, *11*, 2124–2132. [CrossRef] [PubMed]

26. Udayabhaskararao, T.; Kazes, M.; Houben, L.; Lin, H.; Oron, D. Nucleation, growth, and structural transformations of perovskite nanocrystals. *Chem. Mater.* **2017**, *29*, 1302–1308. [CrossRef]

27. Gonzalez-Carrero, S.; Galian, R.E.; Pérez-Prieto, J. Maximizing the emissive properties of $CH_3NH_3PbBr_3$ perovskite nanoparticles. *J. Mater. Chem. A* **2015**, *3*, 9187–9193. [CrossRef]

28. Rodová, M.; Brožek, J.; Knížek, K.; Nitsch, K. Phase transitions in ternary caesium lead bromide. *J. Therm. Anal. Calorim.* **2003**, *71*, 667–673. [CrossRef]
29. Li, G.; Wang, H.; Zhu, Z.; Chang, Y.; Zhang, T.; Song, Z.; Jiang, Y. Shape and phase evolution from $CsPbBr_3$ perovskite nanocubes to tetragonal $CsPb_2Br_5$ nanosheets with an indirect bandgap. *Chem. Commun.* **2016**, *52*, 11296–11299. [CrossRef] [PubMed]
30. Talapin, D.V.; Lee, J.-S.; Kovalenko, M.V.; Shevchenko, E.V. Prospects of colloidal nanocrystals for electronic and optoelectronic applications. *Chem. Rev.* **2009**, *110*, 389–458. [CrossRef] [PubMed]
31. Muller, D.A. Structure and bonding at the atomic scale by scanning transmission electron microscopy. *Nat. Mater.* **2009**, *8*, 263–270. [CrossRef] [PubMed]
32. Hutchison, J.L.; Titchmarsh, J.M.; Cockayne, D.J.; Doole, R.C.; Hetherington, C.J.; Kirkland, A.I.; Sawada, H. A versatile double aberration-corrected, energy filtered HREM/STEM for materials science. *Ultramicroscopy* **2005**, *103*, 7–15. [CrossRef] [PubMed]
33. Allen, L.J.; D'Alfonso, A.J.; Freitag, B.; Klenov, D.O. Chemical mapping at atomic resolution using energy-dispersive X-ray spectroscopy. *MRS Bull.* **2012**, *37*, 47–52. [CrossRef]

crystals

MDPI

Article

CH$_3$NH$_3$Cl Assisted Solvent Engineering for Highly Crystallized and Large Grain Size Mixed-Composition (FAPbI$_3$)$_{0.85}$(MAPbBr$_3$)$_{0.15}$ Perovskites

Yihui Li [1], Taiyang Zhang [1], Feng Xu [1], Yong Wang [1], Ge Li [1], Yang Yang [2,*] and Yixin Zhao [1,*]

[1]   School of Environmental Science and Engineering, Shanghai Jiao Tong University, Shanghai 200240, China; yihuiliwv@sjtu.edu.cn (Y.L.); sun.zhang1988@gmail.com (T.Z.); 1934380351@sjtu.edu.cn (F.X.); wyong9512@163.com (Y.W.); 5111619052@sjtu.edu.cn (G.L.)

[2]   Clean Energy Research Institute, China Huaneng Group, Beijing 102209, China

*   Correspondence: yangyang@qny.chng.com.cn (Y.Y.); yixin.zhao@sjtu.edu.cn (Y.Z.)

Academic Editor: Wei Zhang
Received: 17 July 2017; Accepted: 2 September 2017; Published: 5 September 2017

Abstract: High-quality mixed-cation lead mixed-halide (FAPbI$_3$)$_{0.85}$(MAPbBr$_3$)$_{0.15}$ perovskite films have been prepared using CH$_3$NH$_3$Cl additives via the solvent engineering method. The UV/Vis result shows that the addition of additives leads to enhanced absorptions. XRD and SEM characterizations suggest that compact, pinhole-free and uniform films can be obtained. This is attributable to the crystallization improvement caused by the CH$_3$NH$_3$Cl additives. The power conversion efficiency (PCE) of the F-doped SnO$_2$ (FTO)/compact-TiO$_2$/perovskite/Spiro-OMeTAD/Ag device increases from 15.3% to 16.8% with the help of CH$_3$NH$_3$Cl additive.

Keywords: perovskite; mixed-cation; mixed-halide; additive; CH$_3$NH$_3$Cl; solvent engineering

---

1. Introduction

Organic-inorganic hybrid lead halide perovskites have many attractive features such as tunable band gaps, easy-to-make properties, high optical absorption coefficients and superior charge transport properties. Therefore, they have become one of the most promising materials for optoelectronic applications, and the perovskite solar cells have reached a certified power conversion efficiency (PCE) of 22.1% [1–13]. The typical 3D hybrid lead halide perovskite has the classical APbX$_3$ structure, and the tunable composition for A and X components offers perovskites sufficient freedom in tuning structures and properties [1,3,14–20]. In particular, mixed-cation and mixed-halide perovskites have been an effective approach to optimize the properties of lead halide perovskites [5,21]. For example, the (FAPbI$_3$)$_{0.85}$(MAPbBr$_3$)$_{0.15}$ perovskite is a popular recipe used for solar cells with enhanced performance and stabilities [5]. In spite of the advancement, it is still a challenging task to simultaneously control the film morphology and crystalline quality. It is thus urgent to explore a method to gain high-quality films and improve perovskite crystallization. In previous studies on the MAPbI$_3$ perovskite, the additive method was an effective way to control perovskite crystallization and growth in simple solution chemistry [22–35]. In general, additives can be divided into several types: Organic molecules [25,27–29,36,37], inorganic or ammonium salts [23,30–35], polymers [26] and ionic liquids [22]. Different additives may have different functional mechanisms. For example, the additive can provide homogenous nucleation sites to improve uniformity. It can also coordinate with metal ions to decrease the crystallization rate and enlarge crystals [38–40], and highly efficient semitransparent perovskite solar cells can be achieved [41,42]. In addition, it can change the surface energy to control the crystal growth direction. However, the impact of additives on the mixed composition perovskite is still an area awaiting more explorations.

In this work, we deposit the mixed-cation and mixed-halide $(FAPbI_3)_{0.85}(MAPbBr_3)_{0.15}$ perovskite using CH_3NH_3Cl (MACl) additives. Our results indicate that the MACl additive has little impact on the optical properties of the $(FAPbI_3)_{0.85}(MAPbBr_3)_{0.15}$ perovskite film but can significantly increase crystallinity and crystal grain size.

2. Results and Discussion

These $(FAPbI_3)_{0.85}(MAPbBr_3)_{0.15}$ precursor films with or without MACl additives all turned orange after anti-solvent treatments without thermal annealing (0, 0.15, 0.3 molar ratio MACl with perovskite precursor solution donated as 0 MACl, 0.15 MACl and 0.3 MACl, respectively). The XRD patterns of these stable films before annealing are listed in Figure 1a. Firstly, a typical characteristic peak of perovskite appeared in all these precursor films. Meanwhile, an unknown peak located at ~12° existed in all these precursor films, which might be ascribed to an intermediate phase owing to the complex between the solvent and the perovskite precursor. Interestingly, the peak intensity of this unknown intermediate decreased with the increase of the molar ratio of MACl. The intermediate peak in the precursor film using 0.3 MACl additives became much weaker while the perovskite peak was stronger than in the precursor film without MACl. This indicates that MACl can affect or accelerate the formation of the perovskite phase and inhibit the intermediate phase in these un-annealed films. This indexed intermediate phase is highly likely to be a complex of Pb, I, MA, solvent. The UV/Vis spectra are listed in Figure 1b. All these precursor films showed a characteristic peak at around 510–520 nm without obvious absorption at longer wavelengths although there were some perovskite phases found in these precursor films.

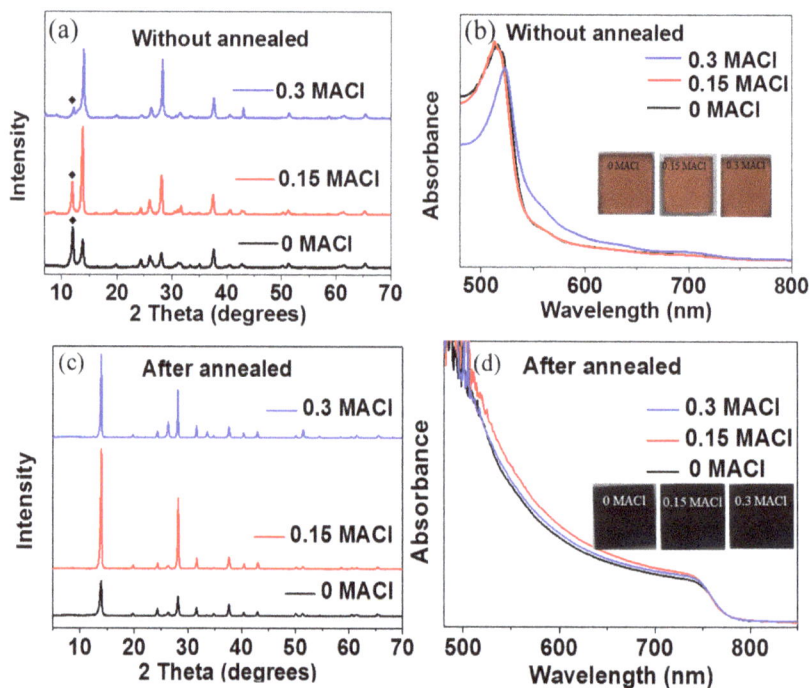

Figure 1. (a) The XRD patterns and (b) UV/Vis spectra of $(FAPbI_3)_{0.85}(MAPbBr_3)_{0.15}$ films using 0, 0.15 and 0.3 MACl additives (**a,b**) without having been annealed and (**c,d**) after having been annealed. Insert: Precursor film images. ♦: the peak of the intermediate.

In order to investigate the film change before and after annealing, the XRD patterns and UV/Vis spectra of $(FAPbI_3)_{0.85}(MAPbBr_3)_{0.15}$ on FTO glasses containing different amounts of MACl are listed in Figure 1c,d. All the orange precursor films turned dark after thermal annealing (see the insert images). The XRD patterns showed that the intermediate peak in all the precursor films disappeared and the samples with MACl additives had the same characteristic peaks of $(FAPbI_3)_{0.85}(MAPbBr_3)_{0.15}$ perovskites as the sample without MACl [43]. The UV/Vis spectra of these samples were almost the same, which is consistent with the XRD patterns. Although these $(FAPbI_3)_{0.85}(MAPbBr_3)_{0.15}$ samples were similar in the phase purity and UV/Vis absorption, the diffraction intensity was considerably enhanced in samples with the MACl additive, especially in the 0.15 MACl sample as shown in Figure 1c. This indicates that the MACl additive can significantly enhance the crystallinity of $(FAPbI_3)_{0.85}(MAPbBr_3)_{0.15}$. The XRD and UV/Vis spectra indicated that MACl greatly improved the perovskite crystallization, especially in the 0.15 MACl sample. It is worth noting that there was no Cl EDX signal found in these annealed 0.15 MACl and 0.3 MACl perovskite films. This means that MACl only functions as a "crystallization improver" affecting the crystal growth process and can be totally removed after annealing.

In order to further understand the perovskite crystallization process with and without MACl, we have investigated these perovskite films annealed for different times. Figure 2 lists the XRD patterns and UV/Vis spectrum evolution of the $(FAPbI_3)_{0.85}(MAPbBr_3)_{0.15}$ perovskite films prepared from 0, 0.15 and 0.3 MACl with different annealing times. The annealing temperature was 150 °C and different annealing times have been tested: 1 min, 3 min, 5 min, 8 min and 12 min (for 0.3 MACl). The above paragraph has suggested that the crystallinity of the perovskite grown from precursors with MACl was better than the corresponding one without MACl. Based on the XRD and UV/Vis spectra results in Figure 2a,b, the crystallization process of the perovskite without MACl additives was that the solvent evaporated and the perovskite crystallization finished in the first 3 min. The perovskite film then exhibited the highest absorbance and XRD intensities. During the period of 5–8 min, the loss of MAI or MABr started to happen in the perovskite without MACl additives. The over-annealing exceeding 5 min then resulted in the appearance of PbI_2. Over-annealing is usually adopted to obtain the PbI_2 passivation effect. According to the XRD patterns and UV/Vis evolution of the perovskites with 0.15 MACl and 0.3 MACl additives, it seems that MACl can retard the crystallization of the $(FAPbI_3)_{0.85}(MAPbBr_3)_{0.15}$ perovskite, although MACl can also help to reduce the intermediate phase in the precursor films. As shown in Figure 2c–f, the 0.15 MACl and 0.3 MACl samples also exhibited the standard $(FAPbI_3)_{0.85}(MAPbBr_3)_{0.15}$ perovskite XRD peaks and reached the maximal absorbance after 1–3 min annealing. However, their XRD intensities kept growing (especially for the 0.3 MACl sample), which might account for the enhanced crystallinity as observed in Figure 1. These MACl additives could also prevent the release of MAI since MACl needed to be expelled before MAI. However, PbI_2 was also observed in both the 0.15 MACl and 0 MACl samples after annealing for 8 min (Figure 2a,c), and PbI_2 appeared at 12 min in the 0.3 MACl sample (Figure 2e). The crystallinity of PbI_2 in the sample without MACl was higher than the one with MACl. Consequently, a weaker PbI_2 XRD peak was found in the 0.15 MACl and 0.3 MACl samples.

From the above-mentioned discussion, the MACl additive can significantly affect the crystallization of the $(FAPbI_3)_{0.85}(MAPbBr_3)_{0.15}$ perovskite. The SEM was taken to further understand the effect of additives on their crystallinity or grain sizes. Figure 3a,c show the SEM images of $(FAPbI_3)_{0.85}(MAPbBr_3)_{0.15}$ perovskite films prepared from 0 MACl, 0.15 MACl and 0.3 MACl. The perovskite crystal sizes in the 0 MACl sample are ~200–300 nm, which is typical for the regular solvent engineering process. Besides the $(FAPbI_3)_{0.85}(MAPbBr_3)_{0.15}$ perovskite crystals, there were also some bright spots and they were probably the PbI_2 residues, which were all formed at the grains size similar to the previous report. With the MACl additive, the $(FAPbI_3)_{0.85}(MAPbBr_3)_{0.15}$ perovskites showed larger crystal grain sizes. The crystal size of the 0.15 MACl sample increased to ~400 nm and that of the 0.3 MACl sample further increased to ~500 nm. They both had some PbI_2 residues existing at the grain boundary. Interestingly, the 0.3 MACl sample seemed to have more PbI_2 residues than the 0.15 MACl sample and also the 0 MACl one. Usually, the large crystal grain can reduce charge recombination and promotes charge transport.

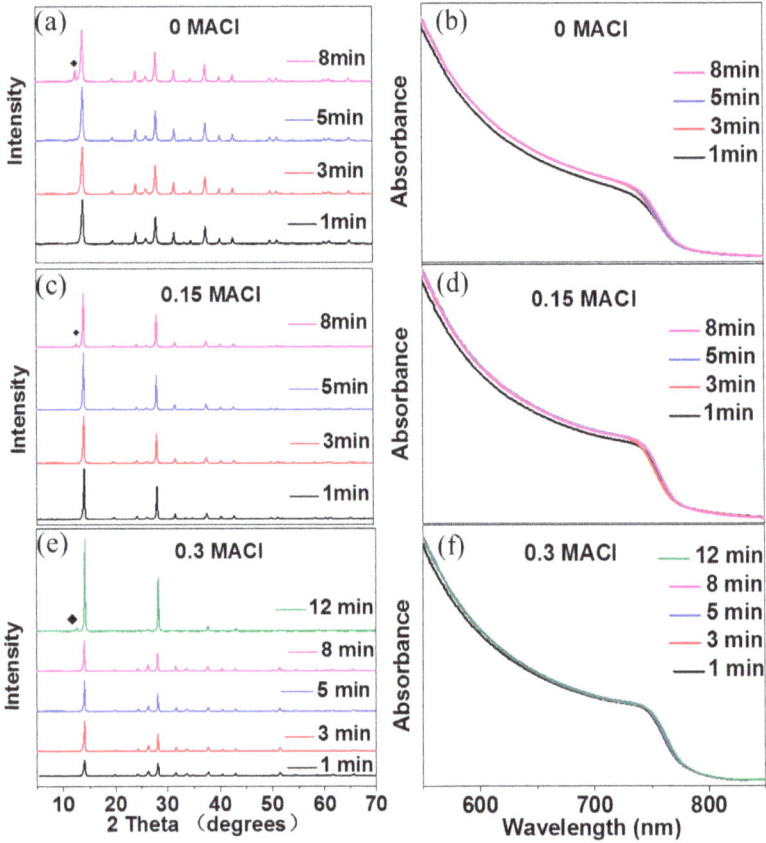

Figure 2. XRD patterns and UV/Vis spectra of $(FAPbI_3)_{0.85}(MAPbBr_3)_{0.15}$ films with (**a,b**) 0, (**c,d**) 0.15 and (**e,f**) 0.3 MACl annealed for 1 min, 3 min, 5 min and 8 min at 150 °C. ◆: PbI_2.

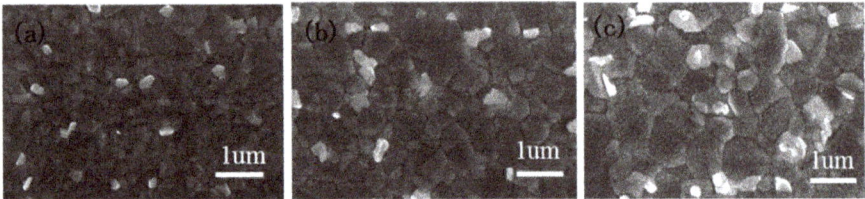

Figure 3. SEM images of $(FAPbI_3)_{0.85}(MAPbBr_3)_{0.15}$ perovskite films using (**a**) 0, (**b**) 0.15 and (**c**) 0.3 MACl additives at 150 °C.

Table 1 lists the photovoltaic parameters of these $(FAPbI_3)_{0.85}(MAPbBr_3)_{0.15}$ perovskite solar cells based on the planar device configuration of $FTO/c\text{-}TiO_2/perovskite/Spiro\text{-}OMeTAD/Ag$. The 0.15 MACl device exhibited improved photovoltaic performance with increased J_{sc}, V_{oc} and

Fill Factor (FF) compared to the 0 MACl device. The efficiency of the 0.3 MACl device was higher than that of the 0 MACl device but lower than that of the 0.15 MACl device. Figure 4 shows the typical J–V curves of these $(FAPbI_3)_{0.85}(MAPbBr_3)_{0.15}$ perovskite solar cells using different molar ratios of MACl additives.

Table 1. Photovoltaic parameters of the FTO/c-TiO$_2$/Perovikite/Spiro-OMeTAD/Ag devices with various molar ratios of MACl additives.

Samples	J_{sc} (mA/cm^2)	V_{oc} (V)	FF (%)	PCE (%)
0.3 MACl	20.6 ± 0.32	1.04 ± 0.01	74 ± 2	15.9 ± 0.4
0.15 MACl	21.3 ± 0.45	1.04 ± 0.01	76 ± 2	16.8 ± 0.3
0 MACl	20.2 ± 0.76	1.01 ± 0.02	75 ± 4	15.3 ± 0.7

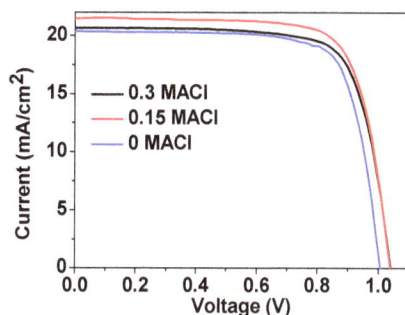

Figure 4. The typical J–V curves of the perovskite solar cells based on $(FAPbI_3)_{0.85}(MAPbBr_3)_{0.15}$ using different molar ratios of MACl additives.

3. Conclusions

In summary, we have developed a MACl-assisted solvent engineering method to deposit the $(FAPbI_3)_{0.85}(MAPbBr_3)_{0.15}$ perovskite film. The MACl additive can impact the crystallization process of the perovskite, which can be used as an additive for improving the thin-film quality of efficient semitransparent perovskite solar cells. The MACl additive can accelerate the crystallization of the precursor film to the perovskite phase but retard the crystallization of the perovskite phase during the thermal annealing process. Consequently, the MACl additive with a suitable MACl content (0.15 MACl) contributes to the formation of homogeneous and large grain perovskite films. The average PCE of the FTO/c-TiO$_2$/perovskite/Spiro-OMeTAD/Ag device increases from 15.3% to 16.8% after utilizing the 0.15 MACl additive. The additive assisted solvent engineering method would be a promising strategy to optimize the deposition of high-quality perovskite films.

4. Materials and Methods

Materials: Patterned FTO glasses were etched with metallic Zn and HCl aqueous solution (2M) and then cleaned. A 20 nm thick compact TiO$_2$ layer deposited on the patterned FTO substrate was prepared by spray pyrolysis of 0.2 M Ti(IV) bis(ethyl acetoacetate)-diisopropoxide in 1-butanol solution at 450 °C, followed by annealing at 450 °C for one hour. MACl was synthesized by reacting methylamine (33 wt % ethanol solution) and 33 wt % hydrocholoride acid with the molar ratio of 1.2:1 in an ice bath for 2 h with stirring, followed by vacuum drying and cleaning with acetonitrile. PbI$_2$, PbBr$_2$, CH$_3$NH$_3$Br(MABr), N,N-Dimethylformamide (DMF), dimethyl sulfoxide (DMSO) and chlorobenzene were purchased from Sigma-Aldrich and FAI was purchased from Shanghai MaterWin New Materials Co., Ltd.

Device preparation: The 1M $(FAPbI_3)_{0.85}(MAPbBr_3)_{0.15}$ perovskite was prepared by mixing stoichiometric FAI, PbI_2, MABr and $PbBr_2$ with the 0.85:0.85:0.15:0.15 molar ratio in DMF/DMSO (4/1 *v/v*). MACl molar ratios of 0, 0.15, 0.3 were added to the precursor solution, respectively. An amount of 80 μL of the perovskite solution was spread on the substrate and spin coated in a two-step program at 1000 and 4000 rpm for 10 and 30 s, respectively. During the second step, 250 μL of chlorobenzene was used as an anti-solvent and dripped on the top of the film 15 s prior to the end of the program. The substrates were then annealed at 150 °C in a glovebox. The hole transport layer was deposited on top of the perovskite film at 4000 rpm for 30 s using the hole transport material (HTM) solution, which consisted of 0.1 M spiro-MeOTAD, 0.035 M bis (trifluoromethane) sulfonamide lithium salt (Li-TFSi), and 0.12 M 4-tert-butylpyridine (tBP) in chlorobenzene/acetonitrile (10:1, *v/v*) solution. Finally, a 150-nm thick Ag film was deposited as a counter electrode using thermal evaporation.

Characterization: The photocurrent-voltage (J–V) characteristic of perovskite solar cells was measured with a 2401 source meter (Keithley, Cleveland, OH, USA) at a scan rate of 0.05 V/s under simulated AM1.5G illumination using Enlitech's 3A light source (Newport Corp., Irvine, CA, USA). The X-ray diffraction (XRD) measurement of perovskite films was performed using an X-ray diffractometer (Rigaku D/Max 2200, Rigaku Corporation, Tokyo, Japan) with Cu Ka radiation. Absorption spectra were measured using a UV/Vis spectrometer (Cary-6000i, Agilent Technologies, Santa Clara, CA, USA). The morphologies of perovskite films were characterized on a FEI Sirion 200 (Hillsboro, OR, USA) scanning electron microscope (SEM).

Acknowledgments: Yixin Zhao acknowledges the support of the NSFC (Grants 51372151 and 21303103) and Houyingdong Grant (151046).

Author Contributions: Yixin Zhao and Yang Yang conceived, designed the experiments and wrote the manuscript; Yihui Li performed the experiments and wrote the paper; Feng Xu and Taiyang Zhang performed the experiments and analyzed the data.

Conflicts of Interest: The authors declare no conflict of interest.

References

1. Stranks, S.D.; Eperon, G.E.; Grancini, G.; Menelaou, C.; Alcocer, M.J.; Leijtens, T.; Herz, L.M.; Petrozza, A.; Snaith, H.J. Electron-hole diffusion lengths exceeding 1 micrometer in an organometal trihalide perovskite absorber. *Science* **2013**, *342*, 341–344. [CrossRef] [PubMed]
2. Nie, W.; Tsai, H.; Asadpour, R.; Blancon, J.C.; Neukirch, A.J.; Gupta, G.; Crochet, J.J.; Chhowalla, M.; Tretiak, S.; Alam, M.A.; et al. High-efficiency solution-processed perovskite solar cells with millimeter-scale grains. *Science* **2015**, *347*, 522–525. [CrossRef] [PubMed]
3. Noh, J.H.; Im, S.H.; Heo, J.H.; Mandal, T.N.; Seok, S.I. Chemical management for colorful, efficient, and stable inorganic-organic hybrid nanostructured solar cells. *Nano Lett.* **2013**, *13*, 1764–1769. [CrossRef] [PubMed]
4. Im, J.-H.; Lee, C.-R.; Lee, J.-W.; Park, S.-W.; Park, N.-G. 6.5% efficient perovskite quantum-dot-sensitized solar cell. *Nanoscale* **2011**, *3*, 4088–4093. [CrossRef] [PubMed]
5. Jeon, N.J.; Noh, J.H.; Yang, W.S.; Kim, Y.C.; Ryu, S.; Seo, J.; Seok, S.I. Compositional engineering of perovskite materials for high-performance solar cells. *Nature* **2015**, *517*, 476–480. [CrossRef] [PubMed]
6. Liu, M.; Johnston, M.B.; Snaith, H.J. Efficient planar heterojunction perovskite solar cells by vapour deposition. *Nature* **2013**, *501*, 395–398. [CrossRef] [PubMed]
7. Zhao, L.C.; Luo, D.Y.; Wu, J.; Hu, Q.; Zhang, W.; Chen, K.; Liu, T.H.; Liu, Y.; Zhang, Y.F.; Liu, F.; et al. High-performance inverted planar heterojunction perovskite solar cells based on lead acetate precursor with efficiency exceeding 18%. *Adv. Funct. Mater.* **2016**, *26*, 3508–3514. [CrossRef]
8. Kim, H.S.; Lee, C.R.; Im, J.H.; Lee, K.B.; Moehl, T.; Marchioro, A.; Moon, S.J.; Humphry-Baker, R.; Yum, J.H.; Moser, J.E.; et al. Lead iodide perovskite sensitized all-solid-state submicron thin film mesoscopic solar cell with efficiency exceeding 9%. *Sci. Rep.* **2012**, *2*, 591. [CrossRef] [PubMed]
9. Kojima, A.; Teshima, K.; Shirai, Y.; Miyasaka, T. Organometal halide perovskites as visible-light sensitizers for photovoltaic cells. *J. Am. Chem. Soc.* **2009**, *131*, 6050–6051. [CrossRef] [PubMed]

10. Burschka, J.; Pellet, N.; Moon, S.J.; Humphry-Baker, R.; Gao, P.; Nazeeruddin, M.K.; Gratzel, M. Sequential deposition as a route to high-performance perovskite-sensitized solar cells. *Nature* **2013**, *499*, 316–319. [CrossRef] [PubMed]

11. Yang, M.; Zhang, T.; Schulz, P.; Li, Z.; Li, G.; Kim, D.H.; Guo, N.; Berry, J.J.; Zhu, K.; Zhao, Y. Facile fabrication of large-grain $CH_3NH_3PbI_{3-x}Br_x$ films for high-efficiency solar cells via CH_3NH_3Br-selective ostwald ripening. *Nat. Commun.* **2016**, *7*, 12305. [CrossRef] [PubMed]

12. Zhao, Y.; Zhu, K. Organic-inorganic hybrid lead halide perovskites for optoelectronic and electronic applications. *Chem. Soc. Rev.* **2016**, *45*, 655–689. [CrossRef] [PubMed]

13. Yin, W.J.; Shi, T.; Yan, Y. Unique properties of halide perovskites as possible origins of the superior solar cell performance. *Adv. Mater.* **2014**, *26*, 4653–4658. [CrossRef] [PubMed]

14. Parrott, E.S.; Milot, R.L.; Stergiopoulos, T.; Snaith, H.J.; Johnston, M.B.; Herz, L.M. Effect of structural phase transition on charge-carrier lifetimes and defects in $CH_3NH_3SnI_3$ perovskite. *J. Phys. Chem. Lett.* **2016**, *7*, 1321–1326. [CrossRef] [PubMed]

15. Chen, Y.; Li, B.; Huang, W.; Gao, D.; Liang, Z. Efficient and reproducible $CH_3NH_3PbI_{3-x}(SCN)_x$ perovskite based planar solar cells. *Chem. Commun.* **2015**, *51*, 11997–11999. [CrossRef] [PubMed]

16. Tai, Q.; You, P.; Sang, H.; Liu, Z.; Hu, C.; Chan, H.L.; Yan, F. Efficient and stable perovskite solar cells prepared in ambient air irrespective of the humidity. *Nat. Commun.* **2016**, *7*, 11105. [CrossRef] [PubMed]

17. Zhou, Y.; Yang, M.; Pang, S.; Zhu, K.; Padture, N.P. Exceptional morphology-preserving evolution of formamidinium lead triiodide perovskite thin films via organic-cation displacement. *J. Am. Chem. Soc.* **2016**, *138*, 5535–5538. [CrossRef] [PubMed]

18. Wang, Z.K.; Li, M.; Yang, Y.G.; Hu, Y.; Ma, H.; Gao, X.Y.; Liao, L.S. High efficiency Pb-In binary metal perovskite solar cells. *Adv. Mater.* **2016**, *28*, 6695–6703. [CrossRef] [PubMed]

19. Saliba, M.; Matsui, T.; Domanski, K.; Seo, J.Y.; Ummadisingu, A.; Zakeeruddin, S.M.; Correa-Baena, J.P.; Tress, W.R.; Abate, A.; Hagfeldt, A.; et al. Incorporation of rubidium cations into perovskite solar cells improves photovoltaic performance. *Science* **2016**, *354*, 206–209. [CrossRef] [PubMed]

20. Nie, Z.; Yin, J.; Zhou, H.; Chai, N.; Chen, B.; Zhang, Y.; Qu, K.; Shen, G.; Ma, H.; Li, Y.; et al. Layered and pb-free organic-inorganic perovskite materials for ultraviolet photoresponse: (010)-Oriented $(CH_3NH_3)_2MnCl_4$ thin film. *ACS Appl. Mater. Interfaces* **2016**, *8*, 28187–28193. [CrossRef] [PubMed]

21. Xu, F.; Zhang, T.; Li, G.; Zhao, Y. Mixed cation hybrid lead halide perovskites with enhanced performance and stability. *J. Mater. Chem. A* **2017**, *5*, 11450–11461. [CrossRef]

22. Wu, Y.; Xie, F.; Chen, H.; Yang, X.; Su, H.; Cai, M.; Zhou, Z.; Noda, T.; Han, L. Thermally stable mapbi3 perovskite solar cells with efficiency of 19.19% and area over 1 cm^2 achieved by additive engineering. *Adv. Mater.* **2017**, 1701073. [CrossRef] [PubMed]

23. Wang, C.L.; Zhao, D.W.; Yu, Y.; Shrestha, N.; Grice, C.R.; Liao, W.Q.; Cimaroli, A.J.; Chen, J.; Ellingson, R.J.; Zhao, X.Z.; et al. Compositional and morphological engineering of mixed cation perovskite films for highly efficient planar and flexible solar cells with reduced hysteresis. *Nano Energy* **2017**, *35*, 223–232. [CrossRef]

24. Li, L.; Chen, Y.H.; Liu, Z.H.; Chen, Q.; Wang, X.D.; Zhou, H.P. The additive coordination effect on hybrids perovskite crystallization and high-performance solar cell. *Adv. Mater.* **2016**, *28*, 9862–9868. [CrossRef] [PubMed]

25. Chueh, C.-C.; Liao, C.-Y.; Zuo, F.; Williams, S.T.; Liang, P.-W.; Jen, A.K.Y. The roles of alkyl halide additives in enhancing perovskite solar cell performance. *J. Mater. Chem. A* **2015**, *3*, 9058–9062. [CrossRef]

26. Chang, C.Y.; Chu, C.Y.; Huang, Y.C.; Huang, C.W.; Chang, S.Y.; Chen, C.A.; Chao, C.Y.; Su, W.F. Tuning perovskite morphology by polymer additive for high efficiency solar cell. *ACS Appl. Mater. Interfaces* **2015**, *7*, 4955–4961. [CrossRef] [PubMed]

27. Liang, P.W.; Liao, C.Y.; Chueh, C.C.; Zuo, F.; Williams, S.T.; Xin, X.K.; Lin, J.; Jen, A.K. Additive enhanced crystallization of solution-processed perovskite for highly efficient planar-heterojunction solar cells. *Adv. Mater.* **2014**, *26*, 3748–3754. [CrossRef] [PubMed]

28. Yantara, N.; Yanan, F.; Shi, C.; Dewi, H.A.; Boix, P.P.; Mhaisalkar, S.G.; Mathews, N. Unravelling the effects of cl addition in single step $CH_3NH_3PbI_3$ perovskite solar cells. *Chem. Mater.* **2015**, *27*, 2309–2314. [CrossRef]

29. Dong, G.H.; Ye, T.L.; Pang, B.Y.; Yang, Y.L.; Sheng, L.; Shi, Y.; Fan, R.Q.; Wei, L.G.; Su, T. $HONH_3Cl$ optimized $CH_3NH_3PbI_3$ films for improving performance of planar heterojunction perovskite solar cells via a one-step route. *Phys. Chem. Chem. Phys.* **2016**, *18*, 26254–26261. [CrossRef] [PubMed]

30. Bi, D.; Gao, P.; Scopelliti, R.; Oveisi, E.; Luo, J.; Gratzel, M.; Hagfeldt, A.; Nazeeruddin, M.K. High-performance perovskite solar cells with enhanced environmental stability based on amphiphile-modified $CH_3NH_3PbI_3$. *Adv. Mater.* **2016**, *28*, 2910–2915. [CrossRef] [PubMed]

31. Ke, W.; Xiao, C.; Wang, C.; Saparov, B.; Duan, H.-S.; Zhao, D.; Xiao, Z.; Schulz, P.; Harvey, S.P.; Liao, W.; et al. Employing lead thiocyanate additive to reduce the hysteresis and boost the fill factor of planar perovskite solar cells. *Adv. Mater.* **2016**, *28*, 5214–5221. [CrossRef] [PubMed]

32. Fei, C.; Guo, L.; Li, B.; Zhang, R.; Fu, H.; Tian, J.; Cao, G. Controlled growth of textured perovskite films towards high performance solar cells. *Nano Energy* **2016**, *27*, 17–26. [CrossRef]

33. Boopathi, K.M.; Mohan, R.; Huang, T.-Y.; Budiawan, W.; Lin, M.-Y.; Lee, C.-H.; Ho, K.-C.; Chu, C.-W. Synergistic improvements in stability and performance of lead iodide perovskite solar cells incorporating salt additives. *J. Mater. Chem. A* **2016**, *4*, 1591–1597. [CrossRef]

34. Tsai, H.; Nie, W.; Cheruku, P.; Mack, N.H.; Xu, P.; Gupta, G.; Mohite, A.D.; Wang, H.-L. Optimizing composition and morphology for large-grain perovskite solar cells via chemical control. *Chem. Mater.* **2015**, *27*, 5570–5576. [CrossRef]

35. Gong, X.; Li, M.; Shi, X.-B.; Ma, H.; Wang, Z.-K.; Liao, L.-S. Controllable perovskite crystallization by water additive for high-performance solar cells. *Adv. Funct. Mater.* **2015**, *25*, 6671–6678. [CrossRef]

36. Zhao, B.G.; Zhu, L.; Zhao, Y.L.; Yang, Y.; Song, J.; Gu, X.Q.; Xing, Z.; Qiang, Y.H. Improved performance of perovskite solar cell by controlling $CH_3NH_3PbI_{3-x}Cl_x$ film morphology with CH_3NH_3Cl-assisted method. *J. Mater. Sci. Mater. Electron.* **2016**, *27*, 10869–10876. [CrossRef]

37. Xie, F.X.; Su, H.; Mao, J.; Wong, K.S.; Choy, W.C.H. Evolution of diffusion length and trap state induced by chloride in perovskite solar cell. *J. Phys. Chem. C* **2016**, *120*, 21248–21253. [CrossRef]

38. Bi, C.; Zheng, X.; Chen, B.; Wei, H.; Huang, J. Spontaneous passivation of hybrid perovskite by sodium ions from glass substrates: Mysterious enhancement of device efficiency revealed. *ACS Energy Lett.* **2017**, *2*, 1400–1406. [CrossRef]

39. Dar, M.I.; Abdi-Jalebi, M.; Arora, N.; Gratzel, M.; Nazeeruddin, M.K. Growth engineering of $CH_3NH_3PbI_3$ structures for high-efficiency solar cells. *Adv. Energy Mater.* **2016**, *6*, 1501358. [CrossRef]

40. Bag, S.; Deneault, J.R.; Durstock, M.F. Aerosol-jet-assisted thin-film growth of $CH_3NH_3PbI_3$ perovskites-a means to achieve high quality, defect-free films for efficient solar cells. *Adv. Energy Mater.* **2017**, 1701151. [CrossRef]

41. Della Gaspera, E.; Peng, Y.; Hou, Q.; Spiccia, L.; Bach, U.; Jasieniak, J.J.; Cheng, Y.-B. Ultra-thin high efficiency semitransparent perovskite solar cells. *Nano Energy* **2015**, *13*, 249–257. [CrossRef]

42. Bag, S.; Durstock, M.F. Efficient semi-transparent planar perovskite solar cells using a 'molecular glue'. *Nano Energy* **2016**, *30*, 542–548. [CrossRef]

43. Sveinbjörnsson, K.; Aitola, K.; Zhang, J.; Johansson, M.B.; Zhang, X.; Correa-Baena, J.-P.; Hagfeldt, A.; Boschloo, G.; Johansson, E.M.J. Ambient air-processed mixed-ion perovskites for high-efficiency solar cells. *J. Mater. Chem. A* **2016**, *4*, 16536–16545. [CrossRef]

crystals

MDPI

Article

Interfacial Kinetics of Efficient Perovskite Solar Cells

Pankaj Yadav [1,*], Daniel Prochowicz [1,2], Michael Saliba [1], Pablo P. Boix [3], Shaik M. Zakeeruddin [1] and Michael Grätzel [1]

[1]   Laboratory of Photonics and Interfaces, Institute of Chemical Sciences and Engineering,
     School of Basic Sciences, Ecole Polytechnique Fédérale de Lausanne (EPFL), CH-1015 Lausanne,
     Switzerland; Daniel.Prochowicz@epfl.ch (D.P.); Michael.Saliba@epfl.ch (M.S.); zakeeruddin@epfl.ch (S.M.Z.);
     Michael.Graetzel@epfl.ch (M.G.)
[2]   Institute of Physical Chemistry, Polish Academy of Sciences, Kasprzaka 44/52, 01-224 Warsaw, Poland
[3]   Instituto de Ciencia Molecular, Universidad de Valencia, C/Catedrático J. Beltrán 2, 46980 Paterna, Spain;
     Pablo.P.Boix@uv.es
*   Correspondence: pankaj.yadav@epfl.ch

Academic Editor: Wei Zhang
Received: 9 June 2017; Accepted: 3 August 2017; Published: 13 August 2017

Abstract: Perovskite solar cells (PSCs) have immense potential for high power conversion efficiency with an ease of fabrication procedure. The fundamental understanding of interfacial kinetics in PSCs is crucial for further improving of their photovoltaic performance. Herein we use the current-voltage (J-V) characteristics and impedance spectroscopy (IS) measurements to probe the interfacial kinetics on efficient MAPbI$_3$ solar cells. We show that series resistance (R_S) of PSCs exhibits an ohmic and non-ohmic behavior that causes a significant voltage drop across it. The Nyquist spectra as a function of applied bias reveal the characteristic features of ion motion and accumulation that is mainly associated with the MA cations in MAPbI$_3$. With these findings, we provide an efficient way to understand the working mechanism of perovskite solar cells.

Keywords: perovskite; MAPbI$_3$; impedance; interfaces

1. Introduction

Organic lead halide-based perovskite solar cells (PSCs) have shown an unprecedented rise in their efficiency from 2% to more than 22% in just a few years [1–4]. This fast increase in efficiency and low temperature fabrication procedure makes perovskites potential candidates for low-cost photovoltaic devices. The most commonly employed PSC structure consists of a mesoporous electron transport layer such as TiO$_2$, followed by perovskite absorber layer and spiro-OMeTAD as a hole transport layer. Various attempts have been made to alter the morphological and structural properties of the hole transporting material (HTM), the absorber material, and electron transporting layer (ETL) to enhance the device performance [5–7]. In spite of the advances in device efficiency and architecture, many aspects of device processes occurring at the bulk (inside) and interfaces are still lagging behind. It has been established that the perovskite absorber material behaves as a mixed electronic-ionic conductor, in which electronic and ionic charges are influenced by applied bias rate, temperature, and illumination [8–11]. Therefore, obtaining exact information on the role of electronic and ionic components as well as their contributions to the device performance is crucial for the further in-depth understanding of the working mechanism of solar cells.

Another aspect that seems to play a significant role in power deliverability is the presence of J-V hysteresis. The surface treatments of ETL, HTL, and the perovskite absorber layer by light metals and carbonaceous materials doping have shown a positive impact in the mitigation of hysteresis behavior [12–18]. Most studies consider only ionic movement as the phenomenon responsible for J-V hysteresis. In fact, under illumination and applied bias, both the components, i.e., electronic and ionic components define the performance of PSCs. It is believed that the presence of ions can change the transport and recombination of electronic charge carriers or vice-versa. In this context, the goal of resolving the responses from ionic and electronic components in relation with the PSC device configuration represents a relevant aspect of the device performance.

Here, we present a semi-quantitative model of a PSC (represented by MAPbI$_3$) that can explain specific issues related to the interfaces and bulk properties of PSCs by considering the coupled responses of electronic and ionic components using impedance spectroscopy (IS). This technique allows us to deconvulate the responses from the both the interfaces and bulk region of the PSC. The well-defined dependencies of Nyquist spectra as a function of applied bias determine the resistive and capacitive parameters, and more importantly resolve the accumulation and movement of ions. Moreover, the presented model permits the quantification of prominent voltage loss sources. On the basis of IS and current-voltage extracted resistive, capacitive, and associated time responses, we postulate that in the investigated solar cell the predominant slow time response is caused by ion motion.

2. Theoretical Background

In general, PSCs consist of a mesoporous or planar electron transport layer, a perovskite absorber layer, a hole transport layer, and gold and fluorine-doped tin oxide (FTO) as a back and front contact, respectively (Figure 1a). Under illumination, photogenerated charges, i.e., electrons and holes, were collected at the FTO and gold contacts via the electron and hole transport layers. It is important to recognize that along with the photo generation, transportation, and recombination of charge carriers, the mobile ions and charge defects can also move and cause the screening of a built-in electric field [19–22]. Recent studies have shown that the accumulation of the ions and vacancies at interfaces changes the interfacial charge concentration and recombination rates, as well as the net photovoltage by setting up an electrostatic potential in addition to the built-in potential [23–25]. Since the electrostatic and built-in potentials have the same polarity (backward scan), the net open circuit voltage (V_{OC}) of PSC devices is the sum of these potentials [24,26].

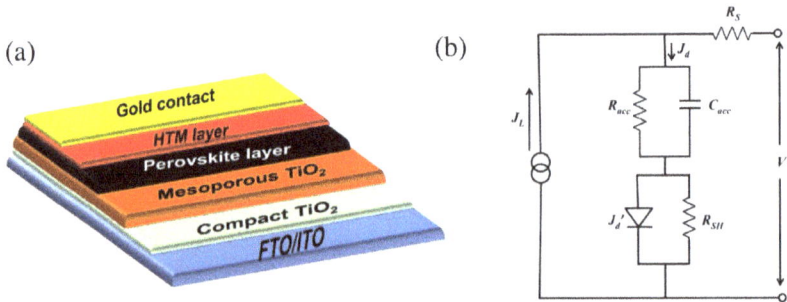

Figure 1. Schematic representation of (**a**) a perovskite solar cell (PSC) and (**b**) a single diode model of a PSC.

A generalized single diode model of a PSC, including the built-in and electrostatic potential formed due to the accumulation of ions and charges, is placed in a model previously discussed by Gottesman and others (Figure 1b) [24,26,27]. In this model, the dark current carrier recombination in the bulk and space charge region (SCR) is represented by a diode equation $J_d = J_0 \left[\left(exp \left(\frac{qV_d}{mk_BT} \right) \right) - 1 \right]$, where q, V_d are the elementary charge and applied voltage, J_0 is the reverse saturation dark current, and k_B and T are the Boltzman constant and cell temperature, respectively. The ideality factor (m) for a single diode falls in the range between $1 \leq m \leq 2$, where m ~ 1 signifies the top of the barrier transport (barrier for holes in ETL and for electrons in HTL) and m ~ 2 signifies the Shockley-Read

Hall (SRH) as a dominant recombination mechanism. In some cases, the ideality factor m is greater than 2, and this behavior is ascribed to the nonlinear shunt resistances across the junctions and/or a significant voltage drop across ETL and HTL [28]. The voltage drop across a diode has the form of $V_d = V + JR_s - V_{acc}$. Under an open circuit voltage condition, $J \approx 0$ and $V \approx VOC = V_d + V_{acc}$ shows that the open-circuit voltage (V_{OC}) of a PSC is defined as the sum of the two potentials in series.

Considering the fact that the accumulation mechanism is due to the electronic and ionic contributions, the accumulation resistance R_{acc} may contain the information about the recombination and ionic components. The alternating current (AC) electrical equivalent circuit shown in Figure 1b is driven through the variable direct current (DC) source in parallel with the AC signal generator. The circuit is also demonstrated as the generalized circuit for the impedance spectra analysis. The low frequency spectra of IS is modeled through R_{acc} and C_{acc} (also denoted as R_{Lf} and C_{Lf} in the IS section). In PSCs, it is well established that the movement of ions can take place across the active layer, and this diffusion is generally defined by the Warburg element (w) depending upon the IS spectra.

3. Results and Discussion

3.1. Device Characterization

The J-V characteristic of the PSC under 1 Sun (AM 1.5 G) at 25 °C in a backward direction scan is shown in Figure 2a (for the film characterization and device fabrication procedure, see supplementary material. The PSC depicts an impressive power conversion efficiency (PCE) of 17.3% with an open circuit voltage (V_{OC}) of 1.07 V, a short circuit current density of 21.6 mA/cm², and a fill factor of 75%. In order to determine the diode characteristic parameters, i.e., (m and J_0), we carried out the V_{OC} vs. illumination (0.1 to 100 mW/cm²) measurements (Figure 2b). In the plot, the symbol represents the experimental data points and the line is defined by using the expression $m = kT/q(V_{OC}/\ln I)$. The value of $m = 1.7$ and $J_0 = 1.4 \times 10^{13}$ A/cm² was obtained using the diode equation. The fabricated PSC also exhibits an efficient charge transport under working conditions, as J_{SC} vs. illumination power law coefficient is almost equal to 1 (see Figure S3).

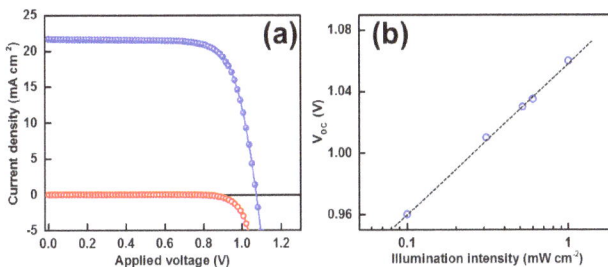

Figure 2. (a) Dark and light J-V characteristics of the MAPbI₃ PSC recorded under standard test conditions and (b) open circuit voltage of the device measured under various illumination levels.

3.2. Impedance Spectroscopy

The electrochemical impedance spectroscopy (IS) of a PSC as a function of applied bias (from 0 V to V_{OC}) under illumination at room temperature is shown in Figure 3. The AC modulation voltage of 20 mV and frequency in the range of 1 MHz to 200 mHz was used. The obtained Nyquist spectra are fitted with the electrical equivalent circuit discussed by Zarazúa et al. [29]. The error in experimental and theoretical data is within the acceptable limit of <1%. The IS spectra in the low forward bias follow the description of the two parallel resistor-capacitor (RC) components connected in series, while the values of each individual pair is different in magnitude. The shown spectra exhibit a maximum value of Z' and Z'' because, in this probed bias range (0–0.4 V), the net response of the diode is mainly dominated by R_{SH} and geometrical capacitance (C_g). In turn, the IS curves at voltage of 0.3 V to 0.6 V are composed of a depressed semicircle in the high frequency region and a straight line in the low frequency region (Figure 3b). When the bias voltages are higher than the knee voltage, the radii of the semicircle in the high frequency region also decreases; however, the straight line in the low frequency region converts into a semicircle arc (Figure 3c). Figure 3e–f demonstrate the general complex Nyquist spectra exhibiting the Warburg features evident by the linear line at 45° at the low frequency region, wherein the dashed line shows the expected behavior of a blocking electrode. Meanwhile, IS spectra exhibiting the double layer characteristics are evident by the semicircular arc present in the low frequency regime.

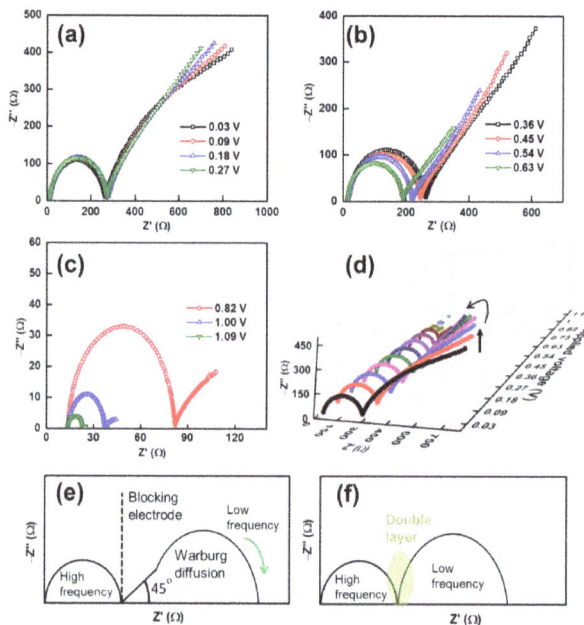

Figure 3. Nyquist spectra of a PSC recorded at various applied biases under illumination: (**a**) at low forward bias where the net photocurrent is constant (see Figure 2a), (**b**) from mid forward bias to knee voltage, and (**c**) beyond knee voltage towards open circuit voltage. (**d**) 3D plot of Nyquist spectra covering the whole range of the applied forward bias. Here, the straight arrow indicates an increase in the low frequency arc at the low forward bias, while the curved arrow shows where the low frequency arc is converted to a straight line at the mid forward bias. (**e**) Typical complex Nyquist spectra exhibiting the Warburg features evident by the linear line at 45° at the low frequency region, wherein the dashed line shows the expected behavior of a blocking electrode. (**f**) Typical complex Nyquist spectra exhibiting the double layer characteristics evident by the semicircular arc in the low frequency regime.

As the applied bias gradually increases beyond the knee voltage, the radii of the high frequency semicircles decreases at a faster rate and the low frequency line is converted to a semicircle arc. The rate of change in the low and high frequency responses under applied bias is shown in Figure 3d (the arrow indicates the transients between semicircles to straight lines). We found that the rate of change in the low frequency arc is faster than the corresponding rate in the high frequency arc. This implies that under the influence of illumination, the coupled influence of charge and ion movement may cause a higher rate of decrease in the low frequency region. In comparison to the electronic conductors such as thin-film solar cells, the low frequency region of IS was defined by the recombination resistance and chemical capacitance. We remark that due to very thin perovskite absorber layer (~200 nm) and ion motion, the chemical capacitance that is usually masked by the electrostatic capacitance arises due to ion accumulation. It is also worth noting that these two kinds of capacitances (i.e., chemical and low frequency capacitance) are totally different in their nature.

The R_S of a PSC extracted from the high frequency intercept on the x-axis as a function of applied bias is shown in Figure 4a. The R_S vs. applied bias plot in the range of 0.0 to 0.4 V depicts a constant value of R_s (13.8 Ω), which may arise from the ohmic components of the PSC. In turn, in the range of 0.4 to 0.8 V, the plot of R_s shows a peak that can be attributed to non-ohmic components of the solar cell. The presence of this peak is commonly observed in silicon and dye-sensitized solar cells [30–32], however, to the best of our knowledge, such behavior has not yet been reported in PSCs. We note that the physical origin of the peak is not fully clear, but can be attributed to the formation of a low Schottky barrier at the perovskite/HTM interface and a voltage drop across the diode that varies with bias, as shown in Figure 4a (blue line). Therefore, it is necessary to develop a proper strategy that allows a reduction of the value of R_S in PSCs.

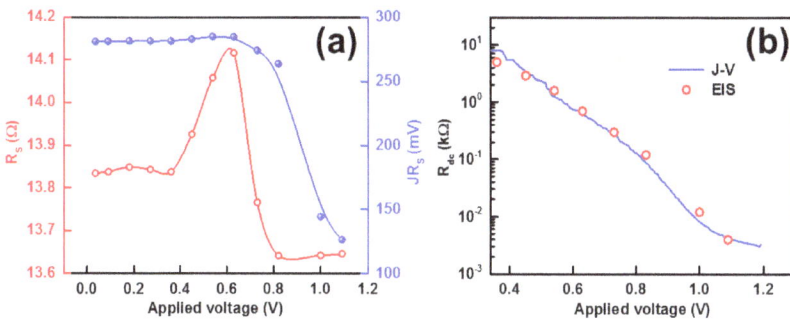

Figure 4. (a) Variation of series resistance R_S depicted from the impedence spectroscopy (IS) measurements, and the voltage drop at R_S under 1 sun illumination as a function of applied bias. (b) A comparative plot of total resistance calculated from IS and differential resistance from *J-V* characteristics measured under 1 sun illumination as a function of applied voltage.

A comparative plot of the total resistance calculated from IS and the differential resistance from *J-V* characteristics measured under 1 sun illumination is shown in Figure 4b. In the plot, the symbol represents the extracted resistance, i.e., $(R_S + R_{Hf} + R_{Lf})$ from IS, and the line represents $\frac{dv}{dJ}$ from the *J-V* curve. We found that the values obtained from both IS and *J-V* exhibit a similar bias dependence. The plot of R_{Hf} and R_{Lf} with respect to $V + JR_S$ is shown in Figure S4 of SI. It should be noted that the magnitude and rate of change for R_{Hf} is lower than that observed for R_{Lf} under the probed range of bias. The value of R_{Hf} varies from 250 Ω to 10 Ω and, at the same time, the corresponding value of R_{Lf} diminishes from 1 KΩ to 1 Ω, signifying the differences in their sources of origin. The value of R_{Lf}, which is associated with the charge accumulation and/or recombination in PSCs, was further examined by the determination of a potential drop across V_{Lf}, where $V_{Lf} \approx J_d R_{Lf}$. It has been found that V_{Lf} is varied in the range of 100 mV to 50 mV, in a bias range from 0.8 V to V_{oc}. More recently, the illumination of PSCs under different time periods and voltage decay periods revealed the formation of electrostatic potential in the range of 100 mV–150 mV in series with built-in potential [24,33]. These results illustrate that the non-ohmic R_S may be associated with the voltage drop across the HTL/perovskite interface. Moreover, the bias dependence of R_{Lf} and the associated V_{Lf} could explain the accumulation of charges and ions at the ETL/perovskite interface, as well as cause a build-up of potential.

As it is outlined in the theoretical section, the high frequency capacitance related to the polarizability of the perovskite absorber layer is defined by geometrical capacitance (C_g). However, the low frequency capacitance is defined by electrostatic capacitance (C_s). The variation of C_g and C_s in the bias range from knee voltage to V_{OC} of a PSC is shown in Figure 5a. In the low forward bias, C_g depicts the constant value in the order of ≈ 1 μF at a frequency of 10 KHz. The C_g plot seems to be constant, while C_s increases exponentially. Considering the parallel plate capacitance as $\frac{\varepsilon_0 \varepsilon}{d}$ with $d = 200$ nm and ε_0 permittivity of the free space, $\varepsilon = 22$ is obtained. The obtained value of ε is comparable with the previously reported values by other authors [9]. On the other hand, C_s exhibits a value in order of mF measured at 1 Hz.

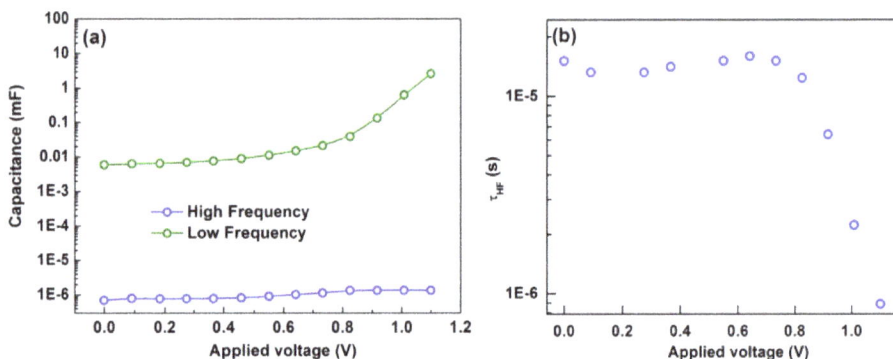

Figure 5. (**a**) The capacitance of a PSC extracted from the low and high frequency Nyquist spectra as a function of applied bias and (**b**) the variation of the high frequency time constant as a function of applied bias for a PSC.

A Mott-schottky (MS) plot of a PSC at the frequency of 10 KHz in the bias range from 0 V to V_{OC} is shown in Figure S5. The MS plot clearly shows the bias dominance of C_g in the low bias range and of C_s beyond the knee voltage. As compared to silicon solar cells, where the MS plot is generally defined by the three distinguishable capacitances, i.e., C_g, C_s, and depletion layer capacitance (C_{dl}), the obtained MS plot for the PSC was only defined by C_g and C_s. The fit of C_s with respect to the bias gives the value of built-in potential (intercept) and defect density (slope). The obtained value of the built-in potential $V_{bi} = 0.1$ V and the defect density of 10^{-17} cm^{-3} are consistent with the reported values for MAPbI$_3$-based solar cells [33].

Time constants corresponding to the high and low frequency spectra as a function of applied bias are shown in Figure 5b and Figure S6. We observed that the high frequency time response (τ_{Hf}) under applied bias follows the decreasing trend of R_{Hf} (not C_{Hf}). Therefore, this time constant cannot be considered as a characteristic response of any physical processes in the PSC. From the low frequency region, a bias independent time constant (τ_{Lf}) in the order of $\sim$1s was obtained. Similar values of slow time constants were also observed by other groups using intensity modulated voltage spectroscopy, open circuit voltage decay, and voltage and current transient techniques [24–27] The mechanism relies on long-lived trapping without recombination, and trap-assisted recombination was assigned for such a long response time, which seems to be anomalously long. It is well known that the presence of traps in a semiconductor accelerates the recombination process that in turn decreases the response time. As the low frequency capacitance has a dielectric nature, the corresponding time constant cannot be considered as a recombination lifetime. In general, IS response under illumination in the frequency range of 1 MHz–200 mHz requires a 3–4 min per voltage scan (considering an average of two data errors for each frequency), which is more than sufficient to accumulate the holes and cations at the TiO$_2$/perovskite interface and to contribute to the electrostatic potential. Considering the fact that the efficiently fast relaxation phenomenon is generally associated with the charge carriers, the low frequency relaxation time (τ_{Lf}) can be assigned to the motion of ions required to reach the equilibrium. A similar observation was also drawn by Gottesman et al. by using the open circuit voltage decay method [24]. The authors demonstrated that V_{OC} in a PSC occurs in two steps, where the fast decay of the built-in potential happens instantly, but the electrostatic potential caused by charge and ion accumulation can take tens of minutes to return its steady-state condition [24]. Moreover, it was also shown that MA$^+$ has a large polarization relaxation time in the range of 1–100 s [34]. Therefore, one can correlate the long τ_{Lf} to the relaxation time of MA$^+$ ions.

4. Conclusions

We have presented a semi-quantitative model that can elucidate the bulk and interfacial characteristics of PSCs. Our results strongly suggest that the low frequency impedance response of a PSC depends on its ionic movements. The ionic movements in the PSC were detected by probing the IS response as a function of applied bias. The presented results also revealed that, in comparison to electronic conductors where the low frequency response of IS was defined by recombination resistance and chemical capacitance, the chemical capacitance in PSCs (mix-conductors) is masked by the electrostatic capacitance that arises from ion accumulation. In addition, we showed that the series resistance (R_S) of a PSC exhibits an ohmic and non-ohmic behavior that causes a significant voltage drop across them. Our conclusion regarding the ion accumulation is supported by the bias dependence of R_{Lf} and the associated V_{Lf} that could explain the accumulation of charges and ions at the ETL/perovskite interface. The slow relaxation time (τ_{Lf}) in a PSC can be assigned to the motion of ions required to reach the equilibrium.

Supplementary Materials: The following are available online at http://www.mdpi.com/2073-4352/7/8/252/s1, Figure S1: UV:Vis absorption spectra of MAPbI$_3$ perovskite absorber layer, Figure S2: (a) Surface and (b) Cross sectional SEM morphology of MAPbI3 perovskite absorber layer, Figure S3: Current Density of the device measured under various illumination levels where the net photocurrent is constant, (b) from mid forward bias to knee voltage, Figure S4: The variation of high and low frequency resistance as a function of V+JR$_S$, Figure S5: A Mott-schottky (MS) plot of PSC at the frequency of 10 KHz in the bias range from 0 V to V$_{OC}$, Figure S6: Variation of low frequency time constant as a function of applied bias for PSC.

Author Contributions: P.Y. conceived and designed the experiments; D.P. and M.S. performed the experiments; P.Y. and P.P.B. analyzed the data; P.P.B. contributed analysis tools; P.Y. and D.P. wrote the paper, S.M.Z and M.Z. supervised the project.

Conflicts of Interest: The authors declare no conflict of interest.

References

1. Chen, W.; Wu, Y.; Yue, Y.; Liu, J.; Zhang, W.; Yang, X.; Chen, H.; Bi, E.; Ashraful, I.; Grätzel, M.; et al. Efficient and stable large-area perovskite solar cells with inorganic charge extraction layers. *Science* **2015**, *350*, 944–948. [CrossRef] [PubMed]
2. Im, J.-H.; Lee, C.-R.; Lee, J.-W.; Park, S.-W.; Park, N.-G. 6.5% efficient perovskite quantum-dot-sensitized solar cell. *Nanoscale* **2011**, *3*, 4088–4093. [CrossRef] [PubMed]
3. Kojima, A.; Teshima, K.; Shirai, Y.; Miyasaka, T. Organometal halide perovskites as visible-light sensitizers for photovoltaic cells. *J. Am. Chem. Soc.* **2009**, *131*, 6050–6051. [CrossRef] [PubMed]
4. Liu, M.; Johnston, M.B.; Snaith, H.J. Efficient planar heterojunction perovskite solar cells by vapour deposition. *Nature* **2013**, *501*, 395–398. [CrossRef] [PubMed]
5. Chang, C.-Y.; Chang, Y.-C.; Huang, W.-K.; Liao, W.-C.; Wang, H.; Yeh, C.; Tsai, B.-C.; Huang, Y.-C.; Tsao, C.-S. Achieving high efficiency and improved stability in large-area ITO-free perovskite solar cells with thiol-functionalized self-assembled monolayers. *J. Mater. Chem. A* **2016**, *4*, 7903–7913. [CrossRef]
6. Feng, S.; Yang, Y.; Li, M.; Wang, J.; Cheng, Z.; Li, J.; Ji, G.; Yin, G.; Song, F.; Wang, Z.; et al. High-Performance Perovskite Solar Cells Engineered by an Ammonia Modified Graphene Oxide Interfacial Layer. *ACS Appl. Mater. Interfaces* **2016**, *8*, 14503–14512. [CrossRef] [PubMed]
7. Bai, Y.; Dong, Q.; Shao, Y.; Deng, Y.; Wang, Q.; Shen, L.; Wang, D.; Wei, W.; Huang, J. Enhancing stability and efficiency of perovskite solar cells with crosslinkable silane-functionalized and doped fullerene. *Nat. Commun.* **2016**, *7*, 12806. [CrossRef] [PubMed]

8. Heo, J.H.; Song, D.H.; Han, H.J.; Kim, S.Y.; Kim, J.H.; Kim, D.; Shin, H.W.; Ahn, T.K.; Wolf, C.; Lee, T.-W.; et al. Planar CH3NH3PbI3 Perovskite Solar Cells with Constant 17.2% Average Power Conversion Efficiency Irrespective of the Scan Rate. *Adv. Mater.* **2015**, *27*, 3424–3430. [CrossRef] [PubMed]

9. Nie, W.; Blancon, J.-C.; Neukirch, A.J.; Appavoo, K.; Tsai, H.; Chhowalla, M.; Alam, M.A.; Sfeir, M.Y.; Katan, C.; Even, J.; et al. Light-activated photocurrent degradation and self-healing in perovskite solar cells. *Nat. Commun.* **2016**, *7*, 11574. [CrossRef] [PubMed]

10. Unger, E.L.; Hoke, E.T.; Bailie, C.D.; Nguyen, W.H.; Bowring, A.R.; Heumuller, T.; Christoforo, M.G.; McGehee, M.D. Hysteresis and transient behavior in current– voltage measurements of hybrid-perovskite absorber solar cells. *Energy Environ. Sci.* **2014**, *7*, 3690–3698. [CrossRef]

11. Heo, J.H.; Han, H.J.; Kim, D.; Ahn, T.K.; Im, S.H. Hysteresis-less inverted CH$_3$NH$_3$PbI$_3$ planar perovskite hybrid solar cells with 18.1% power conversion efficiency. *Energy Environ. Sci.* **2015**, *8*, 1602–1608. [CrossRef]

12. Zhou, Z.; Huang, L.; Mei, X.; Zhao, Y.; Lin, Z.; Zhen, H.; Ling, Q. Highly reproducible and photocurrent hysteresis-less planar perovskite solar cells with a modified solvent annealing method. *Sol. Energy* **2016**, *136*, 210–216. [CrossRef]

13. Kim, J.; Teridi, M.A.M.; Yusoff, A.R.B.M.; Jang, J. Stable and null current hysteresis perovskite solar cells based nitrogen doped graphene oxide nanoribbons hole transport layer. *Sci. Rep.* **2016**, *6*, 27773. [CrossRef] [PubMed]

14. Heo, J.H.; Im, S.H. Highly reproducible, efficient hysteresis-less CH$_3$NH$_3$PbI$_{3-x}$Cl$_x$ planar hybrid solar cells without requiring heat-treatment. *Nanoscale* **2016**, *8*, 2554–2560. [CrossRef] [PubMed]

15. Xu, J.; Buin, A.; Ip, A.H.; Li, W.; Voznyy, O.; Comin, R.; Yuan, M.; Jeon, S.; Ning, Z.; McDowell, J.J.; et al. Perovskite–fullerene hybrid materials suppress hysteresis in planar diodes. *Nat. Commun.* **2015**, *6*, 7081. [CrossRef] [PubMed]

16. Giordano, F.; Abate, A.; Baena, J.P.C.; Saliba, M.; Matsui, T.; Im, S.H.; Zakeeruddin, S.M.; Nazeeruddin, M.K.; Hagfeldt, A.; et al. Enhanced electronic properties in mesoporous TiO2 via lithium doping for high-efficiency perovskite solar cells. *Nat. Commun.* **2016**, *7*, 10379. [CrossRef] [PubMed]

17. Kwon, U.; Kim, B.-G.; Nguyen, D.C.; Park, J.-H.; Ha, N.Y.; Kim, S.-J.; Ko, S.H.; Lee, S.; Lee, D.; Park, H.J. Solution-processible crystalline NiO nanoparticles for high-performance planar perovskite photovoltaic cells. *Sci. Rep.* **2016**, *6*, 30759. [CrossRef] [PubMed]

18. Yang, G.; Wang, C.; Lei, H.; Zheng, X.; Qin, P.; Xiong, L.; Zhao, X.; Yan, Y.; Fang, G. Interface engineering in planar perovskite solar cells: energy level alignment, perovskite morphology control and high performance achievement. *J. Mater. Chem. A* **2017**, *5*, 1658–1666. [CrossRef]

19. Armin, A.; Juska, G.; Philippa, B.W.; Burn, P.L.; Meredith, P.; White, R.D.; Pivrikas, A. Doping-induced screening of the built-in-field in organic solar cells: Effect on charge transport and recombination. *Adv. Energy Mater.* **2013**, *3*, 321–327. [CrossRef]

20. Belisle, R.A.; Nguyen, W.H.; Bowring, A.R.; Calado, P.; Li, X.; Irvine, S.J.C.; McGehee, M.D.; Barnes, P.R.F.; O'Regan, B.C. Interpretation of inverted photocurrent transients in organic lead halide perovskite solar cells: proof of the field screening by mobile ions and determination of the space charge layer widths. *Energy Environ. Sci.* **2017**, *10*, 192–204. [CrossRef]

21. Eames, C.; Frost, J.M.; Barnes, P.R.F.; O'Regan, B.C.; Walsh, A.; Islam, M.S. Ionic transport in hybrid lead iodide perovskite solar cells. *Nat. Commun.* **2015**, *6*, 7497. [CrossRef] [PubMed]

22. Calado, P.; Telford, A.M.; Bryant, D.; Li, X.; Nelson, J.; O'Regan, B.C.; Barnes, P.R.F. Evidence for ion migration in hybrid perovskite solar cells with minimal hysteresis. *Nat. Commun.* **2016**, *7*, 13831. [CrossRef] [PubMed]

23. Yu, H.; Lu, H.; Xie, F.; Zhou, S.; Zhao, N. Native defect-induced hysteresis behavior in organolead iodide perovskite solar cells. *Adv. Funct. Mater.* **2016**, *26*, 1411–1419. [CrossRef]

24. Gottesman, R.; Lopez-Varo, P.; Gouda, L.; Jimenez-Tejada, J.A.; Hu, J.; Tirosh, S.; Zaban, A.; Bisquert, J. Dynamic phenomena at perovskite/electron-selective contact interface as interpreted from photovoltage decays. *Chem* **2016**, *1*, 776–789. [CrossRef]

25. Domanski, K.; Roose, B.; Matsui, T.; Saliba, M.; Turren-Cruz, S.-H.; Correa-Baena, J.-P.; Carmona, C.R.; Richardson, G.; Foster, J.M.; de Angelis, F.; et al. Migration of cations induces reversible performance losses over day/night cycling in perovskite solar cells. *Energy Environ. Sci.* **2017**, *10*, 604–613. [CrossRef]

26. Ravishankar, S.; Almora, O.; Echeverría-Arrondo, C.; Ghahremanirad, E.; Aranda, C.; Guerrero, A.; Fabregat-Santiago, F.; Zaban, A.; Garcia-Belmonte, G.; Bisquert, J. Surface polarization model for the dynamic hysteresis of perovskite solar cells. *J. Phys. Chem. Lett.* **2017**, *8*, 915–921. [CrossRef] [PubMed]

27. Ludmila, C.; Satoshi, U.; Piyankarage, V.V.J.; Shoji, K.; Yasutake, T.; Jotaro, N.; Takaya, K.; Hiroshi, S. Simulation of current–voltage curves for inverted planar structure perovskite solar cells using equivalent circuit model with inductance. *Appl. Phys. Express* **2017**, *10*, 025701.

28. Agarwal, S.; Seetharaman, M.; Kumawat, N.K.; Subbiah, A.S.; Sarkar, S.K.; Kabra, D.; Namboothiry, M.A.G.; Nair, P.R. On the uniqueness of ideality factor and voltage exponent of perovskite-based solar cells. *J. Phys. Chem. Lett.* **2014**, *5*, 4115–4121. [CrossRef] [PubMed]

29. Zarazua, I.; Han, G.; Boix, P.P.; Mhaisalkar, S.; Fabregat-Santiago, F.; Mora-Seró, I.; Bisquert, J.; Garcia-Belmonte, G. Surface Recombination and Collection Efficiency in Perovskite Solar Cells from Impedance Analysis. *J. Phys. Chem. Lett.* **2016**, *7*, 5105–5113. [CrossRef] [PubMed]

30. Yadav, P.; Tripathi, B.; Pandey, K.; Kumar, M. Recombination kinetics in a silicon solar cell at low concentration: electro-analytical characterization of space-charge and quasi-neutral regions. *Phys. Chem. Chem. Phys.* **2014**, *16*, 15469–15476. [CrossRef] [PubMed]

31. Yadav, P.; Pandey, K.; Tripathi, B.; Kumar, C.M.; Srivastava, S.K.; Singh, P.K.; Kumar, M. An effective way to analyse the performance limiting parameters of poly-crystalline silicon solar cell fabricated in the production line. *Sol. Energy* **2015**, *122*, 1–10. [CrossRef]

32. Garland, J.E.; Crain, D.J.; Roy, D. Utilization of electrochemical impedance spectroscopy for experimental characterization of the diode features of charge recombination in a dye sensitized solar cell. *Electrochim. Acta* **2014**, *148*, 62–72. [CrossRef]

33. Almora, O.; Aranda, C.; Mas-Marzá, E.; Garcia-Belmonte, G. On Mott-Schottky analysis interpretation of capacitance measurements in organometal perovskite solar cells. *Appl. Phys. Lett.* **2016**, *109*, 173903. [CrossRef]

34. Leguy, A.M.A.; Frost, J.M.; McMahon, A.P.; Sakai, V.G.; Kockelmann, W.; Law, C.; Li, X.; Foglia, F.; Walsh, A.; O'Regan, B.C.; et al. The dynamics of methylammonium ions in hybrid organic–inorganic perovskite solar cells. *Nat. Commun.* **2015**, *6*, 7124. [CrossRef] [PubMed]

crystals

MDPI

Review

Analysing the Prospects of Perovskite Solar Cells within the Purview of Recent Scientific Advancements

Aakash Bhat [1], Bhanu Pratap Dhamaniya [2], Priyanka Chhillar [2], Tulja Bhavani Korukonda [2], Gaurav Rawat [2] and Sandeep K. Pathak [2,*]

[1] Department of Physics, Friedrich-Alexander-Universität, Erlangen-Nürnberg, 91054 Erlangen, Germany;
    aakashbhat7@gmail.com
[2] Center for Energy Studies, Indian Institute of Technology Delhi, Delhi 110016, India;
    pratapbhanu92@gmail.com (B.P.D.); priyankachhillar3@gmail.com (P.C.);
    tulja.bhavani058@gmail.com (T.B.K.); gauravr343@gmail.com (G.R.)
*   Correspondence: spathak@iitd.ac.in

Received: 6 April 2018; Accepted: 16 May 2018; Published: 6 June 2018

Abstract: For any given technology to be successful, its ability to compete with the other existing technologies is the key. Over the last five years, perovskite solar cells have entered the research spectrum with tremendous market prospects. These cells provide easy and low cost processability and are an efficient alternative to the existing solar cell technologies in the market. In this review article, we first go over the innovation and the scientific findings that have been going on in the field of perovskite solar cells (PSCs) and then present a short case study of perovskite solar cells based on their energy payback time. Our review aims to be comprehensive, considering the cost, the efficiency, and the stability of the PSCs. Later, we suggest areas for improvement in the field, and how the future might be shaped.

Keywords: Perovskite Solar Cells; review; solar cells; Payback time; efficiency; stability; cost analysis

1. Introduction

In the past few years, perovskite based solar cells have come into mainstream academic discussions, with research prospects increasing exponentially in the past three years. Perovskite was first used as an absorber for a wide spectrum of solar energy [1,2]. However, in the past few years, perovskite has been researched solely for efficiency improvements and stability issues. The power conversion efficiency has reached as high as 22.7% [3] which makes it a legitimate competitor for other cells such as polycrystalline silicon solar cells. Scientists have begun using perovskites for tandem solar cell applications, which would further increase the efficiency of solar cells to 25–30% [4]. For instance, CsFA ([HC(NH$_2$)$_2$]$_{0.83}$Cs$_{0.17}$Pb(I$_{0.83}$Br$_{0.17}$)$_3$) based PSCs have been reported having an efficiency of 23.6% when placed on top of infrared-tuned silicon heterojunction bottom cell.

Recently, crystalline silicon heterojunction solar cells have achieved a PCE greater than 25% [5] when an interdigitated back contact structure was used to reduce optical loss. A heterojunction between amorphous silicon and crystalline silicon, with intrinsic thin layer solar cell has shown an efficiency of 24.7% with a cell thickness of 98 μm [6]. Meanwhile, black silicon solar cell has been a breakthrough in the field of efficient and stable solar cells with an efficiency of 22.1% attributed to interdigitated back contacts [7] and wide absorption in solar spectrum. On the other hand, new technologies such as IBSC (Intermediate Band Solar Cells), organic solar cells, and low cost CdTe solar cells have also boomed, which require intense research before being commercialised.

Organic–inorganic hybrid perovskite has a high absorption coefficient (~10^5 cm^{-1}) [8], wide range absorption in solar spectrum (300–900 nm) [9], low excitonic binding energy (19–50 meV) [10], long electron/hole diffusion length (100–1000 nm) [11], and a tuneable band gap [12]. Because of these properties it has attracted considerable attention nowadays.

Unlike most conventional technologies, the biggest problem faced by perovskite solar cells is stability in the operating environment. However, its low-cost fabrication advantage has driven the research so far despite the competitive market. The fabrication methods involve solution processable techniques such as spin-coating and dip coating. There exist other approaches such as chemical vapour deposition (CVD) [13], atomic layer deposition (ALD) [14], plasma enhanced CVD (PECVD) [15], etc., which are costlier compared to solution processable methods.

In this article, we first go over studies on efficiency of PSCs. After that, we discuss other aspects of perovskite solar cells which are being researched: stability, cost, and energy payback time. We also discuss tandem solar cells and later, the future of perovskites as energy solutions at low cost.

2. Discussion

Within the context of this paper, it is important to first define the relationship between efficiency, stability, and cost, and the overall effect of the three on any solar cell technology. The most basic view that is be sufficient for now can be summarized as follows:

A solar cell, essentially, is made of a light absorbing material sandwiched between charge extraction layers and connected to electrodes. While identifying an effective absorber material which can harvest a significant portion of visible spectrum plays a crucial role in photo-charge generation, the electron and hole transport layers (ETL and HTL, respectively) aid the charge transport and collection. Quantizing the quality of the solar cell can be effectively done by efficiency measurements. However, commercialization demands a trade-off between efficiency and the production cost. In the case of PSCs, although the efficiencies were not comparable to Si-solar cells during the early days, economic lab scale processing and cheap material inventory lead to intensified research. Although the efficiencies leapt to new heights in less than a decade, PSCs constantly suffered in stability and reproducibility. Perovskite is highly sensitive to environment unlike conventional absorber materials and requires extra efforts to achieve a stable performing cell. Thereby the research groups strive to address the trio: (1) cost effectiveness; (2) high efficiency; and (3) outdoor/field stability. It is implicit that the three are interlinked; the key to commercialization is to positively tweak the balance between this trio as much as possible.

Furthermore, research in PSCs involves the optimization of the three issues stated above, whereas work on established Si-cells is often carried out to reduce the production cost. With an aim to resolve the vague idea about the efficiency of the PSC, we first tried to define the theoretical limitations on efficiency.

Theoretical Limitations in the Performance of Perovskite Solar Cell

When there are a high number of research groups involved in efforts of increasing the power conversion efficiency of perovskite solar cells, it is important to know the maximum amount of photon energy conversion that can be attained. Earlier, in 1961, Shockley and Queisser proposed a detailed balanced limit method for a single junction single absorber layer, which gave an idea of limiting values. However, this theoretical efficiency limit is restricted by several parameters and takes only the ideal cases into consideration such as: photo recycling and pondering upon the optical band gap of the absorber having only radiative recombination [16,17] (in the theoretical limit). Hence, the necessity arises to determine the practical limit for the efficiency along with the basic facts mentioned in the S-Q model [16].

Pabitra et al. (2014) [18] made a realistic approach to explain the limit of perovskite solar cell technology in terms of basic cell parameters such as short circuit current, open circuit voltage, fill factor, etc. Their calculations demonstrated that the ratio of maximum Jsc attained until date by a laboratory cell to the maximum theoretical current was 79% for mixed halide case and 71% for iodide-based perovskite. One can infer from it that there is a possibility of improvement in the Jsc.

Similar calculations were made for Voc values and reported overpotential was 0.73 eV for mixed halide perovskites. The reason for this might be the losses at the interfaces.

Seki et al. reported that the maximum PCE that can be achieved in perovskite solar cell is 26–27%, which can be obtained at an optical band gap of around 1.4 eV and considering the voltage loss of 0.2 eV between TiO_2 and $CH_3NH_3PbI_3$ layer. They calculated that, if the voltage loss can be reduced to 0.1 eV, then efficiency limit can be extended to 30% [19]. Later, another study conducted by Oscar et al. calculated the efficiency limit as 25–27%. In the observations, they mentioned that intrinsic properties of the material are not the limiting factor of conversion efficiency, but the proper alignment of ETL and HTL are [20]. Recently, Wei et al. reported that the maximum limit for conversion efficiency is 31%. They have also certified that the efficiency limit of perovskite can be improved with light trapping through h texturing [21].

With the studies performed on the maximum efficiency limit, it is clear, that there is much more scope for performance improvement. Numerous researchers have investigated the field which will be expounded in the next section.

3. Scientific Progress Made So Far in Perovskite Solar Cell

3.1. Photo Conversion Efficiency

As discussed earlier, PSCs are proven to be low cost solar cells with easy and flexible fabrication technologies.

Edward J. W. Crossland et al. [22] in 2013 showed that mesoporous single crystals compiled with perovskite sensitizer layer delivered higher efficiencies than normal solar cells at low temperatures. Since then, there has been a race for the highest efficiency. The efficiency limit of PSC has been predicted to be about 31% [17], nearly reaching the Shockley–Queisser limit of 33%, according to the detailed balance method [23].

The performance of Perovskite solar cells depends on many factors, including the restrictions of stability and cost. For example, Mesoporous TiO_2 usage has shown the highest efficiency but it requires a temperature of 450 °C to be fabricated. On the other hand, planar heterojunction cells are simpler but have hysteresis losses in efficiency. Although devices with efficiencies more than 20% have been fabricated, the limit can be overstepped by focussing on the imperfections involved in the fabrication process of the cell. The major research is centred on different components of the perovskite solar cell which can be tweaked to achieve better results. Therefore, the discussion on efficiency can be divided into the following basic areas according to their functioning.

3.1.1. Improvement in Perovskite Film Quality

A good quality absorber layer is the foremost requirement to attain highly efficient devices. Perovskite film fabrication method generally involves solution processable techniques such as one step deposition, sequential deposition method, etc. The polycrystalline perovskite absorber developed via these techniques is generally associated with incomplete film coverage and defect sites at grain boundaries. These crystallographic imperfections and charge trap centres favour non-radiative recombination and hence deteriorate the electrical output of the device. Various research groups have facilitated numerous techniques to improve the film quality by means of improving the microstructures and passivating the defect states. A few of them are covered in this section.

Improvement in Microstructure

Film morphology of the substrate is highly dependent on the precursor used in the fabrication of the film. Snaith's group in their article made a comparison in the perovskite film quality fabricated when taking $PbCl_2$, PbI_2, and $Pb(Ac)_2$ as the lead sources in the precursor. They observed that the film prepared from $Pb(Ac)_2$ precursor have smaller grains but dense and uniform film as compared to the film made from $PbCl_2$ and PbI_2 sources. However, $PbCl_2$ based film give larger grains with less number of nucleation sites but the film quality is not up to the mark due to incomplete coverage. Another advantage mentioned is the shorter annealing time required for crystallization of the Pb(Ac)$_2$ based film when compared to the others. Overall, they concluded that using Pb(Ac)$_2$ in the precursor gives faster crystallization with smooth film and relatively better surface coverage along with low energy requirements [24]. In the sequence of achieving uniform perovskite film, Nie et al. (2015) [25] tried coating the perovskite film on hot substrate with pre-heated precursor solution. They analysed that increasing the substrate temperature (>130) leads to larger grains. With substrate temperature of 190 °C, grain size up to 1 mm was achieved. When the substrate is heated at temperatures above the perovskite crystallisation temperature, there is sufficient amount of solvent available and it will reduce the number of crystallisation sites, resulting in bigger grains. With the method reported, they were able to produce a device with 18% PCE efficiency.

Saliba et al. [26] incorporated Rb cation in the perovskite which helps in stabilizing the black phase of perovskite. The addition of this cation supports the crystallization of photoactive phase giving sharp PL emission. With the RbCsFAMA based perovskite device, they got the best cell efficiency of 21.6%.

Yun et al. in 2017 demonstrated a new method to improve the perovskite film morphology. The added NiO nanotubes in the precursor solution of perovskite to support large grain formation by acting as a framework at the time of annealing. With this the film obtained had bigger grains (average size 1100 nm) as compared to controlled film (700 nm). The best efficiency achieved by this modification was 19.3% for planer devices [27]. In the research work conducted by Yongzhen et al. using ionic liquids and molecular additives such as methylammonium acetate and thiosemicarbazide respectively in the precursor solution enhanced the film quality, providing smooth dense and larger grain films. With this chemical engineering, a PCE of around 19% was achieved with the advantage of improved stability. They noted that the device showed an improved stability by retaining 80% of its initial efficiency after 500 h while subjected to a temperature of 85 °C [28].

Yang Bai et al. used a crosslink structure of C60-SAM and silane. The modification provides a surface passivation giving high efficiency of around 19%, as well as the hydrophobic CF$_3$ group in the material improves the stability by retaining 90% of the initial PCE even after 30 days of ambient exposure [29].

Zhou et al. (2014) reported that a better morphology can be achieved using controlled humidity environment. At the same time, they used Y-doped TiO$_2$ as the ETL for increased conductivity. Overall modifications result in the device efficiency of 19.3% for the champion cell [30].

Yang et al. (2016) reported an Ostwald ripening phenomenon in CH$_3$NH$_3$PbI$_3$ films when treated with MABr in IPA solution. With the optimum concentration of MABr (2 mg/mL), compact film with larger grains was fabricated which leads to stabilised PCE of around 19% [31]. Similar kind of MABr surface treatment was also employed by Han et al. with 2% absolute improvement in the PCE. Similarly, Jose at al. reported MACl treatment to enhance the optoelectronic properties of controlled perovskite film [32]. Bert et al. treated perovskite layer with MA and got smooth pin hole free compact layer which improved the photo-physical properties and hence the performance [33].

Additives in precursor: Liu et al. (2017) added V$_2$O$_x$ solution in the precursor which helped in obtaining the large perovskite grains. Existence of V$_2$O$_x$ in the precursor forms hydrogen bonding with the CH$_3$NH$_3$PbI$_x$Cl$_{3-x}$ which helps at the time of crystal growth. With 2.5 wt % of V$_2$O$_x$, they achieved an efficiency of 16.14%. In addition, the additive helps in retaining 70% of its initial efficiency after 1000 h of air exposure [34].

In the two-step method, Wu and co-workers (2017) added DMF in the FA/MA/IPA solution and found better perovskite conversion with improved optical properties and less defect centres. Steady state power output of 20.1% was achieved with 2% DMF content [35].

Zhang et al. (2018) reported that adding 1 vol % graphene oxide in the perovskite precursor solution improves the grain size from 100 nm (for pristine film) to 200 nm. With the improved quality they achieved a PCE of 17.59% [36]. Similarly, Faraji et al. have used benzoquinone (BQ) as additive to get the higher efficiency [37]. Xie et al. have reported cyclic urea as additive in the precursor solution for better performance [38].

Wang et al. demonstrated vapour deposition solution hybrid method (VSHM) to achieve smooth and pin hole free perovskite film. In the method reported, they developed $CH_3NH_3PbI_3$ perovskite film by solution deposition method, after which they have fabricated another layer of perovskite by vacuum deposition. The vacuum deposited perovskite layer followed a two-step approach fabricating $PbCl_2$ layer followed by MAI deposition [39].

Reduction in Non-Radiative Losses

Defect sites arise in the perovskite layer at the time of fabrication and are the major cause of non-radiative recombination. These trap sites reduce the optoelectronic properties of perovskite layer thereby lowering the device efficiency. Several groups have tried to reduce these losses in the absorber material by employing various modifications/passivation to the perovskite layer. Zhang et al. used hypo-phosphorous acid (HPA) in the perovskite precursor solution which improved its morphology by helping in the formation of large diffused grains. The addition of HPA also reduces the possibilities of iodine oxidation at the time of crystallization of the film, thus lowering the bulk defects [40]. Yang et al. reported a new mechanism to stabilize the iodine deficiencies in the perovskite layer. In the two-step approach of perovskite formation, they used an iodide ion rich solution in the deposition of the second step. This iodide ion-based solution was prepared by dissolving solid iodine at 80 °C in isopropyl alcohol. With this defect engineering, they achieved a PSC device with a certified efficiency of 22.1% [1].

Henry's group in 2014 reported treatment of perovskite surface with Lewis bases such as pyridine and thiophene. The treatment resulted in the passivation of unsaturated Pb-bonds available on the surface of the perovskite and lead to increased luminescence and efficiency [41].

Non-radiative losses arising in the perovskite film can also be a cause of strain in the crystal. Wang. et al. employed Al^{3+} doping in the perovskite by supplementing Al-actylacetonate in the precursor solution. A trade off in the optoelectronic properties of the perovskite has been observed by tuning the concentration of the dopant material. They concluded that adding 0.15 mol% of Al dopant can lower the strain in the film, eventually minimizing the electronic disorders. With 0.15 mol % doping, they achieved a PCE of 19.1% with the additional benefit of negligible hysteresis [42]. Doping of alkali metal such as Na^+ and K^+ was also reported by Wangen et al. (2018) to improve the surface smoothness and to reduce the defects sites at grain boundaries [43]. The inherent property of these cations is that they cannot be oxidised or reduced, therefore making them appropriate for the purpose. Zhao et al. were able to passivate the grain boundary defects states by treatment of diammonium iodine C8 without altering the 3D structure of the perovskite. C8 treatment lead to filling the spaces between the grain boundaries and reduced the roughness of the film, consequently improving the electronic quality of the active layer [44].

Ming et al. processed the perovskite surface with small molecules (named as boron subphthalocyanine chloride (SubPc)) solution. The molecules of SubPc intrude in the grain boundary regions and suppress the trap states associated with the grain boundaries leading to enhanced device performance [45]. Jamaludin et al. incorporated tetraethylammonium bromide cation on the surface of the perovskite for the defect state passivation [46]. Jiang and co-workers introduced a polymeric 2D material g-C_3N_4 in the precursor solution of the $CH_3NH_3PbI_3$. At the time of film fabrication, these g-C_3N_4 molecules slow down the crystallisation process of perovskite and reduce the possibility of charge recombination at interfaces. With the modification, a notable efficiency of 19.49% was acquired [47].

3.1.2. Improvement in Hole Collection Efficiency

Spiro-OMeTAD remains the most widely used hole transporting layer in PSCs, and the material with which other HTMs are compared. Its usage however precedes perovskite as it has previously been used as an HTL in dye sensitized solar cells [48]. Nam Joong Jeon et al. [49] studied three Spiro-OMeTAD derivatives and observed the optical and electronic properties by changing the positions of the two methoxy substituents in each quadrant. The cell performance was found to be dependent upon the position of the methoxy groups with o-OMe substituents showing the highest efficiency of 16.7%. In a similar manner, Dong Shi et al. [50] grew single crystals of Spiro-OMeTAD and determined its crystal structure, appearing contrary to what was previously known, finding that Spiro-OMeTAD's single-crystal structure has a hole mobility that is three orders of magnitude greater than that of its thin-film counterpart. They used methanol as an anti-solvent to crystallise the Spiro-OMeTAD crystals showing further optimization in the same direction that could help in boosting the solar cell efficiency.

Nazeeruddin and group invented a new method to passivate the defects on the perovskite surface at HTM side. They introduced a thin layer of another perovskite $FAPbBr_{3-x}I_x$ between existing perovskite and hole transporting material. This passivation was achieved by spin casting FABr on the primary layer. It was expected that the unreacted PbI_2 in the primary layer may react with the FABr and form another perovskite, as stated earlier. With the XRD pattern, they proved that the partially reacted PbI_2 was converted into perovskite, suppressing the defects and enhancing the Voc. This modification resulted in an efficiency of 21.3% for the best cell [51].

Recently, Jae Choul Yu and co-workers reported 18% efficient perovskite devices in the inverted architecture using PEDOT:GO as the hole transport layer. This increased device performance can be attributed to superior morphology of the film with reduced contact potential between HTL/perovskite. Use of this PEDOT and GO composite instead of conventional PEDOT:PSS film improves the hysteresis loss in the device [52].

NiO_x has emerged as a new HTL material due to its advantages such as robustness to the environment and preferable energy level alignments. Seongrok et al. used NiO_x based HTL in their inverted perovskite architecture which exhibited an efficiency of 16.4% [53]. Park et al. also fabricated NiO_x as HTL layer by electrochemical method and reported a power conversion efficiency of 17% for 1 cm^2 area devices [54]. Furthermore, many groups have suggested doping of a transition metal in NiO_x to enhance the electrical properties and morphology of the layer. Recently, Wei Chen et al. reported Cs-doping in NiO_x layers in inverted perovskites which facilitates deep valence band leading to an efficiency gain of 19.35% [55]. Similarly, doping of Co, Li-Cu, Cu, and Ag have also been reported [56–59].

A small subset of improving the hole transport layer is changing the layer itself, which often allows for an inverted structure of the perovskite to be used.

Dewei Zhao et al. in 2014 studied the inverse perovskite solar cell architecture, placing the hole transport layer on the transparent electrode. Because of problems with PEDOT:PSS in the inverted case, the researchers chose poly[N,N'-bis(4-butylphenyl)-N,N'-bis(phenyl)benzidine] (poly-TPD) as the hole transport layer and electron block layer. The device showed an efficiency of 13.8% on an average with a maximum of 15.3%. Poly TPD devices performed much better than PEDOT:PSS based devices owing to large crystallites in the former case [60].

Another case of a Spiro-structured HTM is the spiro[fluorene-9,9'-xanthene] (SFX) based HTM-X60 (Figure 1) which has been shown to give an efficiency of 19.84%. X60 is comparatively easily synthesizable and therefore might be a promising candidate for large-scale industrial production [61]. Knowing that Spiro-OMeTAD is expensive, there are other materials which have been studied for HTM purposes. However, none of them is probably as well known as H101 based on 3,4-ethylenedioxythiopene (Figure 2). H101 is the first heterocycle containing material which achieved an efficiency greater than 10%, therefore being comparable to Spiro-OMeTAD. Additionally, it is also cheaper and easier to synthesise [62].

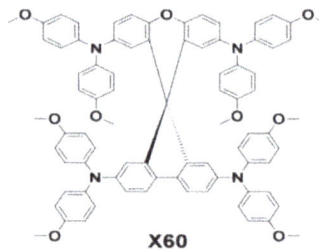

Figure 1. The structure of X60 as used by Xu et al. [61].

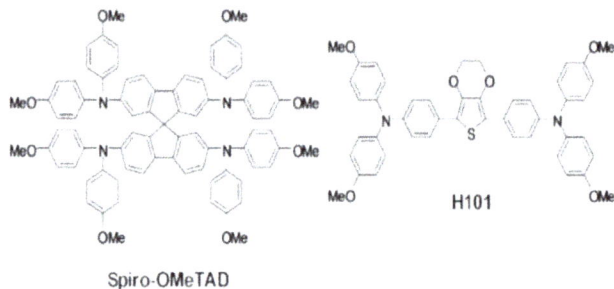

Figure 2. Spiro-OMeTAD and H101 structure comparison; H101 being easier to synthesise. (from Li et al. [62]).

Triazatruxene-based Hole transporting materials with 18.3% PCE [63], PTAA with max PCE of 12% [64], poly (3,4-ethylenedioxythiopene) PEDOT with 17% PCE [65], Spiro-Phenylpyrazole/Fluorine with 14.2% [66], conjugated polyelectrolytes with 12%+ [67], and MEH-PPV and P3HT with 9.65% [68] are other hole-transporting materials which have been successfully investigated.

3.1.3. Improvement in the Electron Collection Efficiency

Efficiency of the overall device is greatly dependent on how well we can extract the photo-induced charges generated in the absorber layer. Inefficient transport of charges from perovskite to electrodes results in the dropdown of power conversion efficiencies. The term "inefficient charge transport" refers to recombination losses at the interface due to existence of trap states and losses due to defects in the charge collecting material itself. For collection of electrons a commonly reported ETL is c-TiO_2 or mesoporous TiO_2. This material is associated with the surface defects generated from the oxygen vacancies in the near conduction band region due to unsaturated Ti(IV) valancies. At the time of charge transfer from absorber layer to ETL, there is a high probability that the electron can be trapped through the oxygen vacancies at the interface and lead to non-radiative recombination losses.

Three basic ways can be identified to mitigate the recombination losses in the charge transport from perovskite to ETL.

Overcoming the Defects Present in the Material Itself

Doping in TiO_2 has been proven as an effective way to reduce the defect densities. In 2014, Pathak et al. reported Al doping TiO_2 for reducing the deep sub band gap trap sites in TiO_2 for efficient charge collection in DSSC and perovskite solar cells [69]. Later, Chen et al. reported that Nb doped TiO_2 improves the PCE of perovskite by improving the carrier density and electron injection probability [70]. Michael Graetzel's group in 2016 reported an improved PCE of 19% with Li doped TiO_2 [71]. Jun et al. [72] and Xiangling et al. [73] also analysed the effect of Indium and Fe^{3+} doping in TiO$_2$, respectively, and found that the device performance was upgraded as compared to that of pristine TiO$_2$. Doping in TiO$_2$ will lead to passivation in the oxygen deficiency and/or the low-level trap states generated at the time of material synthesis due to unsaturated Ti(III), and hence boost the performance.

Minimizing the Interface Recombination by Passivating Layer/Element between ETL and Perovskite

Various research groups have tried and succeeded in the modification of the TiO$_2$ based ETL for effective charge extraction. Seo et al. in their work used mesoporous TiO$_2$ based architecture with the TiO$_2$ nanoparticles of 200 m^2/g surface area which is much larger than what have been used till now in the conventional mesoporous based devices (60 m^2/g). The large area facilitates better contact with the perovskite giving efficient extraction. In addition, they modified the nanoparticle surface with the doping of Cs-halide. The doping with Cs based halide saturates the oxygen vacancies present on the surface by reducing the Ti(IV) into Ti(III). By the modification they were able to achieve a PCE of 21% [74]. Snaith et al. [75] also implemented the modification of the surface of c-TiO$_2$ by Caesium Bromide (CsBr) coating, inhibiting the UV induced degradation. It also enhances the efficiency by providing improved electron transfer rate. In a similar approach, Sun et al. reported silane modification on the TiO$_2$ surface which absorbs UV radiation as well as improves the film quality [76].

Another study performed to passivate the interface defect sites by Hairen et al. showed the improved efficiency of 20.1% for small area devices using Cl capped TiO$_2$ nanocrystals in mixed cation (FA and MA) halide (I and Br) based perovskites. In the figures of charge recombination lifetimes collected by them, Cl-TiO$_2$ NCs shows almost double the lifetime (145 μs) when compared to pure TiO$_2$ (65 μs). Cl capping in the TiO$_2$ helps in the surface passivation of the perovskite film by inhibiting the trap state formation leading to reduced recombination losses. By employing the same ETL to CsMAFA based perovskite, a high efficiency of 21.4% was achieved for small area devices [77].

Hao li and co-workers demonstrated a new carbon quantum dot and TiO$_2$ based electron transport layer for better charge extraction. Introduction of Carbon QDs in TiO$_2$ gives rise in the current values indicating efficient charge transport from perovskite to TiO$_2$. They observed that the increase in the wt % of CQDs in the TiO$_2$ prompts the charge transport but overall device performance will be improved with 10% (by weight) CQD incorporation in TiO$_2$. However, they did not notice any chemical bonding between quantum dots and TiO$_2$ in XPS studies inferring that CQDs acts as spacer between them. The best cell efficiency obtained was 18.9% with the Jsc and Voc values of 21.36 mA/cm^2 and 1.14 V respectively [78]. In the same direction, Jun et al. used PMMA:PCBM ultrathin layer at ETL/perovskite interface in mesoporous based perovskite device. They identified that the use of PMMA passivates the dangling bonds at the interface but simultaneously lowers the conductivity of the device. Addition of PCBM is used to balance this conductivity loss. Correct ratio of PMMA:PCBM (1:3) mixture showed the best cell efficiency of 20.4% [79].

Wenzhe Li et al. showed that inter-facial adjustments could lead to increased efficiency (and resilience) in perovskite solar cells, since the layers have small surface areas. They used a Caesium Bromide layer between the electron collection layer and the absorber layer of perovskite. This lead to an increased efficiency and stability of the solar cell: Firstly, this layer reduces the work function of TiO$_2$. Secondly, it inhibits the degradation of the perovskite layer because of UV rays. Thirdly, it increases the electron transfer rate from perovskite to c-TiO$_2$ [75].

Severin et al. summed up that *t*BP (4-*tert*-butylpyridine) adsorbs to the TiO$_2$ surface increasing the V$_{oc}$ [80]. There is a negative shift in the energy bands of TiO$_2$ reducing the recombination rate which implies long charge carrier lifetime, and hence the efficiency is increased. Song et al. demonstrated that when 40 nm thick A-TiO$_2$ (prepared by a potentio-static anodization method) is deposited on FTO glass, the efficiency is maximum [81], a 22% increase from 12.5% to 15.2%.

Replace the Existing TiO$_2$ Layer with New ETLs

TiO$_2$ is widely used but it is associated with the flaw of high temperature annealing which is one of the limiting factors in its commercialization. In addition, the device instability under UV irradiation for longer times has been a critical parameter. Snaith's group in 2011 reported the advantages of tin oxide as an electron collecting layer as compared to TiO$_2$ [82]. SnO$_2$ provides the benefits of low temperature annealing, relatively higher band gap of 3.2 eV, and has better electron mobility (100 times faster than TiO$_2$). Its deep level conduction edge facilitates faster charge transport from absorber layer to ETL. Qi Jiang and co-workers reported a PSC device of 20% efficiency with SnO$_2$ based ETL while showing that the hysteresis can be lowered to a much greater extent in SnO$_2$ based device [83]. The device could be produced at under 150 °C and utilized a PbI$_2$ passivation phase in the perovskite layer, thus showing that meso-porosity can be replaced for high efficiency devices. In an approach Zhu et al. researched the use of SnO$_2$ Nanocrystals as an ETL which provides good efficiency while retaining 90% of the initial PCE after 30 days of exposure to ambient conditions with RH~70% [84]. Recently, they also synthesized a new ETL material named c-HATNA. Utilizing this ETL, they were able to achieve high device performance with a PCE of 18.21% [85]. Similarly, a new bilayer ETL has been proposed by Escobar et al. in which they have reported a combination of ZrO$_2$ nanoparticles and TiO$_2$ compact layer. The concept reduces any possibility of short circuiting between perovskite and TiO$_2$. With the combination, they witnessed a stabilized power output of 17.0% with almost no hysteresis [86]. Due to the high temperature requirements in the fabrication of TiO$_2$, organic charge-transport materials were employed as ETL. PCBM was reported as a possible replacement for TiO$_2$ providing the advantage of room-temperature fabrication [87]. Jangwon and co-workers [88] reported a 14.1% efficient single unit cell device with PCBM as electron transport layer using a p-i-n architecture. A buffer layer was also introduced between PCBM and Al electrode which helps in better performance of the device by reducing the energy barrier between them. The PCE dropped to 8.7% when the architecture was employed on a large substrate of 10 × 10 cm^2 area. Low temperature PSCs utilizing nanocomposites of graphene and TiO$_2$ nanoparticles were reported with an efficiency of 15.6% to counter the sintering requirement (by using a solution-based deposition method) [89].

3.1.4. Miscellaneous Studies

Hysteresis

A major component in the energy loss of PSC, which leads to a lower efficiency is the hysteresis loop. The studies conducted by Dualeh et al. (2013) [90] and snaith et al. (2014) [91] show that hysteresis is dependent on the scan rates used during current–voltage measurements, although the relationship found was contradictory in the two cases. Hysteresis loss is a function of pinhole formation and other faults in perovskite solar cells. Majority of these losses in the PSC are dependent on the morphology of the film, which is difficult to control. This has been attributed to slow transient capacitive current, dynamic trapping and de-trapping processes, and band bending due to ion migrations or ferroelectric polarization, as reviewed by Bo Chen et al. [92]. Rafael S. Sanchez et al. [93] detected slow dynamic processes in lead halide PSCs with a low frequency characteristic time which cannot be attributed to recombination processes in the cell and are correlated with J-V hysteresis. Wu et al. proposed the presence of a majority carrier bottleneck at either one or at both the perovskite/transport layer interfaces, combined with recombination at that interface, as the cause of hysteresis [94]. Shao et al. (2014) [95] showed that photocurrent hysteresis in the perovskite device could be eliminated by fullerene passivation using PCBM. According to their studies, trap sites on the surface and at grain boundaries of the perovskite material can be the reason behind the hysteresis. However, it has been posited that further research is required for a complete understanding of the science behind the hysteresis; therefore, controlled experiments have been suggested to probe the interstitial and interface causes of hysteresis [96].

Role of 4-*tert*-ButylPyridine

Habisreutinger et al. [80] investigated the role of 4-*tert*-ButylPyridine in perovskite solar cells. They concluded that the inclusion of tBP in perovskite solar cells leads to an increase in the steady-state efficiency of the cell. By deconstructing the solar cell, it was found that tBP interacts with the perovskite to make it more selective for holes, thus the entire study becomes independent of the hole-transporting material used. Since tBP is used on *p*-type side of the cells, the solar cells can be made more efficient by studying the tBP and perhaps similar chemicals which could make Perovskite more selective to hole selection.

Graded Band-Gap PSC

A new concept of graded band-gap perovskite solar cells has been added to the field by Ergen et al. They achieved this based on an architecture of two perovskite layers ($CH_3NH_3SnI_3$ and $CH_3NH_3PbI_{3-x}Br_x$) incorporating GaN, Boron Nitride, and graphene aerogel, reporting an efficiency maximum of 21.7% [97].

Method of Film Fabrication

There have been several modifications in the method of perovskite film fabrication itself. Most recently, Ayi et al. [98] used a two-step spin-coating method for the preparation of perovskite film with varied spinning speed, spinning time, and temperature. They created a pin-hole free film with large grain size at room temperature with spinning speed 1000 rpm for 20 s which was annealed at 100 °C for 300 s. In 2015, Nie et al. [25] used a hot casting technique to grow continuous, pinhole-free perovskite films with millimetre-scale crystalline grains. The hot casting was done at a temperature of 180 °C which resulted in large crystalline grains. The group reported an efficiency of up to 18% with roughly 2% variability and no hysteresis loss, which they attributed to the superb morphology of the perovskite film formed. Thus, the two processes have shown the difference in perovskite formation and the utility of perovskite as a substance is liable to constant change. For example, recently, Fei Han et al. [99] used a dissolution-recrystallisation method (first time reported) to improve device performance. They tested four combinations of DRM and found DRM-2 (4 μL DSMO and 96 μL chlorobenzene) to give the most efficient cell with an efficiency of 16.76% compared with 13.81% for only chloro-benzene treated devices. There was a high level of reproducibility as well, with most devices showing a PCE of 15–17%. The SEM scans revealed lesser defects and grain boundaries thus improving the TiO_2/perovskite interface as well.

Transitioning from Lead to Other Elements

Because of the toxic nature of lead, one of the challenges for perovskite researchers is to utilize other elements in the place of lead while keeping the perovskite structure and properties identical. The choice of element is dependent theoretically on the Goldschmidt tolerance factor [100] and practically on the formation of black perovskite [101]. Yao et al. [102] tested the optoelectronic properties of a Lead substituted Strontium based perovskite solar cell for different amount of strontium substitution. They found a positive effect on the thermal stability and output voltage (1.11 V as opposed to 1.07 V in the device without doping) with a = 0.05 amount of Strontium substitution, where "a" represents the stoichiometric coefficient of strontium in the material. The maximum efficiency achieved was 16.3%. In another study, Eperon et al. [103] reported a method to form pure black phase $FAPbI_3$ infiltrated into mesoporous scaffolds while also enhancing the performance of single cation solar cells. Their study focused on treating $MAPbI_3$/$FAPbI_3$ with solutions of formamidinium or methylammonium iodide to tune the band gap of perovskite to 1.48–1.57 eV.

Crystals 2018, 8, 242

Inverted Structure of Perovskite

Apart from the above discussion, another important aspect is the use of inverted Perovskite cell structure where the positions of ETL and HTL are interchanged. This structure has been derived from organic solar cells and has been a successful transfer [104]. Due to the low hysteresis and simple fabrication method involved, inverted PSC (*p*-type layer at front) is worthy of investigation [105]. Recently, Wu et al. [106] reported a PCE of 18.21% within an aperture area of 1.022 cm^2, for an inverted PSC which had a perovskite-fullerene graded heterojunction. The concept of perovskite-fullerene bulk hetero-junction to enhance photo-electron collection was employed though the PCBM distributed within the perovskite in a gradient. Perovskite-fullerene bulk hetero-junction has previously been reported to enhance performance [107].

3.1.5. Perovskite-Tandem Solar Cell

Presently, researchers are looking at ways to fabricate high efficiency tandem solar cells (TSCs). These solar cells are usually created by an interconnection between a top perovskite cell and a bottom silicon (or other inorganic material) solar cell. Perovskite has been shown to be an ideal tandem material because it matches the three required criteria: solution processable, large bandgap, low energy losses [108]. There are different possible configurations in tandem solar cells (Figure 3): series tandem, module tandem, and four-terminal tandem; among these the four-terminal tandem is expected to have the highest efficiency. The architecture can also be divided into three types, as shown in Figure 4. Limiting efficiency for TSCs has been shown to be around 68.2% for an infinite number of sub-cells (under one sun illumination) [109], while efficiencies of 45% have been shown for the three configurations [110].

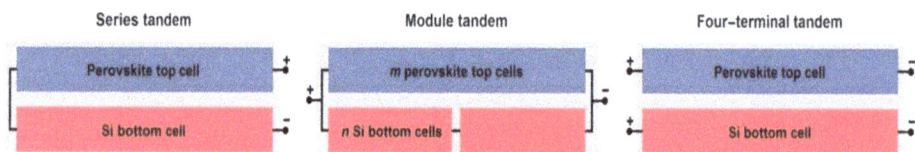

Figure 3. Si-Perovskite Tandem configurations, characterised by the number and connection of Perovskite and Silicon solar cell. Figure Taken from Futscher et al. [100].

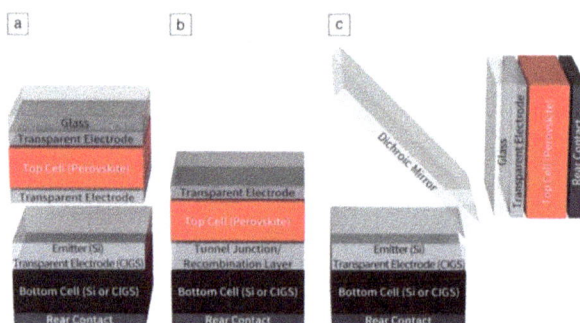

Figure 4. Tandem architecture in terms of internal stacking: (**a**) mechanically Stacked; (**b**) monolithically integrated; and (**c**) spectrally split. Taken from Bailie et al. [108].

The maximum efficiency for tandem solar cells has been reported to be around 28% utilizing an optical splitting system which splits light of different wavelengths and transports it to respective cells [111]. Hence, the splitting makes possible the usage of different cells in their respective bandgaps.

100

TSCs include perovskite along with Kesterite (Cu$_2$ZnSn(S,Se)$_4$) [112], crystalline silicon [113–116], Copper Indium Gallium Selenide [117], and recently even perovskite with perovskite [118] where it was pointed out that there was need for high-performance low bandgap perovskite solar cells. A recent field which is also progressing is polymer-based organic photovoltaic cells which enable solution-processable tandem solar cells [119].

In 2015, Mailoa et al. [115] fabricated a two-terminal perovskite/silicon multi-junction solar cell, but it was only able to reach an efficiency of 13.7%, signalling the need for more research so that the potential could be tapped into, which was concluded to be dependent on more efficient silicon and perovskite cell layers. Later, Fan Fu et al. [120] fabricated a CH$_3$NH$_3$PbI$_3$ layer over ~100 nm compact ZnO layer and ~50 nm PCBM. This electron transport layer was low-cost, lightweight, and compatibile with monolithic tandem fabrication. Further, Spiro-OMeTAD was spin coated onto that followed by deposition of 60 nm gold contact by thermal evaporation. In addition, PCBM below perovskite smoothened the surface and eradicated the J-V hysteresis. The device showed an open-circuit voltage of 1.101 V, short-circuit current of 17.6 mA cm^{-2} and a fill-factor of 74.9%, resulting in a PCE of 14.5%. When perovskite was mechanically stacked over the CIGS solar cell, the efficiency jumped to 20.5%. The J$_{SC}$ of the CIGS cell was found to decrease because of reduced light intensity.

Bush and co-workers [4] fabricated monolithic perovskite/silicon tandem solar cells to reach an efficiency of 23.6% with zero hysteresis. They improved upon previous tandem solar cells by combining caesium formamidinium lead halide perovskite (band-gap of 1.63 eV) with an infrared-tuned silicon hetero-junction bottom cell (which had a rear-reflector of silicon nanoparticles). As reported, the final solar cell had a V$_{oc}$ of 1.65 V, a J$_{sc}$ of 18.1 mA/cm^2, and a fill factor of 79% with a 1 cm^2 aperture area and no hysteresis. Furthermore, the stability of the tandem cells also increased because of the ITO employed which trapped the volatile methylammonium cation, thus reducing its degradation. The authors tested the tandem cell for more than 1000 h in a temperature of 85 °C and 85% relative humidity. It was also able to sustain maximum power for 30 min under constant illumination. They concluded that performance could be increased by widening the band-gap of the perovskite and reducing front-surface reflections.

David P. McMeekin et al. [121] used a metal halide perovskite cell in tandem with a silicon bottom cell to show that four terminal tandem cells of efficiencies greater than 25% may be obtained. The main perovskite cell employed caesium instead of the formamidinium cation to get a tunability of band-gap to around 1.75 eV (eliminating the instability region of iodine to bromine phase change) which had a PCE of over 19%. While greater efficiencies have been reported in single terminal perovskite cells, this solar cell serves better in the case of tandem solar cells by increasing the band-gap from the conventional 1.55 eV to 1.75 eV.

3.2. Long Term Device Stability

3.2.1. Degradation Mechanisms

Moisture Degradation

It is evident that stability of the perovskite solar cell is one of the major factors that needs to be addressed before industrial application of this technology begins. Perovskite solar cells are inherently very sensitive to ambient operating conditions. Their performance starts deteriorating once they are exposed to moisture, UV radiation, and an increase in temperature. Degradation of the device is attributed mainly to the degradation of the perovskite layer itself or the degradation of interfacial layers. Among all factors, humidity is the biggest felon in the degradation of hygroscopic perovskite film. Exposure of perovskite film to humid environment leads to its hydration eventually resulting in the loss of performance. Leguy et al. found that hydration leads to formation of monohydrate and dihydrate phase of perovskite [122]. The hydrate formation mechanism can be understood by the following equations:

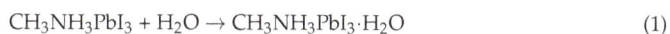

$$CH_3NH_3PbI_3 + H_2O \rightarrow CH_3NH_3PbI_3 \cdot H_2O \tag{1}$$

$$4(CH_3NH_3PbI_3 \cdot H_2O) \rightarrow (CH_3NH_3)_4PbI_6 \cdot 2H_2O \tag{2}$$

Upon prolonged exposure to moisture, excess amount of water leads to formation of lead iodide by dissolving the methylammonium iodide part.

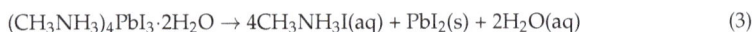

$$(CH_3NH_3)_4PbI_3 \cdot 2H_2O \rightarrow 4CH_3NH_3I(aq) + PbI_2(s) + 2H_2O(aq) \tag{3}$$

However, Neu et al. in their report had shown the direct decomposition of perovskite into MAI and lead iodide on exposure to UV illumination coupled with air (containing oxygen and moisture), without any intermediate hydrated phase [123].

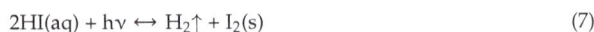

$$CH_3NH_3PbI_3 \leftrightarrow CH_3NH_3I(aq) + PbI_2(s) \tag{4}$$

$$CH_3NH_3I(aq) \leftrightarrow CH_3NH_2(aq) + HI(aq) \tag{5}$$

$$4HI(aq) + O_2 \leftrightarrow 2I_2(s) + 2H_2O \tag{6}$$

$$2HI(aq) + h\nu \leftrightarrow H_2\uparrow + I_2(s) \tag{7}$$

Decomposition of MAI gives CH_3NH_2 and HI [124]. HI further decomposes into H_2, I_2 and H_2O on exposure to oxygen and illumination [125].

Light Induced Degradation

Another factor influencing the stability of perovskite solar cell device is UV light exposure. Charge collection efficiency of the device is severely limited with UV irradiation. The reason for this deterioration is the presence of deep charge trap states on the TiO_2 surface. These sites adsorb the oxygen in the atmosphere and act as a recombination site for the hole generated upon shining of ultraviolet light. Desorption of oxygen from these sites leads to trapping of the electron from the conduction band of TiO_2. The trapped electrons on these deep defect sites can easily recombine with holes available in the HTM and perovskite layer thereby resulting in a drastic drop in charge carrier collection efficiency [126].

In principal, passivation of these trap states and avoiding UV to reach to TiO_2 are the basic solution to suppress the UV light degradation. Thus far, there are three methods to bypass the problem: either pacify the trap sites or avoid UV light from reaching the TiO_2 or replace the TiO_2 with another material.

Thermal Degradation

Increased temperature during operational conditions also causes instability of the perovskite film. High temperature exposure causes decomposition of $CH_3NH_3PbI_3$ into its constituents [127]. Perovskite decomposes to give PbI_2 which accumulates as a yellow phase between the layers of the cell and is orthorhombic and non-perovskite in structure. This decomposition of the perovskite is accelerated when temperature is increased further and is observed even in an inert environment (with argon). The PbI_2 is stable within the layers of perovskite solar cell, but it leads to a decrease in overall stability. Conings et al. demonstrated the decomposition of perovskite after heating at 85 °C for 24 h [128]. It has been observed that under exposure to much lower temperatures MAPbI3 converts to PbI_2. According to recent studies performed by Kim and his co-workers on exposure to temperatures greater than 100 °C, MAPbI3 decomposes into PbI_2, CH_3I, and NH_3. PbI_2 remains whereas CH_3I, and NH_3 evaporate [129]. They also elucidated that prolonged exposure to 80 °C results in degradation of perovskite. They established the fact that organic MA cation plays a critical role in the thermal induced degradation since a change in the orientation as well as decomposition of MA on exposure to thermal stress is observed. Thus, replacing it with some inorganic cation can be beneficial in improving thermal stability.

3.2.2. Improving the Stability of Perovskite Solar Cells

Globally, researchers are trying hard to achieve highly efficient and at the same time highly stable perovskite solar cells. Many attempts have been made to impart stability to the perovskite solar cell. Thus far, researchers have succeeded to achieve the stability of more than one year in perovskite solar cells [130]. The various possible ways discovered to impart stability to the perovskite solar cells are: (i) by altering the perovskite material since MAPbI₃ is more prone to degradation; (ii) by providing hydrophobic HTL; (iii) by altering ETL so that UV induced degradation can be avoided; and (iv) by proper encapsulation of the device to avoid moisture ingress.

By Altering the Perovskite Material

Alteration of halide composition has proven to be an efficient way of improving stability. Tuning of divalent cation A and halide X in perovskite composition can make it more stable to the ambient environment. Mixed halide perovskite compositions generally incorporating Br in MAPbI₃ have proven to be more stable in humid environments. Noh et al. showed that incorporation of Br in the MAPbI₃ perovskite results in enhanced stability. [131] They have shown that when compared to the MAPbI₃ perovskite degrading in a couple of hours, MAPb(I_{1-x}Br$_x$)$_3$ with x = 0.2 was stable up to 20 days. According to them replacement of large Iodine ions with smaller Br ions results in cubic symmetry, which leads to enhanced stability. Jeon et al. also proved that Br incorporation into MAPbI₃ results in an enhanced efficiency of 16.2% along with improved stability [132]. Although researchers have shown an enhancement in the moisture stability and stabilized crystal structure on incorporation of Br, a study by Hoke and co-workers elucidated upon reversible light-induced phase transition in Br incorporated mixed halide perovskite. They believe that light soaking results in the generation of minority iodine-enriched and majority bromide-enriched phases, though dark exposure treatment can revert to the initial structure [133].

Some authors [134–136] have realized the use of pseudo-halogen for stability enhancement in MAPbI₃ perovskite solar cell. Tai et al. [137] and Chen et al. showed that incorporation of lead (II) thiocyanate (Pb(SCN)₂) in the precursor leads to enhanced stability and performance. High-quality CH₃NH₃PbI₃ₓ(SCN)ₓ can be formed in humidity of more than 70% and with efficiency up to 15%, maintaining 85% of the efficiency for more than 500 h without any encapsulation. This enhanced stability is attributed to the strong interaction between Pb^{2+} cation and the SCN⁻ anion.

Stability can be enhanced by replacing the MA hygroscopic part with some other organic or inorganic cation with lesser sensitivity towards moisture. Larger formamidinium (FA) cation has shown to be more resistant to moisture and thermal degradation as compared to MA [138]. FAPbI₃ perovskites have slightly lower bandgap (1.48 eV) as compared to MAPbI₃ (1.55 eV) providing even wider absorption, rendering it suitable for photovoltaic application [139]. FAPbI₃ is reported to be unstable in the black α-FAPbI₃ phase at room temperature and a transition to yellow δ-FAPbI₃ photo-inactive phase is observed [140]. Tolerance factor of FAPbI₃ is greater than 1, making the perovskite phase unstable and transition to yellow phase is quite favourable. This issue has been resolved by incorporation of smaller cation such as Cs and Rb into FA based perovskite to make the tolerance factor lie in the range 0.9–1 [141]. Due to entropic stabilisation, caesium assists the crystallization of black phase in FA based perovskite. Jeon et al. demonstrated that a combination of FAPbI₃ and MAPbBr₃, (FAPbI₃)₁₋ₓ(MAPbBr₃)ₓ results in phase stabilization and enhanced efficiency of around 18% [142]. Yang et al. also elucidated that a mixture of FA and MA can yield highly efficient (>19%) and structurally stable perovskite [143]. Another possible way of countering the instability issue could be the replacement of organic MA part with some inorganic cation. Caesium (Cs) is detected to be an acceptable alternative to organic MA cation by various research groups with enhanced moisture and thermal stability (stable up to 460 °C) [2] and suitable bandgap (1.73 eV) required for solar cell applications [144]. In the initial studies, it was observed that the photoactive δ-CsPbI₃ phase is not stable below 315 °C, as there is a transition from black cubic δ-CsPbI₃ phase to yellow orthorhombic α-CsPbI₃ phase. Till date various approaches have been discovered to stabilize the cubic δ-CsPbI₃ phase. Eperon et al. showed that addition of hydroiodic in the $CsPbI_3$ precursor results in formation of stable black δ-$CsPbI_3$ phase on annealing at 100 °C but with 1.7% PCE [145]. Recently, it has been observed that incorporation of sulfobetaine zwitterion (1.5 wt %) in $CsPbI_3$ precursor solution results in the formation of room temperature stable δ-$CsPbI_3$ phase. They found that plasma treated $CsPb(I_{0.98}Cl_{0.02})_3$ device leads to a stabilized PCE of 11.4%, better than $CsPbI_3$ device [144].

Further studies on Cs based perovskites, hunting for higher efficiencies, have also been positive. Saliba et al. presented that a triple Cs/MA/FA cation mixture perovskite can achieve an efficiency of 21.1% and maintains efficiency of about 18% under operational conditions after 250 h. Adding Cs cation to MA/FA mixture suppresses yellow phase PbI_2 impurities. In recent papers [4,121] CsFA ($Cs_{0.17}FA_{0.83}$ $Pb(Br_{0.17}I_{0.83})_3$) perovskite tandem cell was operated at humidity 40% and temperature 35 °C. Even then the Device performed for over 1000 h of testing with minimum degradation. Caesium (5%) induces more uniform grains thereby enhancing the better charge transport resulting in a higher fill factor [146]. Sun and co-workers showed that incorporation of $Pb(SCN)_2$ in triple cation mixed halide perovskite $FA_{0.7}MA_{0.2}Cs_{0.1}Pb(I_{5/6}Br_{1/6})_3$ results in highly efficient and stable perovskite, and without the requirement of thermal annealing [147]. They have realized a room temperature flexible solar cell with an efficiency of 10.55%. This opens a path for the realization of low temperature, highly efficient, stable, and flexible perovskite solar cells. Bu and co-workers successfully demonstrated a quadruple cation perovskite ($K_x(Cs_{0.05}(FA_{0.85}MA_{0.15})_{0.95}Pb(I_{0.85}Br_{0.15})_3$ with enhanced efficiency and stability [148]. Partial substitution of the B site with Bi^{3+} has also proven to be an efficient way of stabilizing α-$CsPbI_3$ phase, retaining 68% of initial PCE after 168 h of exposure to ambient conditions without encapsulation [149].

Many 2D perovskites solar cells have been realized with better moisture stability but which compromise on the PCE [150]. Grancini et al. showed that 2D/3D perovskite junction interface engineering results in an extraordinary stability of more than one year [130]. They showed that an HTM free solar cell can be realized by substituting Spiro-OMeTAD and Au contact with carbon matrix in the proposed 2D/3D ($HOOC(CH_2)_4NH_3)_2PbI_4$/$MAPbI_3$ perovskite junction solar cell. They elucidated that 2D perovskite layer functions as a protective window against moisture for the underlying efficient 3D perovskite layer, at the same time acting as barrier for electron recombination. In a recent study, Wang and co-workers also investigated a 2D/3D heterostructure by incorporating *n*-butylammonium cations into $FA_{0.83}Cs_{0.17}Pb(I_yBr_{1-y})_3$ perovskite leading to enhanced efficiency and stability [151]. They found an average PCE of $17.5 \pm 1.3\%$ and retention of 80% PCE in ambient environment for 1000 and 4000 h with encapsulation.

By Altering the ETL

Usually, TiO_2 is used as an electron transport layer (ETL) in perovskite solar cells. Batmunkh and co-workers modified the TiO_2 ETL by incorporating Single-Walled Carbon Nano-Tubes (SWCNTs) in the TiO_2 NPs and sandwiching this layer between compact TiO_2 and perovskite layer [152]. They witnessed an increase in the light stability because SWCNT are highly conducting and they provide alternative paths for electrons to travel. This way they do not get trapped in the deep trap states developed in TiO_2 on light exposure. Air stability is also enhanced as CNTs are hydrophobic so they do not allow the adsorption of moisture onto TiO_2 surface, which might lead to degradation of on grown perovskite film. These surface modifications improve the stability under UV light as well as enhance the efficiency of the device through passivation. Pathak et al. demonstrated that Neodymium doped TiO_2 has increased conductivity [153]. Only 0.3% doping of Nd in TiO_2 passivates all the non-stoichiometry defects resulting in increased device performance. PSC prepared with Nd-doped TiO_2 electrodes have higher early-time stability without affecting the short-circuit current. Lifetime of the encapsulated devices is improved with neodymium doping.

Replacement of TiO_2 with ZnO had also been tried by Yang et al. and it turned out be more thermally unstable than TiO_2 [154]. The reason behind it is that ZnO consists of deep trap states caused by electrons hopping from perovskite LUMO and might cause trap assisted recombination, resulting in performance loss. Mahmud et al. presented a method of modifying the surface of ZnO with Caesium Acetate (CA) and Caesium Carbonate (CC) showing improved stability and device performance [155]. They have elucidated that CA modified ETL in mixed organic cation $MA_{0.6}FA_{0.4}PbI_3$ perovskite film shows a PCE of 16.45% and higher stability than CC modified one, when stored in controlled N_2 ambient for 30 days under RH of 35–40%. This is because CC ETL is more sensitive to humidity, causing easy moisture assisted degradation. Cao el al. studied modification of ZnO with a thin passivation layer of MgO and protonated ethanolamine (EA) with PCE of about 21.1% and stable up to 300 h [156]. MgO eliminates charge recombination and EA helps in efficient charge transport. M. Arafat Mahmud et al. also investigated the effect of adsorbed carbon derivatives, fullerene (C60) and $PC_{71}BM$ on the triple cation $MA_{0.57}FA_{0.38}Rb_{0.05}PbI_3$ perovskite and ZnO ETL interface [157]. They found that C60 modified device turned out to be more efficient and stable holding back 94% of initial PCE after one month of studies in N_2 filled glove box without encapsulation. This enhancement is because of less hydrophilicity of C60 modified ZnO and passivation of trap states by intercalation of oxygen in the interstitial trap states of ZnO. Pang and co-workers in a recent study have shown highly mesoporous $Zn_2Ti_3O_8$ (m-ZTO) as an efficient ETL which retarded the interfacial recombination resulting in high efficiency [158]. The devices so formed displayed a decrease of just 12% on 100 days of exposure to 10% RH and ambient temperature.

Wang and co-workers introduced a new ETL layer of cerium oxide. They reported that this CeO_x (x = 1.87) based electron transport material can be processed at low temperatures and is equally efficient for electron extraction from perovskite absorber layer. Stability test showed that this material is more robust as compared to conventional TiO_2 based ETL [159]. They also improvised a buffer layer of $PC_{61}BM$ between ETL and absorber to enhance the UV stability. Shin et al. testified replacement of TiO_2 with Lanthanum (La)-doped $BaSnO_3$ with enhanced efficiency of about 21.2% and retaining 93% of initial PCE after 1000 h of sun illumination [160].

Recently, Lee and co-workers tried replacing TiO_2 ETL with Triton X-100 surface modified phenyl-C61-butyric-acid-methyl ester (PCBM) where NiO_x nanocrystals served the purpose of HTL [161]. They observed an increase in the average PCE from 10.76% to 15.68% along with an increase in the stability. PSC preserved 83.8% of the PCE after 800 h of exposure to ambient conditions. Jin Heo and co-workers investigated an inverted planar $ITO/PEDOT:PSS/MAPbI_3/PCBM/Au$ device with enhanced efficiency of about 18.1% along with air and moisture stability [105]. This improvement in stability is attributed to the top layer being exposed to PCBM ETL which is hydrophobic. Kim et al. demonstrated the use of an edged-selectively fluorine(F) functionalized graphene nanoplatelets (EFGnPs-F) layer between PCBM and Al electrode, exhibiting improved stability; being stable for 30 days at 50% humidity [162]. This is because of the hydrophobic nature of C-F bonding protecting the perovskite layer against moisture. Recently, Kim et al. showed high performance and stability with naphthalene diimide (NDI)-based polymer with dicyanothiophene (P(NDI2DT-TTCN)) as ETL [163]. P(NDI2DT-TTCN) improves the charge transfer and at the same time passivates the surface and increases moisture stability.

By Providing Hydrophobic HTL

Spiro-OMeTAD is usually used as a hole transporting material. This is doped with chemicals such as *tert*-butyl pyridine (tBP) and Li-bis(trifluoromethanesulfonyl)imide (Li–TFSI) for better conductivity. In the absence of tBP, the film formation suffers from de-wetting and poor homogeneity [82]. These dopants increase the device performance and at the same time tBP absorbs moisture and dissolves $MAPbI_3$, thus rendering it unstable in the ambient condition [164]. Many attempts have been made to design devices with dopant free HTL. Many other dopant free HTLs have been realized, e.g., triazatruxene, 5,10,15-tribenzyl-5*H*-diindolo[3,2-*a*:3',2'-*c*]-carbazole (TBDI) and other carbazole, thiophene [165], and triarylamine derivatives, thus avoiding the possibility of dopant induced degradation [166].

Li et al. recently proposed a cell structure employing a double-layer HTM configuration comprising of CuSCN layer and Spiro-OMeTAD layer for enhanced stability and efficiency [167]. Existence of pin holes in the as-deposited Spiro-OMeTAD layer have been observed by Hawash et al. which may provide sites for the migration of ions from the perovskite layer towards Spiro-OMeTAD, resulting in the degradation of device [168]. Thus, the CuSCN inorganic layer sandwiched between the perovskite and Spiro-OMeTAD blocks methyl amine ion migration into Spiro-OMeTAD. Kim et al. proposed a mixed hole transporting layer comprising of Spiro-OMeTAD and poly(3-hexylthiophene) (P3HT), exhibiting 18.9% PCE and enhanced stability [169]. Ginting et al. showed that when MAPbI$_3$ is decomposed, HI and CH$_3$NH$_2$ gases are released in the air [170].

Due to moisture/O$_2$, distribution of iodine concentration is affected within the MAPbI$_3$ layer. This leads to the diffusion of iodine towards the Spiro-OMeTAD layer and it gets accumulated near the Spiro-OMeTAD/MAPbI$_3$ interface. Short circuit current and fill factor are adversely affected due to reduced carrier mobility. To address this issue, Snaith et al. [171] demonstrated a method to form HTL free perovskite solar cell by sequential deposition and observed that when single walled nanotube (SWNT)-PMMA is deposited sequentially on the perovskite film, SWNT forms a dense and interconnected network. PMMA matrix fills in all the holes between the nanotube mesh and protects the underneath perovskite film from moisture degradation. It has been concluded that the thickness of PMMA layer should be around 300 nm. This is because thicker layers increase series resistance and reduce the charge collection efficiency, whereas thinner layers lead to an increased charge recombination and reduced shunt resistance. Similarly, Luo et al. studied usage of Cross-stacked super-aligned carbon nanotubes (CSCNTs) for an efficient and stable HTL free device [172]. Mei and co-workers fabricated (5-AVA)$_x$(MA)$_{1-x}$PbI$_3$ based HTL free perovskite device using mesoporous TiO$_2$, ZrO$_2$, and porous carbon tri-layer [173]. They demonstrated stability for >1000 h in ambient air and illumination. Replacement of Spiro-OMeTAD with other equally efficient HTLs has been realized worldwide. According to a recent study performed by Chang et al. a *p*-type conjugated polymer poly[(2,5-bis(2-hexyldecyloxy)-phenylene)-alt-(5,6-difluoro-4,7-di(thiop-hen-2-yl)benzo[c]-[1,2,5]-thiadiazole)] (PPDT2FBT) doped with a non-hygroscopic Lewis acid, tris(penta-fluorophenyl)borane (BCF, 2−6 wt %) has been proven to be a much more efficient HTL than standard Spiro-OMeTAD [174]. They showed PCE up to 17.7% with standard MAPbI$_3$ PSC and retaining 60% of the primary PCE upon exposure to 85% humidity for 500 h without any encapsulation.

Another commonly used HTL is poly(3,4-ethylenedioxythiophene):poly(styrenesulfonate) (PEDOT:PSS) but its acidic nature corrodes the ITO layer, and its hygroscopic nature causes the degradation of perovskite film. [171] These limitations combined with UV instability result in the irreversible deterioration of perovskite/PEDOT:PSS interface causing device degradation. Wang et al. [175] tried to solve this limitation of PEDOT:PSS by forming a bilayer HTL using Vanadium pentoxide (V$_2$O$_5$) and PEDOT:PSS, showing a PCE of 15% (20% more than when using conventional PEDOT:PSS HTL). After 18 days of stability testing, it was observed that the bilayer HTL based device retains 95% of its initial PCE whereas PEDOT:PSS HTL device exhibits a retentivity of only 55%. Recently, Shuang and co-workers investigated a mixture of hydrophobic polymer Nafion and PEDOT:PSS for better efficiency and highly enhanced stability than its PEDOT:PSS counterpart [176]. It acts as an electron blocking layer and avoids recombination hence increasing performance, and, at the same time, its chemical inertness and hydrophobicity enhance the stability in air.

Yang et al. studied the replacement of PEDOT:PSS with 4,4′-cyclohexylidenebis[*N,N*-bis(4-methylphenyl) benzenamine] (TAPC) as an efficient HTL in p-i-n MAPbI$_3$ perovskite device with better PCE (18.80%) and stability (retaining 6% PCE after 30 days exposure to RH between 50 and 85%) without encapsulation as compared to a conventional PEDOT:PSS device [177]. They have optimized the annealing temperature required and found it to be 120 °C for better performance of the device.

Lee et al. demonstrated the use of a hydrophobic HTL, poly(*N,N*′-bis(4-butylphenyl)-*N,N*′-bis (phenyl)benzidine) (PTPD), along with a thin hydrophilic layer of poly[(9,9-bis(3′-(*N,N*-dimethylamino) propyl)-2,7-fluorene)-alt-2,7-(9,9-dioctylfluorene)](PFN) for enhanced stability retaining ~80% PCE after 16 days of ambient exposure [178]. In this study they elucidated how the thin PFN layer, though hydrophilic, does not have any detrimental effect, but it results in a good morphology of the film, acting as a compatibilizer between perovskite and PTPD. PTPD is a hydrophobic HTL, so it restricts moisture interaction with perovskite film. Wang and his co-workers showed an inorganic NiO_x nanoparticles based HTL with para-bromobenzoic acid showing good stability, while maintaining 80% of its initial PCE after 15 days of exposure to 30% humidity [179]. This is accounted for the enhanced film morphology and passivation of defect states in NiO_x by para-bromobenzoic acid. NiO as HTL has also proven to enhance thermal stability [180].

Petrozza et al. elucidated the enhancement in the stability of the solar cell when Al_2O_3 nanoparticles were put in between perovskite and HTL [181]. Addition of buffer layer improved the device efficiency as it reduces the thickness of HTL, decreasing the series resistance and as a result the Fill Factor increased. Similarly, Koushik et al. introduced an atomic layer deposited (ALD) ultra-thin layer of Al_2O_3 between the mixed halide $CH_3NH_3PbI_{3-x}Cl_x$ perovskite film and HTL for stability enhancement (up to 60 days) and an improved PCE of 18% [182]. Hydrophobic ALD Al_2O_3 not only restricts the reach of moisture to the perovskite film but also passivates the defect states on the perovskite, making it more stable in moisture and oxygen.

By Proper Device Encapsulation

Owing to the moisture sensitivity of perovskite solar cells, proper encapsulation of the device is critical. Researchers worldwide have demonstrated many protective coatings to suppress the ingression of moisture into the perovskite layer. Depositing a protective layer of epoxy/Ag paint, CNTs, Al_2O_3, Cr_2O_3, and ZnO has proven to enhance the perovskite stability to a considerable amount [183–186]. Hwang et al. used a hydrophobic polymer Polytetrafluoroethylene as a perovskite shielding layer, showing negligible PCE loss after 30 days of ambient exposure [187]. Bush et al. used an ethylene-vinyl acetate (EVA) coated glass cover as an encapsulant, which is generally used in commercial silicon devices [4]. They sandwiched the perovskite layer between two glass sheets with EVA and butyl rubber edge seal to prevent moisture exposure. These Devices withstood damp heat test for 1008 h without any loss in PCE. Cheacharoen et al. compared EVA and Surlyn based encapsulation, and their tests proved EVA to be better because of its high mechanical stability due to low elastic modulus [188]. Liu et al. reported polydimethylsiloxane as an efficient encapsulant for HTL free perovskite device employed with carbon electrode, exhibiting stability for about 3000 h [189]. Matteocci et al. demonstrated a Kapton film covered with silicon based adhesive and UV curable methacrylate glue used for gluing glass-glass as an efficient encapsulant [190]. Han and his co-workers studied the comparison between two encapsulations: UV-curable epoxy and glass cover along with HG desiccant sheet in between [191]. They found the latter to be more stable under light and moisture exposure. Similarly, Dong and his co-workers testified various encapsulants and demonstrated UV-curable epoxy along with desiccant and an SiO_2 protective layer as an efficient encapsulant [192]. Shi et al. used glass/polyisobutylene (PIB)/glass encapsulation, and the device remained for 540 h under damp heat without any loss in PCE [193]. PIB tape is used as a sealant and it protects moisture ingression as well as retarding the decomposition generated gaseous escape. This is also non-reactive to perovskite solar cell dissimilar to the UV-curable epoxy and EVA. Some other sophisticated encapsulants may open the scope for even better stability.

Electrodes

Conventionally, the PSC relies on Au, Ag, and Al based electrodes. However, due to Ag [191] and Al [192] not being as favoured because of stability issues, PSCs utilise more costly Au electrodes. Recent studies however have demonstrated the use of other materials such as MoO_x/Al electrodes that inhibit the decomposition of perovskite films [194], multi-walled carbon nanotubes that are more cost effective and stable than gold [195], and even Cu which has a predicted stability lifetime of over 22 years at the nominal operating cell temperature [196]. Kaltenbrunner et al. proposed Cr_2O_3/Cr-Au metal contact for enhanced stability by shielding the Au contact from oxidation and chemical etching due to MAPbI$_3$ released Iodine [185]. These studies are based on finding cost effective electrode materials which either do not degrade the perovskite film or do not trap electrons at the surface.

In recent years, however, the impact of stability improvements is also being seen a little in the marketplace, although it has yet to reach any public modules. Australian PSC developer Dyesol claimed a significant breakthrough in perovskite stability for commercial use [197]. Similarly, a few other companies such as Saule Technologies and Oxford Photovoltaics are also unveiling plans for future perovskite production at the commercial level [198].

3.3. Cost of Fabricating Perovskite Solar Cell

When it comes to overall solar and renewable energy, the world has seen a steep decrease in cost and it has been predicted that the costs will drop 66% by 2040 [199,200]. These predictions, however, are based on the performance of silicon based solar cells, as shown in Figure 5a,b. While it is well known and generally accepted that PSCs are low cost, very few techno-economic studies have been performed. However, recently, according to Cai's comparative cost-performance analysis on high efficiency and moderate efficiency PSCs, the Levelized Cost of Electricity (LCOE: defined as the ratio of total life cycle cost to total lifetime energy production) was 3.5–4.9 US cents/kWh with an efficiency of 12% and lifetime of 15 years (Figure 6) and the counterpart multi-Si cells (with an efficiency >18.4%) are estimated to have an LCOE of $0.23 per watt [201–203].

According to calculations done by Maniarasu et al. [204] in the fabrication of one metre square perovskite solar cell with 70% active area, 52% of the total raw material cost is taken up by the hole transporting material (Spiro-OMeTAD—604 USD) and counter electrode (gold—330 USD), and 43% by FTO glass (766 USD). On the other hand, Perovskite layer takes only 0.0038% of the total cost. Therefore, any improvement in cost requires change of HTM and Electrode to cheaper alternatives, for example PCBM and Al, respectively. Pertinent to this, Petrus et al. suggested the use of azomethine-based HTMs—Diazo-OMeTPA [205] and EDOT-OMeTPA—for low cost and highly efficient PSCs. The studies revealed that Schiff-base condensation chemical synthesis for these HTMs, which is eco-friendly, attributes to the reduced cost in contradiction to coupling reactions (conventionally used to synthesize HTMs) [206]. Wu et al. developed electron rich HTMs with thiophene pi-linker claiming to reduce the synthesis cost by tenfold, in comparison with Spiro-OMeTAD, due to its shorter synthesis route [207]. Arylamine derivatives with pyrene core [208], two-dimensional triazatruxene-based derivatives [63], phenothiazine-based molecules [209], biphenyl based [210], bi- and tetra-thiophene based molecules [209,211], triazine based [212] HTMs could be a prospective replacement for Spiro-OMeTAD. Although polymeric HTMs such as PE-DOT:PSS and P3HT are proven to be cost effective, their polydispersity results in variation in the molecular weight for batch to batch production, hindering their commercialization [213]. Chemically stable inorganic p-type semiconductors-NiO [214], CuSCN [215], and CuI [216] are often cheap due to inexpensive precursors and act as hole-extracting materials in PSCs, and yet they are least explored. Despite noble metals being widely used as counter electrodes for state-of-art devices, use of carbon-based materials can cut down the material cost for PSCs. Carbon in various forms, CNTs, graphitic sheets/flakes, fullerenes, and soot have been explored for this purpose. These materials offer flexibility of printing as well as solution-based processing, paving the path to scale up production. Composites of p-type semiconductors bent with carbon materials, for example NiO in carbon matrix [180,217], P3HT and PMMA blended with SWCNTs [218] exhibited a great potential as HTMs for low cost PSCs.

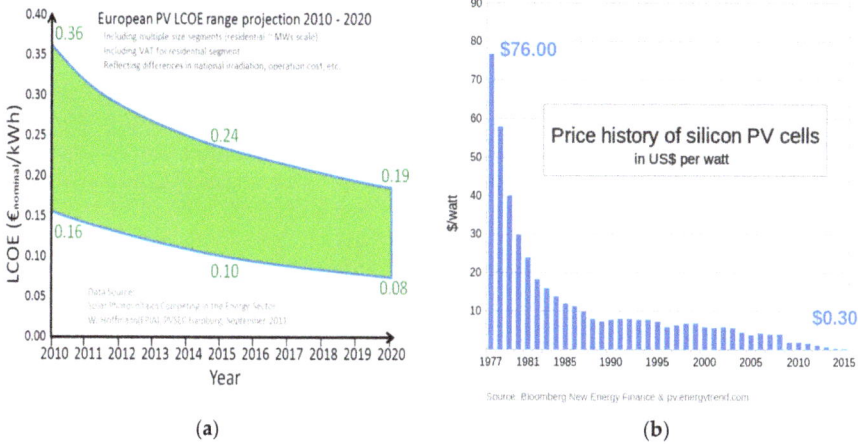

Figure 5. (a) LCOE range of photovoltaics (Europe); and (b) price history of silicon photovoltaics (Wikipedia.com).

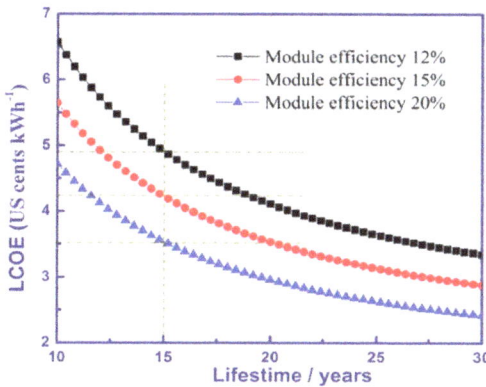

Figure 6. Effect of lifetime on LCOE, as a function of efficiency. As the lifetime increases, the LCOE goes drastically down but should reach a saturation point at the end (taken from Cai et al. [201]).

4. Economic Feasibility of Perovskite Solar Cell

4.1. Energy Payback Time and Energy Return on Energy Invested

An economical way of looking at solar cells is by comparing the energy payback time of different solar cells [219]. The EPBT determines the duration of time a PV system must operate before it recovers the amount of energy invested throughout its lifetime. Using the formula given by Bhandari et al. we have:

$$EPBT \text{ (years)} = \frac{\text{Embeded primary energy (MJm}-2)}{\text{Annual primary energy generated by the system(MJ m}-2 \text{ year}-1)} = \frac{W_1}{(I \times \eta \times PR/\varepsilon)} \quad (8)$$

where W_1 is the embedded primary energy; ε is the electricity to primary energy conversion factor (0.35 usually); I is the total solar insolation incident on unit surface per year; η is the average module efficiency; and PR is the system performance ratio.

According to the study carried out by Celik et al. [220] in 2016, only four Life cycle assessment studies had been done before them, where a life cycle study determines the energy usage of a solar cell during the entire process from creation to disposal. Using estimations and ideas about which type of perovskite solar cell would be used commercially in the future, Celik et al. did an ex-ante life cycle assessment including the EPBT of perovskite solar cells. The efficiency assumed was 15%. The studies are shown in the graphic below along with the data on harmonized EPBT values of other solar cells. The fourth bar in Figure 7 combines the data of the previous studies with data for PCBM that we calculated, while the fifth bar has data for single walled carbon nanotubes based solar cells which have an estimated theoretical efficiency of 28%. Gong et al. found an EPBT of 0.299 and 0.221 if TiO_2 and ZnO are replaced by PCBM in the cell [88], and an EPBT of 0.964 and 0.721 if SnO_2 is replaced by PCBM in the case of solution processed cell and HTL free cell thought of by Celik et al. The latter case is perhaps more interesting, as Gong's study does not utilise much electricity due to the processes involved (no annealing is done), which can further effect both the efficiency and the lifetime of the solar cells. Due to the lower amount of PCBM material used as compared to SnO_2 (0.154 gm as opposed to nearly 500 gm of Sn) and the low electricity requirement for the former, PCBM lowers the EPBT by nearly 25%, assuming the same efficiency of the cell in the case of CuSCN HTL. On the other hand, when we have an HTL free device, the EPBT is even lower. Because Spiro-OMeTAD takes more energy than CuSCN, the latter is a better alternative, unless the cell is HTL free. The data for Spiro-OMeTAD and PCBM are from Espinosa et al. who did an LCA (Life-cycle assessment) study at the lab level [221] and from Annick Anctil [222].

Another way of looking at solar cell is to look at the EROI, or the energy return on energy invested, which is a function of EPBT and lifetime, given as [219]:

$$EROI = \frac{\text{Lifetime(years)}}{\text{EPBT(years)}} \qquad (9)$$

The EROI is presently set as a 3:1 ratio for any solar cell technology to be considered viable. Therefore, we can calculate the minimum requirement of lifetime for the solar cells used in the study to be considered as viable contenders by using the EPBT and studies in Figure 7. The resulting lifetimes when the EROI is set as 3 can be seen in Figure 8.

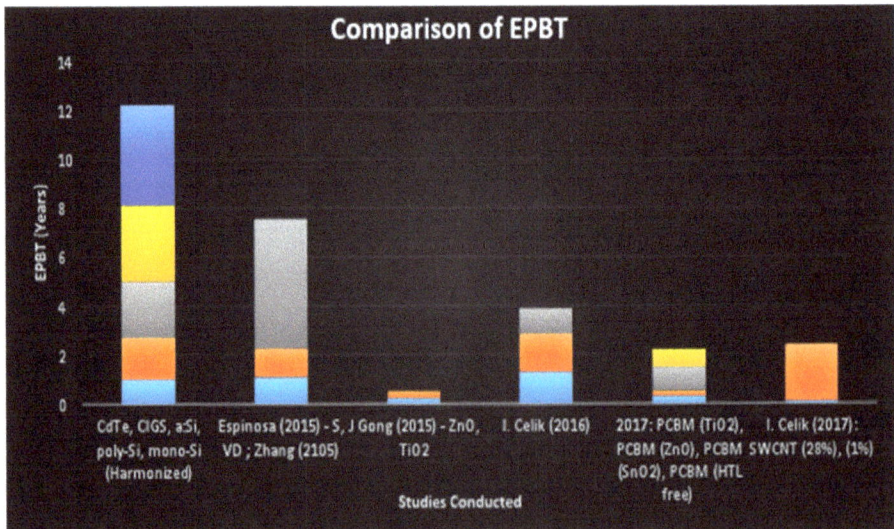

Figure 7. Comparison of the EPBT of different solar cells. Calculations have been done over time and the studies have been collected here. The figure shows the clear difference in EPBT based on materials used and calculation methods.

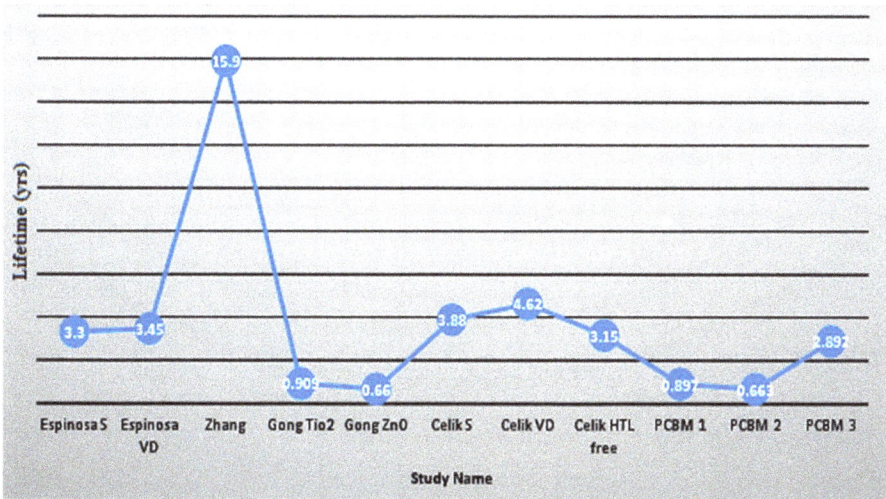

Figure 8. A look at the viability of commercial solar cells. Figure 7 provides a reference.

4.2. Global Consideration of EPBT

Apart from the architecture of the cell, the energy embedded in the manufacturing of materials is another important consideration in the solar insolation, which affects the EPBT. Thus far, most of the above studies have taken the average of 1700 KWh/m^2/year, except for Gong who took 1960 KWh/m^2/year to consider areas in the United States which get more heat. However, for emerging energy markets, especially in terms of competition, perhaps India and China need to be considered separately [223–226].

Consider south, central, and west India for example. The daily average for these places (including major states of Kerala, Karnataka, Madhya Pradesh, and especially Rajasthan) is greater than 5 KWh/m^2/year as shown in the Figure 9. This translates to more than 1825 KWh/m^2/year of solar insolation in such places on average.

About 1,400,000 km^2 of land is available with an annual solar insolation of 5–5.5 KWh/m^2/day which is around 1825–2007 KWh/m^2/year, while 650,000 km^2 has a solar insolation of 5.5–6 KWh/m^2/day translating into 2007–2190 KWh/m^2/year, and this rounds up to 2,000,000 km^2 of area (62.5% of total Indian area) that is available. Considering that 10% of the land is utilised, solar PV systems installation with perovskite solar cells (PCE = 15% and performance ratio of 75%; a conservative estimate), solar insolation average of 1970 KWh/m^2/year, and assuming the conversion factor (mentioned above) of India to also be 0.35, we arrive at:

$$Energy\ Produced\left(\frac{KWh}{Year}\right) = \frac{1970 \times 0.15 \times 0.75}{0.35} \times 200,000 \qquad (10)$$

which is 1266.4 × 10^5 KWh/year, or 4.55 × 10^{14} MJ/year. That is 455 TJ of energy per year, giving roughly 0.000035 GJ of the per capita energy, or 0.0015% of the total per capita energy consumption of India (22.042 GJ) [227] and 0.0086% of the per capita electrical energy consumption (1122 KWh) [228]. With such a low EPBT and manufacturing time, perovskite solar cells provide a much-needed cost effective and energy intensive solution to the problem of solar energy.

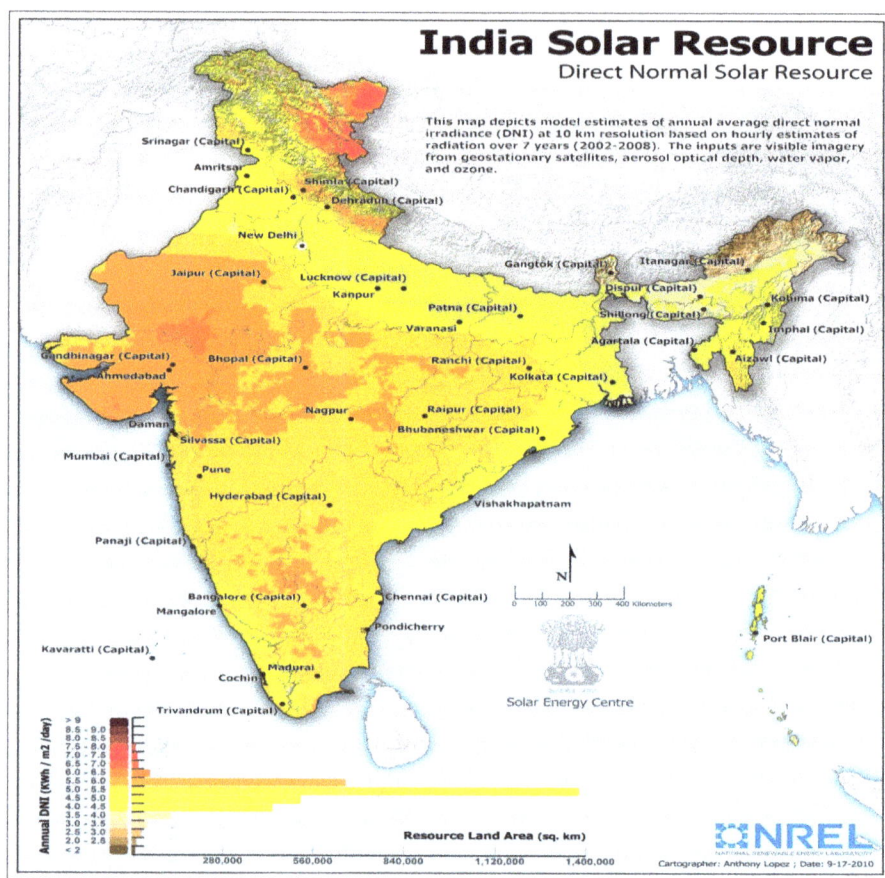

Figure 9. Amount of solar insolation in India on average showing areas of extreme heat which may be considered in the calculations of energy payback time. (Taken from NREL solar radiation database).

5. Perovskite versus the Rest

It is safe to assume that the only competition PSCs will face in the future is from the very technology it is trying to overtake. To achieve the level of success silicon based solar cells have achieved, PSCs first need to achieve a stability of at least a few years (2–5). However, thus far, such a goal seems rather far off considering that this stability threshold has not been reached by laboratory cells yet. A low stability device not only affects the efficiency of the solar cell but also increases the cost since cell modules must be replaced, thus requiring human resources and cost for instalment of modules in a commercial setting.

When it comes to efficiency, PSCs are on their way to match Silicon solar cells and other alternatives. Lab based cells (1 cm^2) should soon reach efficiency records of Silicon based cells. While cell modules (>1 m^2) are far from such an efficiency limit, an efficiency threshold of 15–20% can be considered viable. Obviously, a higher efficiency level is desired, but even within such a range, perovskites will outdo silicon cells in terms of energy to cost ratio.

Lastly, the one area where PSCs can be said to have overtaken silicon cells is the cost of fabrication (and the low energy requirement). This can be said to be the primary motivator of research in the field. However, much can still be done about the cost in the case of PSCs, especially about components such as electrodes and HTM. While Silicon cells have achieved their most desirable configuration, PSCs still require further research.

Therefore, to outdo silicon-based cells, a conservative estimate of a 1 m^2 PSC module can be predicted as:

- **Stability:** 3 years (with the cell retaining at least 95–99% of its original efficiency)
- **Efficiency:** 15%
- **Cost of Fabrication:** 1300–1500 USD

6. Conclusions

In this review article, we have presented the state of perovskite solar cells over the past few years. In being comprehensive, we have studied the problems present within different aspects associated with the cells, namely stability, efficiency, and cost of fabrication/energy of fabrication, and have also provided a detailed review of the research (both past and ongoing) which has been done to evolve the state of perovskite solar cells to the next level. If PSCs were to be thought of as caterpillars in the year 2013, now, it is safe to say that they are undergoing a metamorphosis; they are in the developmental state known as a chrysalis.

We have shown the achievements of the past few years, the problems overcome, and new challenges which have yet to be conquered, keeping not just a scientific mindset but also an economical and therefore a competitive mindset with respect to silicon solar cells.

PSCs have reached efficiencies greater than 22%, which is an exponential increase considering they started at a meagre 4%. Research to increase the efficiency has focused on the film quality, hole and electron transport layers which form the second most important parts of the cell, the electrodes used, and even tandem solar cells which utilize either two PSCs or a perovskite coupled with a silicon cell or other alternatives. The use of tandems is one to look out for, especially since tandems often utilise a greater spectrum of the sunlight, thereby having a much higher efficiency. Research has also been focused in areas utilising more than two components of the cell, which include the interaction of HTM/ETM and perovskite layers; utilising buffer layers to increase the overall efficiency.

Following efficiency, the most important aspect of the solar cells is stability. As we have reported, this is also the aspect which must be researched the most if PSCs are to compete in the market. The research in this area has been focused on the interaction of PSCs with the environment which leads to a decay of the layers. The cells fair badly with moisture, heat, and sunlight itself. Methods to counter this include utilisation of buffer layers, changing the materials of the perovskite cells to those which do not decay, and better techniques of encapsulation. Future research needs to strive for both longer stability times with high efficiency (for lab cells) and a higher stability for bigger albeit low efficiency modules.

The third significant factor is the cost of fabrication, in terms of both cost of materials and energy utilised, and it also forms a good measuring factor for the PSCs based on EPBT and EROI calculations. Here, we have also collected data and added a bit of our own to show that PSCs (assuming they reach a required level of stability and efficiency) will outdo silicon cells when it comes to energy and cost requirements. Perovskite solar cells are possibly the only technology which gives a much faster payback than Silicon cells while also being the easiest to fabricate. Finally, extending the scope of EPBT measurements, we have included a small section that talks about the use of PSCs in areas other than the USA and Europe, focusing on India and China, although we have stayed away from the geo-political conditions which often accompany the spread of any new technology.

In conclusion, the research which has been ongoing for the past five years has achieved significant results. However, if PSCs need to move out from the confines of labs into the sunlit world, especially in an era dominated by energy, the research needs to accelerate further. With the emergence of private companies vying for the future market, it appears perovskites should soon be everywhere; however, practically, that moment is probably far away. To answer the question of the future, this will suffice for now: theoretically, perovskites are already a star; practically, they have yet to leave the womb.

Author Contributions: A.B. wrote the first draft of the paper (including EPBT studies) with contribution from G.R. B.P.D., P.C., and T.B.K. provided major inputs in efficiency and stability studies along with fact-checking the rest of paper. Their contributions gave depth and made the paper more comprehensive. S.K.P. oversaw and guided the entire study, structuring and editing the paper and providing the rest with resources when necessary.

Conflicts of Interest: The authors declare no conflict of interest.

References

1. Yang, W.S.; Park, B.W.; Jung, E.H.; Jeon, N.J.; Kim, Y.C.; Lee, D.U.; Shin, S.S.; Jangwon, S.; Kim, E.K.; Noh, J.H.; et al. Iodide management in formamidinium-lead-halide–based perovskite layers for efficient solar cells. *Science* **2017**, *356*, 1376–1379. [CrossRef] [PubMed]
2. Burschka, J.; Pellet, N.; Moon, S.-J.; Humphry-Baker, R.; Gao, P.; Nazeeruddin, M.K.; Grätzel, M. Sequential Deposition as a Route to High-Performance Perovskite-Sensitized Solar Cells. *Nature* **2013**, *499*, 316. [CrossRef] [PubMed]
3. Im, J.-H.; Lee, C.-R.; Lee, J.-W.; Park, S.-W.; Park, N.-G. 6.5% Efficient Perovskite Quantum-Dot-Sensitized Solar Cell. *Nanoscale* **2011**, *3*, 4088–4093. [CrossRef] [PubMed]
4. Bush, K.A.; Palmstrom, A.F.; Yu, Z.J.; Boccard, M.; Rongrong Cheacharoen, J.P.M.; McMeekin, D.P.; Hoye, R.L.Z.; Bailie, C.D.; Leijtens, T.; Peters, I.M.; et al. 23.6%-Efficient Monolithic Perovskite/Silicon Tandem Solar Cells with Improved Stability. *Nat. Energy* **2017**, *2*, 17009. [CrossRef]
5. Masuko, K.; Shigematsu, M.; Hashiguchi, T.; Fujishima, D.; Kai, M.; Yoshimura, N.; Yamaguchi, T.; Ichihashi, Y.; Mishima, T.; Matsubara, N.; et al. Achievement of More Than 25% Conversion Efficiency with Crystalline Silicon Heterojunction Solar Cell. *IEEE J. Photovolt.* **2014**, *4*, 1433–1435. [CrossRef]
6. Taguchi, M.; Yano, A.; Tohoda, S.; Matsuyama, K.; Nakamura, Y.; Nishiwaki, T.; Fujita, K.; Maruyama, E. 24.7% Record Efficiency HIT Solar Cell on Thin Silicon Wafer. *IEEE J. Photovolt.* **2014**, *4*, 96–99. [CrossRef]
7. Savin, H.; Repo, P.; von Gastrow, G.; Ortega, P.; Calle, E.; Garín, M.; Alcubilla, R. Black Silicon Solar Cells with Interdigitated Back-Contacts Achieve 22.1% Efficiency. *Nat. Nanotechnol.* **2015**, *10*, 624. [CrossRef] [PubMed]
8. Ma, S.; Zhang, H.; Zhao, N.; Cheng, Y.; Wang, M.; Shen, Y.; Tu, G. Spiro-Thiophene Derivatives as Hole-Transport Materials for Perovskite Solar Cells. *J. Mater. Chem. A* **2015**, *3*, 12139–12144. [CrossRef]
9. Zheng, L.; Ma, Y.; Chu, S.; Wang, S.; Qu, B.; Xiao, L.; Chen, Z.; Gong, Q.; Wu, Z.; Hou, X. Improved Light Absorption and Charge Transport for Perovskite Solar Cells with Rough Interfaces by Sequential Deposition. *Nanoscale* **2014**, *6*, 8171–8176. [CrossRef] [PubMed]
10. Sum, T.C.; Mathews, N. Advancements in Perovskite Solar Cells: Photophysics behind the Photovoltaics. *Energy Environ. Sci.* **2014**, *7*, 2518–2534. [CrossRef]
11. Xing, G.; Mathews, N.; Sun, S.; Lim, S.S.; Lam, Y.M.; Grätzel, M.; Mhaisalkar, S.; Sum, T.C. Long-Range Balanced Electron- and Hole-Transport Lengths in Organic-Inorganic $CH_3NH_3PbI_3$. *Science* **2013**, *342*, 344–347. [CrossRef] [PubMed]
12. Stoumpos, C.C.; Malliakas, C.D.; Kanatzidis, M.G. Semiconducting Tin and Lead Iodide Perovskites with Organic Cations: Phase Transitions, High Mobilities, and Near-Infrared Photoluminescent Properties. *Inorg. Chem.* **2013**, *52*, 9019–9038. [CrossRef] [PubMed]
13. Leyden, M.R.; Ono, L.K.; Raga, S.R.; Kato, Y.; Wang, S.; Qi, Y. High Performance Perovskite Solar Cells by Hybrid Chemical Vapor Deposition. *J. Mater. Chem. A* **2014**, *2*, 18742–18745. [CrossRef]
14. Zardetto, V.; Williams, B.L.; Perrotta, A.; Di Giacomo, F.; Verheijen, M.A.; Andriessen, R.; Creatore, M. Atomic layer deposition for perovskite solar cells: Research status, opportunities and challenges. *Sustain. Energy Fuels* **2017**, *1*, 30–55. [CrossRef]
15. Wang, K.; Zhao, W.; Liu, J.; Niu, J.; Liu, Y.; Ren, X.; Feng, J.; Liu, Z.; Sun, J.; Wang, D.; et al. CO_2 Plasma-Treated TiO_2 Film as an Effective Electron Transport Layer for High-Performance Planar Perovskite Solar Cells. *ACS Appl. Mater. Interfaces* **2017**, *9*, 33989–33996. [CrossRef] [PubMed]
16. Shockley, W.; Quiesser, H.J. Detailed Balance Limit of Efficiency of *p-n* Junction Solar Cells. *J. Appl. Phys.* **1961**, *32*, 510. [CrossRef]

17. Sha, W.E.I.; Ren, X.; Chen, L.; Choy, W.C.H. The Efficiency Limit of $CH_3NH_3PbI_3$ Perovskite Solar Cells. *Appl. Phys. Lett.* **2015**, *106*, 221104. [CrossRef]

18. Nayak, P.K.; Cahen, D. Updated Assessment of Possibilities and Limits for Solar Cells. *Adv. Mater.* **2013**, *26*, 1622–1628. [CrossRef] [PubMed]

19. Yoshida, K.S.; Furube, A.; Yoshida, Y. Theoretical Limit of Power Conversion Efficiency for Organic and Hybrid Halide Perovskite Photovoltaics. *Jpn. J. Appl. Phys.* **2015**, *54*, 08KF04.

20. Grånäs, O.; Vinichenko, D.; Kaxiras, E. Establishing the Limits of Efficiency of Perovskite Solar Cells from First Principles Modeling. *Sci. Rep.* **2016**, *6*, 36108. [CrossRef] [PubMed]

21. Sha, W.E.I.; Zhang, H.; Wang, Z.S.; Zhu, H.L.; Ren, X.G.; Lin, F.; Jen, A.K.-Y.; Choy, W.C.H. Quantifying Efficiency Loss of Perovskite Solar Cells by a Modified Detailed Balance Model. *Adv. Energy Mater.* **2017**, *8*, 1701586. [CrossRef]

22. Crossland, E.J.W.; Noel, N.; Sivaram, V.; Leijtens, T.; Alexander-Webber, J.A.; Snaith, H.J. Mesoporous TiO_2 Single Crystals Delivering Enhanced Mobility and Optoelectronic Device Performance. *Nature* **2013**, *495*, 215–219. [CrossRef] [PubMed]

23. Miller, O.D.; Yablonovitch, E.; Kurtz, S.R. Strong Internal and External Luminescence as Solar Cells Approach the Shockley-Queisser Limit. *IEEE J. Photovolt.* **2012**, *2*, 303–311. [CrossRef]

24. Zhang, W.; Saliba, M.; Moore, D.T.; Pathak, S.K.; Hörantner, M.T.; Stergiopoulos, T.; Stranks, S.D.; Eperon, G.E.; Alexander-Webber, J.A.; Abate, A.; et al. Ultrasmooth Organic-Inorganic Perovskite Thin-Film Formation and Crystallization for Efficient Planar Heterojunction Solar Cells. *Nat. Commun.* **2015**, *6*, 6142. [CrossRef] [PubMed]

25. Nie, W.; Tsai, H.; Asadpour, R.; Blancon, J.-C.; Neukirch, A.J.; Gupta, G.; Crochet, J.J.; Chhowalla, M.; Tretiak, S.; Alam, M.A.; et al. High-Efficiency Solution-Processed Perovskite Solar Cells with Millimeter-Scale Grains. *Science* **2015**, *347*, 522–525. [CrossRef] [PubMed]

26. Saliba, M.; Matsui, T.; Domanski, K.; Seo, J.-Y.; Ummadisingu, A.; Zakeeruddin, S.M.; Correa-Baena, J.-P.; Tress, W.R.; Abate, A.; Hagfeldt, A.; et al. Incorporation of Rubidium Cations into Perovskite Solar Cells Improves Photovoltaic Performance. *Science* **2016**, *354*, 206. [CrossRef] [PubMed]

27. Yun, J.; Jun, J.; Yu, H.; Lee, K.; Ryu, J.; Lee, J.; Jang, J. Highly Efficient Perovskite Solar Cells Incorporating NiO Nanotubes: Increased Grain Size and Enhanced Charge Extraction. *J. Mater. Chem. A* **2017**, *5*, 21750–21756. [CrossRef]

28. Wu, Y.; Xie, F.; Chen, H.; Yang, X.; Su, H.; Cai, M.; Zhou, Z.; Noda, T.; Han, L. Thermally Stable $MAPbI_3$ Perovskite Solar Cells with Efficiency of 19.19% and Area over 1 cm^2 Achieved by Additive Engineering. *Adv. Mater.* **2017**, *29*, 1701073. [CrossRef] [PubMed]

29. Bai, Y.; Dong, Q.; Shao, Y.; Deng, Y.; Wang, Q.; Shen, L.; Wang, D.; Wei, W.; Huang, J. Enhancing Stability and Efficiency of Perovskite Solar Cells with Crosslinkable Silane-Functionalized and Doped Fullerene. *Nat. Commun.* **2016**, *7*, 12806. [CrossRef] [PubMed]

30. Zhou, H.P.; Chen, Q.; Li, G.; Luo, S.; Song, T.B.; Duan, H.-S.S.; Hong, Z.R.; You, J.B.; Liu, Y.S.; Yang, Y. Interface Engineering of Highly Efficient Perovskite Solar Cells. *Science* **2014**, *345*, 542–546. [CrossRef] [PubMed]

31. Yang, M.; Zhang, T.; Schulz, P.; Li, Z.; Li, G.; Kim, D.H.; Guo, N.; Berry, J.J.; Zhu, K.; Zhao, Y. Facile Fabrication of Large-Grain $CH_3NH_3PbI_{3-x}Br_x$ Films for High-Efficiency Solar Cells via CH_3NH_3Br-Selective Ostwald Ripening. *Nat. Commun.* **2016**, *7*, 12305. [CrossRef] [PubMed]

32. Uribe, J.I.; Ciro, J.; Montoya, J.F.; Osorio, J.; Jaramillo, F. Enhancement of Morphological and Optoelectronic Properties of Perovskite Films by CH_3NH_3Cl Treatment for Efficient Solar Minimodules. *ACS Appl. Energy Mater.* **2018**, *1*, 1047–1052. [CrossRef]

33. Conings, B.; Bretschneider, S.A.; Babayigit, A.; Gauquelin, N.; Cardinaletti, I.; Manca, J.V.; Verbeeck, J.; Snaith, H.J.; Boyen, H.-G. Structure–Property Relations of Methylamine Vapor Treated Hybrid Perovskite $CH_3NH_3PbI_3$ Films and Solar Cells. *ACS Appl. Mater. Interfaces* **2017**, *9*, 8092–8099. [CrossRef] [PubMed]

34. Liu, Z.; He, T.; Liu, K.; Zhi, Q.; Yuan, M. Solution Processed Double-Decked V_2O_x/PEDOT:PSS Film Serves as the Hole Transport Layer of an Inverted Planar Perovskite Solar Cell with High Performance. *RSC Adv.* **2017**, *7*, 26202–26210. [CrossRef]

35. Wu, J.; Xu, X.; Zhao, Y.; Shi, J.; Xu, Y.; Luo, Y.; Li, D.; Wu, H.; Meng, Q. DMF as an Additive in a Two-Step Spin-Coating Method for 20% Conversion Efficiency in Perovskite Solar Cells. *ACS Appl. Mater. Interfaces* **2017**, *9*, 26937–26947. [CrossRef] [PubMed]

36. Zhang, X.; Ji, G.; Xiong, D.; Su, Z.; Zhao, B.; Shen, K.; Yang, Y.; Gao, X. Graphene Oxide as an Additive to Improve Perovskite Film Crystallization and Morphology for High-Efficiency Solar Cells. *RSC Adv.* **2018**, *8*, 987–993. [CrossRef]

37. Faraji, N.; Qin, C.; Matsushima, T.; Adachi, C.; Seidel, J. Grain Boundary Engineering of Halide Perovskite $CH_3NH_3PbI_3$ Solar Cells with Photochemically-Active Additives. *J. Phys. Chem. C* **2018**, *122*, 4817–4821. [CrossRef]

38. Xie, L.; Cho, A.-N.; Park, N.-G.; Kim, K. Efficient and Reproducible $CH_3NH_3PbI_3$ Perovskite Layer Prepared Using a Binary Solvent Containing a Cyclic Urea Additive. *ACS Appl. Mater. Interfaces* **2018**, *10*, 9390–9397. [CrossRef] [PubMed]

39. Wang, J.-F.; Zhu, L.; Zhao, B.-G.; Zhao, Y.-L.; Song, J.; Gu, X.-Q.; Qiang, Y.-H. Surface Engineering of Perovskite Films for Efficient Solar Cells. *Sci. Rep.* **2017**, *7*, 14478. [CrossRef] [PubMed]

40. Zhang, W.; Pathak, S.; Sakai, N.; Stergiopoulos, T.; Nayak, P.K.; Noel, N.K.; Haghighirad, A.A.; Burlakov, V.M.; deQuilettes, D.W.; Sadhanala, A.; et al. Enhanced Optoelectronic Quality of Perovskite Thin Films with Hypophosphorous Acid for Planar Heterojunction Solar Cells. *Nat. Commun.* **2015**, *6*, 10030. [CrossRef] [PubMed]

41. Noel, N.K.; Abate, A.; Stranks, S.D.; Parrott, E.S.; Burlakov, V.M.; Goriely, A.; Snaith, H.J. Enhanced Photoluminescence and Solar Cell Performance via Lewis Base Passivation of Organic–Inorganic Lead Halide Perovskites. *ACS Nano* **2014**, *8*, 9815–9821. [CrossRef] [PubMed]

42. Wang, J.T.-W.; Wang, Z.; Pathak, S.; Zhang, W.; deQuilettes, D.W.; Wisnivesky-Rocca-Rivarola, F.; Huang, J.; Nayak, P.K.; Patel, J.B.; Mohd Yusof, H.A.; et al. Efficient Perovskite Solar Cells by Metal Ion Doping. *Energy Environ. Sci.* **2016**, *9*, 2892–2901. [CrossRef]

43. Zhao, W.; Yao, Z.; Yu, F.; Yang, D.; Liu, S.F. Alkali Metal Doping for Improved $CH_3NH_3PbI_3$ Perovskite Solar Cells. *Adv. Sci.* **2018**, *5*, 1700131. [CrossRef] [PubMed]

44. Zhao, T.; Chueh, C.-C.; Chen, Q.; Rajagopal, A.; Jen, A.K.-Y. Defect Passivation of Organic–Inorganic Hybrid Perovskites by Diammonium Iodide toward High-Performance Photovoltaic Devices. *ACS Energy Lett.* **2016**, *1*, 757–763. [CrossRef]

45. Xu, M.; Feng, J.; Ou, X.L.; Zhang, Z.Y.; Zhang, Y.F.; Wang, H.Y.; Sun, H.B. Surface Passivation of Perovskite Film by Small Molecule Infiltration for Improved Efficiency of Perovskite Solar Cells. *IEEE Photonics J.* **2016**, *8*, 1–7. [CrossRef]

46. Jamaludin, N.F.; Yantara, N.; Ng, Y.F.; Li, M.; Goh, T.W.; Thirumal, K.; Sum, T.C.; Mathews, N.; Soci, C.; Mhaisalkar, S. Grain Size Modulation and Interfacial Engineering of $CH_3NH_3PbBr_3$ Emitter Films through Incorporation of Tetraethylammonium Bromide. *ChemPhysChem* **2018**, *19*, 1075. [CrossRef] [PubMed]

47. Jiang, L.L.; Wang, Z.K.; Li, M.; Zhang, C.C.; Ye, Q.Q.; Hu, K.H.; Lu, D.Z.; Fang, P.F.; Liao, L.S. Passivated Perovskite Crystallisation via g-C_3N_4 for High-Performance Solar Cells. *Adv. Funct. Mater.* **2018**, *28*, 1705875. [CrossRef]

48. Seo, J.; Park, S.; Kim, Y.C.; Jeon, N.J.; Noh, J.H.; Yoon, S.C.; Seok, S.I. Benefits of Very Thin PCBM and LiF Layers for Solution-Processed p–i–n Perovskite Solar Cells. *Energy Environ. Sci.* **2014**, *7*, 2642–2646. [CrossRef]

49. Jeon, N.J.; Lee, H.G.; Kim, Y.C.; Seo, J.; Noh, J.H.; Lee, J.; Seok, S.I. O-Methoxy Substituents in Spiro-OMeTAD for Efficient Inorganic–Organic Hybrid Perovskite Solar Cells. *J. Am. Chem. Soc.* **2014**, *136*, 7837–7840. [CrossRef] [PubMed]

50. Shi, D.; Qin, X.; Li, Y.; He, Y.; Zhong, C.; Pan, J.; Dong, H.; Xu, W.; Li, T.; Hu, W.; et al. Spiro-OMeTAD Single Crystals: Remarkably Enhanced Charge-Carrier Transport via Mesoscale Ordering. *Sci. Adv.* **2016**, *2*, e1501491. [CrossRef] [PubMed]

51. Cho, K.T.; Paek, S.; Grancini, G.; Roldan-Carmona, C.; Gao, P.; Lee, Y.; Nazeeruddin, M.K. Highly Efficient Perovskite Solar Cells with a Compositionally Engineered Perovskite/Hole Transporting Material Interface. *Energy Environ. Sci.* **2017**, *10*, 621–627. [CrossRef]

52. Yu, J.C.; Hong, J.A.; Jung, E.D.; Kim, D.B.; Baek, S.-M.; Lee, S.; Song, M.H. Highly efficient and stable inverted perovskite solar cell employing PEDOT:GO composite layer as a hole transport layer. *Sci. Rep.* **2018**, *8*, 1070. [CrossRef] [PubMed]

53. Seo, S.; Park, I.J.; Kim, M.; Lee, S.; Bae, C.; Jung, H.S.; Shin, H. An ultra-thin, un-doped NiO hole transporting layer of highly efficient (16.4%) organic-inorganic hybrid perovskite solar cells. *Nanoscale* **2016**, *8*, 11403–11412. [CrossRef] [PubMed]

54. Park, I.J.; Kang, G.; Park, M.A.; Kim, J.S.; Seo, S.W.; Kim, D.H.; Zhu, K.; Park, T.; Kim, J.Y. Highly Efficient and Uniform 1 cm^2 Perovskite Solar Cells with an Electrochemically Deposited NiO$_x$ Hole-Extraction Layer. *ChemSusChem* **2017**, *10*, 2660–2667. [CrossRef] [PubMed]

55. Chen, W.; Liu, F.Z.; Feng, X.Y.; Djurišić, A.B.; Chan, W.K.; He, Z.B. Cesium Doped NiO$_x$ as an Efficient Hole Extraction Layer for Inverted Planar Perovskite Solar Cells. *Adv. Energy Mater.* **2017**, *7*, 1700722. [CrossRef]

56. Huang, A.B.; Zhu, J.T.; Zheng, J.Y.; Yu, Y.; Liu, Y.; Yang, S.W.; Bao, S.H.; Lei, L.; Jin, P. Achieving High-Performance Planar Perovskite Solar Cells with Co-Sputtered Co-Doping NiO$_x$ Hole Transport Layers by Efficient Extraction and Enhanced Mobility. *J. Mater. Chem. C* **2016**, *4*, 10839–10846. [CrossRef]

57. Liu, M.H.; Zhou, Z.J.; Zhang, P.P.; Tian, Q.W.; Zhou, W.H.; Kou, D.X.; Wu, S.X. p-type Li, Cu-codoped NiO$_x$ hole-transporting layer for efficient planar perovskite solar cells. *Opt. Express* **2016**, *24*, 128–132. [CrossRef] [PubMed]

58. Kim, J.H.; Liang, P.W.; Williams, S.T.; Cho, N.; Chueh, C.C.; Glaz, M.S.; Ginger, D.S.; Jen, A.K.Y. High-Performance and Environmentally Stable Planar Heterojunction Perovskite Solar Cells Based on a Solution-Processed Copper-Doped Nickel Oxide Hole-Transporting Layer. *Adv. Mater.* **2015**, *27*, 695–701. [CrossRef] [PubMed]

59. Zheng, J.; Hu, L.; Yun, J.S.; Zhang, M.; Lau, C.-F.J.; Bing, J.; Deng, X.; Ma, Q.; Cho, Y.; Fu, W.-F.; et al. Solution-Processed, Silver-Doped NiOx as Hole Transporting Layer for High Efficiency Inverted Perovskite Solar Cells. *ACS Appl. Energy Mater.* **2018**, *1*, 561–570. [CrossRef]

60. Zhao, D.; Sexton, M.; Park, H.-Y.; Baure, G.; Nino, J.C.; So, F. High-Efficiency Solution-Processed Planar Perovskite Solar Cells with a Polymer Hole Transport Layer. *Adv. Energy Mater.* **2015**, *5*, 1401855. [CrossRef]

61. Xu, B.; Bi, D.; Hua, Y.; Liu, P.; Cheng, M.; Gratzel, M.; Sun, L. A low-cost spiro[fluorene-9,9'-xanthene]-based hole transport material for highly efficient solid-state dye-sensitized solar cells and perovskite solar cells. *Energy Environ. Sci.* **2016**, *9*, 873–877. [CrossRef]

62. Li, H.; Fu, K.; Hagfeldt, A.; Grätzel, M.; Mhaisalkar, S.G.; Grimsdale, A.C. A simple 3,4-ethylenedioxythiophene based hole-transporting material for perovskite solar cells. *Chem.-Int. Ed.* **2014**, *53*, 4085–4088. [CrossRef] [PubMed]

63. Rakstys, K.; Abate, A.; Dar, M.I.; Gao, P.; Jankauskas, V.; Jacopin, G.; Kamarauskas, E.; Kazim, S.; Ahmad, S.; Grätzel, M.; et al. Triazatruxene-Based Hole Transporting Materials for Highly Efficient Perovskite Solar Cells. *J. Am. Chem. Soc.* **2015**, *137*, 16172–16178. [CrossRef] [PubMed]

64. Heo, J.H.; Im, S.H.; Noh, J.H.; Mandal, T.N.; Lim, C.-S.; Chang, J.A.; Lee, Y.H.; Kim, H.J.; Sarkar, A.; Nazeeruddin, M.K.; et al. Efficient inorganic-organic hybrid heterojunction solar cells containing perovskite compound and polymeric hole conductors. *Nat. Photon.* **2013**, *7*, 486–491. [CrossRef]

65. Jiang, X.; Yu, Z.; Zhang, Y.; Lai, J.; Li, J.; Gurzadyan, G.G.; Sun, L. High-Performance Regular Perovskite Solar Cells Employing Low-Cost Poly (ethylenedioxythiophene) as a Hole-Transporting Material. *Sci. Rep.* **2017**, *7*, 42564. [CrossRef] [PubMed]

66. Wang, Y.; Su, T.-S.; Tsai, H.-Y.; Wei, T.-C.; Chi, Y. Spiro-Phenylpyrazole/Fluorene as Hole-Transporting Material for Perovskite Solar Cells. *Sci. Rep.* **2017**, *7*, 7859. [CrossRef] [PubMed]

67. Choi, H.; Mai, C.-K.; Kim, H.-B.; Jeong, J.; Song, S.; Bazan, G.C.; Heeger, A.J. Conjugated polyelectrolyte hole transport layer for inverted-type perovskite solar cells. *Nat. Commun.* **2015**, *6*, 7348. [CrossRef] [PubMed]

68. Chen, H.-W.; Huang, T.-Y.; Chang, T.-H.; Sanehira, Y.; Kung, C.-W.; Chu, C.-W.; Ikegami, M.; Miyasaka, T.; Ho, K.-C. Efficiency Enhancement of Hybrid Perovskite Solar Cells with MEH-PPV Hole-Transporting Layers. *Sci. Rep.* **2016**, *6*, 34319. [CrossRef] [PubMed]

69. Pathak, S.K.; Abate, A.; Ruckdeschel, P.; Roose, B.; Gödel, K.C.; Vaynzof, Y.; Santhala, A.; Watanabe, S.; Hollman, D.J.; Noel, N.; et al. Performance and Stability Enhancement of Dye-Sensitized and Perovskite Solar Cells by Al Doping of TiO$_2$. *Adv. Funct. Mater.* **2014**, *24*, 6046–6055. [CrossRef]

70. Chen, B.X.; Rao, H.-S.; Li, W.-G.; Xu, Y.-F.; Chen, H.-Y.; Kuang, D.-B.; Su, C.-Y. Achieving high-performance planar perovskite solar cell with Nb-doped TiO$_2$ compact layer by enhanced electron injection and efficient charge extraction. *J. Mater. Chem. A* **2016**, *4*, 5647–5653. [CrossRef]

71. Giordano, F.; Abate, A.; Correa Baena, J.P.; Saliba, M.; Matsui, T.; Im, S.H.; Graetzel, M. Enhanced electronic properties in mesoporous TiO$_2$ via lithium doping for high-efficiency perovskite solar cells. *Nat. Commun.* **2016**, *7*, 10379. [CrossRef] [PubMed]

72. Peng, J.; Duong, T.; Zhou, X.; Shen, H.; Wu, Y.; Mulmudi, H.K.; Wan, Y.; Zhong, D.; Li, J.; Tsuzuki, T.; et al. Efficient Indium-Doped TiO$_x$ Electron Transport Layers for High Performance Perovskite Solar Cells and Perovskite-Silicon Tandems. *Adv. Energy Mater.* **2017**, *7*, 1601768. [CrossRef]
73. Gu, X.; Wang, Y.; Zhang, T.; Liu, D.; Zhang, R.; Zhang, P.; Li, S. Enhanced electronic transport in Fe^{3+}-doped TiO$_2$ for high efficiency perovskite solar cells. *J. Mater. Chem. C* **2017**, *5*, 10754–10760. [CrossRef]
74. Seo, J.Y.; Uchida, R.; Kim, H.S.; Saygili, Y.; Luo, J.; Moore, C.; Kerrod, J.; Wagstaff, A.; Eklund, M.; McIntyre, R.; et al. Boosting the Efficiency of Perovskite Solar Cells with CsBr-Modified Mesoporous TiO$_2$ Beads as Electron-Selective Contact. *Adv. Funct. Mater.* **2017**, *28*, 1705763. [CrossRef]
75. Li, W.; Zhang, W.; Van Reenen, S.; Sutton, R.J.; Fan, J.; Haghighirad, A.; Johnston, M.; Wang, L.; Snaith, H. Enhanced UV-Light Stability of Planar Heterojunction Perovskite Solar Cells with Caesium Bromide Interface Modification. *Energy Environ. Sci.* **2016**, *9*, 490–498. [CrossRef]
76. Sun, Y.; Fang, X.; Ma, Z.; Xu, L.; Lu, Y.; Yu, Q.; Yuan, N.; Ding, J. Enhanced UV-Light Stability of Organometal Halide Perovskite Solar Cells with Interface Modification and a UV Absorption Layer. *J. Mater. Chem. C* **2017**, *5*, 8682–8687. [CrossRef]
77. Tan, H.; Jain, A.; Voznyy, O.; Lan, X.; de Arquer, F.P.G.; Fan, J.Z.; Quintero-Bermudez, R.; Yuan, M.; Zhang, B.; Zhao, Y.; et al. Efficient and stable solution-processed planar perovskite solar cells via contact passivation. *Science* **2017**, *355*, 722–726. [CrossRef] [PubMed]
78. Li, H.; Shi, W.; Huang, W.; Yao, E.-P.; Han, J.; Chen, Z.; Yang, Y. Carbon Quantum Dots/TiOx Electron Transport Layer Boosts Efficiency of Planar Heterojunction Perovskite Solar Cells to 19%. *Nano Lett.* **2017**, *17*, 2328–2335. [CrossRef] [PubMed]
79. Peng, J.; Wu, Y.; Ye, W.; Jacobs, D.A.; Shen, H.; Fu, X.; Wan, Y.; Duong, T.; Wu, N.; Barugkin, C.; et al. Interface Passivation Using Ultrathin Polymer-Fullerene Films for High-Efficiency Perovskite Solar Cells with Negligible Hysteresis. *Energy Environ. Sci.* **2017**, *10*, 1792–1800. [CrossRef]
80. Habisreutinger, S.N.; Noel, N.K.; Snaith, H.J.; Nicholas, R.J. Investigating the Role of 4-Tert Butylpyridine in Perovskite Solar Cells. *Adv. Energy Mater.* **2017**, *7*, 1–8. [CrossRef]
81. Choi, J.; Song, S.; Hörantner, M.T.; Snaith, H.J.; Park, T. Well-Defined Nanostructured, Single-Crystalline TiO$_2$ Electron Transport Layer for Efficient Planar Perovskite Solar Cells. *ACS Nano* **2016**, *10*, 6029–6036. [CrossRef] [PubMed]
82. Tiwana, P.; Docampo, P.; Johnston, M.B.; Snaith, H.J.; Herz, L.M. Electron Mobility and Injection Dynamics in Mesoporous ZnO, SnO$_2$, and TiO$_2$ Films Used in Dye-Sensitized Solar Cells. *ACS Nano* **2011**, *5*, 5158–5166. [CrossRef] [PubMed]
83. Jiang, Q.; Zhang, L.; Wang, H.; Yang, X.; Meng, J.; Liu, H.; Yin, Z.; Wu, J.; Zhang, X.; You, J.; et al. Enhanced Electron Extraction Using SnO$_2$ for High-Efficiency Planar-Structure HC(NH$_2$)$_2$PbI$_3$-Based Perovskite Solar Cells. *Nat. Energy* **2016**, *1*, 16177. [CrossRef]
84. Zhu, Z.; Bai, Y.; Liu, X.; Chueh, C.-C.; Yang, S.; Jen, A.K.-Y. Enhanced Efficiency and Stability of Inverted Perovskite Solar Cells Using Highly Crystalline SnO$_2$ Nanocrystals as the Robust Electron-Transporting Layer. *Adv. Mater.* **2016**, *28*, 6478–6484. [CrossRef] [PubMed]
85. Zhu, Z.; Zhao, D.; Chueh, C.-C.; Shi, X.; Li, Z.; Jen, A.K.-Y. Highly Efficient and Stable Perovskite Solar Cells Enabled by All-Crosslinked Charge Transporting Layers. *Joule* **2018**, *2*, 168–183. [CrossRef]
86. Mejía Escobar, M.A.; Pathak, S.; Liu, J.; Snaith, H.J.; Jaramillo, F. ZrO$_2$/TiO$_2$ Electron Collection Layer for Efficient Meso-Superstructured Hybrid Perovskite Solar Cells. *ACS Appl. Mater. Interfaces* **2017**, *9*, 2342–2349. [CrossRef] [PubMed]
87. Malinkiewicz, O.; Yella, A.; Lee, Y.H.; Espallargas, G.M.; Graetzel, M.; Nazeeruddin, M.K.; Bolink, H.J. Perovskite Solar Cells Employing Organic Charge-Transport Layers. *Nat. Photon.* **2013**, *8*, 128–132. [CrossRef]
88. Gong, J.; Darling, S.B.; You, F. Perovskite Photovoltaics: Life-Cycle Assessment of Energy and Environmental Impacts. *Energy Environ. Sci.* **2015**, *8*, 1953–1968. [CrossRef]
89. Wang, J.T.-W.; Ball, J.M.; Barea, E.M.; Abate, A.; Alexander-Webber, J.A.; Huang, J.; Saliba, M.; Mora-Sero, I.; Bisquert, J.; Snaith, H.J.; et al. Low-Temperature Processed Electron Collection Layers of Graphene/TiO$_2$ Nanocomposites in Thin Film Perovskite Solar Cells. *Nano Lett.* **2014**, *14*, 724–730. [CrossRef] [PubMed]
90. Dualeh, A.; Moehl, T.; Tétreault, N.; Teuscher, J.; Gao, P.; Nazeeruddin, M.K.; Grätzel, M. Impedance Spectroscopic Analysis of Lead Iodide Perovskite-Sensitized Solid-State Solar Cells. *ACS Nano* **2014**, *8*, 362–373. [CrossRef] [PubMed]

91. Snaith, H.J.; Abate, A.; Ball, J.M.; Eperon, G.E.; Leijtens, T.; Noel, N.K.; Stranks, S.D.; Wang, J.T.-W.; Wojciechowski, K.; Zhang, W. Anomalous Hysteresis in Perovskite Solar Cells. *J. Phys. Chem. Lett.* **2014**, *5*, 1511–1515. [CrossRef] [PubMed]

92. Chen, B.; Yang, M.; Priya, S.; Zhu, K. Origin of J-V Hysteresis in Perovskite Solar Cells. *J. Phys. Chem. Lett.* **2016**, *7*, 905–917. [CrossRef] [PubMed]

93. Sanchez, R.S.; Gonzalez-Pedro, V.; Lee, J.W.; Park, N.G.; Kang, Y.S.; Mora-Sero, I.; Bisquert, J. Slow Dynamic Processes in Lead Halide Perovskite Solar Cells. Characteristic Times and Hysteresis. *J. Phys. Chem. Lett.* **2014**, *5*, 2357–2363. [CrossRef] [PubMed]

94. Wu, Y.; Shen, H.; Walter, D.; Jacobs, D.; Duong, T.; Peng, J.; Jiang, L.; Cheng, Y.B.; Weber, K. On the Origin of Hysteresis in Perovskite Solar Cells. *Adv. Funct. Mater.* **2016**, *26*, 6807–6813. [CrossRef]

95. Shao, Y.; Xiao, Z.; Bi, C.; Yuan, Y.; Huang, J. Origin and Elimination of Photocurrent Hysteresis by Fullerene Passivation in $CH_3NH_3PbI_3$ Planar Heterojunction Solar Cells. *Nat. Commun.* **2014**, *5*, 1–7. [CrossRef] [PubMed]

96. Ono, L.K.; Qi, Y. Surface and Interface Aspects of Organometal Halide Perovskite Materials and Solar Cells. *J. Phys. Chem. Lett.* **2016**, *7*, 4764–4794. [CrossRef] [PubMed]

97. Ergen, O.; Gilbert, S.M.; Pham, T.; Turner, S.J.; Tan, M.T.Z.; Worsley, M.A.; Zettl, A. Graded Bandgap Perovskite Solar Cells. *Nat Mater* **2017**, *16*, 522–525. [CrossRef] [PubMed]

98. Bahtiar, A.; Rahmanita, S.; Inayatie, Y.D. Pin-Hole Free Perovskite Film for Solar Cells Application Prepared by Controlled Two-Step Spin-Coating Method. *IOP Conf. Ser. Mater. Sci. Eng.* **2017**, *196*, 012037. [CrossRef]

99. Han, F.; Luo, J.; Wan, Z.; Liu, X.; Jia, C. Dissolution-Recrystallization Method for High Efficiency Perovskite Solar Cells. *Appl. Surf. Sci.* **2017**, *408*, 34–37. [CrossRef]

100. Li, Z.; Yang, M.; Park, J.-S.; Wei, S.H.; Berry, J.J.; Zhu, K. Stabilizing Perovskite Structures by Tuning Tolerance Factor: Formation of Formamidinium and Cesium Lead Iodide Solid-State Alloys. *Chem. Mater.* **2016**, *28*, 284–292. [CrossRef]

101. Yi, C.; Luo, J.; Meloni, S.; Boziki, A.; Ashari-Astani, N.; Gratzel, C.; Zakeeruddin, S.M.; Rothlisberger, U.; Gratzel, M. Entropic Stabilization of Mixed A-Cation ABX_3 Metal Halide Perovskites for High Performance Perovskite Solar Cells. *Energy Environ. Sci.* **2016**, *9*, 656–662. [CrossRef]

102. Yao, E.P.; Sun, P.; Huang, W.; Yao, E.P.; Yang, Y.; Wang, M. Efficient Planar Perovskite Solar Cells Using Halide Sr-Substituted Pb Perovskite. *Nano Energy* **2017**, *36*, 213–222.

103. Eperon, G.E.; Beck, C.E.; Snaith, H.J. Cation Exchange for Thin Film Lead Iodide Perovskite Interconversion. *Mater. Horiz.* **2016**, *3*, 63–71. [CrossRef]

104. Meng, L.; You, J.; Guo, T.-F.F.; Yang, Y. Recent Advances in the Inverted Planar Structure of Perovskite Solar Cells. *Acc. Chem. Res.* **2015**, *49*, 155–165. [CrossRef] [PubMed]

105. Heo, J.H.; Han, H.J.; Kim, D.; Ahn, T.K.; Im, S.H. Hysteresis-Less Inverted $CH_3NH_3PbI_3$ Planar Perovskite Hybrid Solar Cells with 18.1% Power Conversion Efficiency. *Energy Environ. Sci.* **2015**, *8*, 1602–1608. [CrossRef]

106. Wu, Y.; Yang, X.; Chen, W.; Yue, Y.; Cai, M.; Xie, F.; Bi, E.; Islam, A.; Han, L. Perovskite Solar Cells with 18.21% Efficiency and Area over $1 Cm^2$ Fabricated by Heterojunction Engineering. *Nat. Energy* **2016**, *1*, 16148. [CrossRef]

107. Chiang, C.-H.; Wu, C.-G. Bulk Heterojunction Perovskite–PCBM Solar Cells with High Fill Factor. *Nat. Photon.* **2016**, *10*, 196. [CrossRef]

108. Bailie, C.D.; McGehee, M.D. High-Efficiency Tandem Perovskite Solar Cells. *MRS Bull.* **2015**, *40*, 681–686. [CrossRef]

109. De Vos, A. Detailed Balance Limit of the Efficiency of Tandem Solar Cells. *J. Phys. D Appl. Phys.* **1980**, *13*, 839. [CrossRef]

110. Futscher, M.H.; Ehrler, B. Efficiency Limit of Perovskite/Si Tandem Solar Cells. *ACS Energy Lett.* **2016**, 2–7. [CrossRef]

111. Uzu, H.; Ichikawa, M.; Hino, M.; Nakano, K.; Meguro, T.; Hernández, J.L.; Kim, H.-S.; Park, N.-G.; Yamamoto, K. High Efficiency Solar Cells Combining a Perovskite and a Silicon Heterojunction Solar Cells via an Optical Splitting System. *Appl. Phys. Lett.* **2015**, *106*, 13506. [CrossRef]

112. Todorov, T.; Gershon, T.; Gunawan, O.; Sturdevant, C.; Guha, S. Perovskite-Kesterite Monolithic Tandem Solar Cells with High Open-Circuit Voltage. *Appl. Phys. Lett.* **2014**, *105*, 173902. [CrossRef]

113. Loper, P.; Moon, S.-J.; Martin de Nicolas, S.; Niesen, B.; Ledinsky, M.; Nicolay, S.; Bailat, J.; Yum, J.-H.; De Wolf, S.; Ballif, C. Organic-Inorganic Halide Perovskite/Crystalline Silicon Four-Terminal Tandem Solar Cells. *Phys. Chem. Chem. Phys.* **2015**, *17*, 1619–1629. [CrossRef] [PubMed]

114. Lang, F.; Gluba, M.A.; Albrecht, S.; Rappich, J.; Korte, L.; Rech, B.; Nickel, N.H. Perovskite Solar Cells with Large-Area CVD-Graphene for Tandem Solar Cells. *J. Phys. Chem. Lett.* **2015**, *6*, 2745–2750. [CrossRef] [PubMed]

115. Mailoa, J.P.; Bailie, C.D.; Johlin, E.C.; Hoke, E.T.; Akey, A.J.; Nguyen, W.H.; McGehee, M.D.; Buonassisi, T. A 2-Terminal Perovskite/Silicon Multijunction Solar Cell Enabled by a Silicon Tunnel Junction. *Appl. Phys. Lett.* **2015**, *106*. [CrossRef]

116. Werner, J.; Dubuis, G.; Walter, A.; Löper, P.; Moon, S.-J.; Nicolay, S.; Morales-Masis, M.; De Wolf, S.; Niesen, B.; Ballif, C. Sputtered Rear Electrode with Broadband Transparency for Perovskite Solar Cells. *Sol. Energy Mater. Sol. Cells* **2015**, *141*, 407–413. [CrossRef]

117. Kranz, L.; Abate, A.; Feurer, T.; Fu, F.; Avancini, E.; Löckinger, J.; Reinhard, P.; Zakeeruddin, S.M.; Grätzel, M.; Buecheler, S.; et al. High-Efficiency Polycrystalline Thin Film Tandem Solar Cells. *J. Phys. Chem. Lett.* **2015**, *6*, 2676–2681. [CrossRef] [PubMed]

118. Zhao, D.; Yu, Y.; Wang, C.; Liao, W.; Shrestha, N.; Grice, C.R.; Cimaroli, A.J.; Guan, L.; Ellingson, R.J.; Zhu, K.; et al. Low-Bandgap Mixed Tin–lead Iodide Perovskite Absorbers with Long Carrier Lifetimes for All-Perovskite Tandem Solar Cells. *Nat. Energy* **2017**, *2*, 17018. [CrossRef]

119. Li, G.; Chang, W.-H.; Yang, Y. Low-Bandgap Conjugated Polymers Enabling Solution-Processable Tandem Solar Cells. *Nat. Rev. Mat.* **2017**, *2*, 17043. [CrossRef]

120. Fu, F.; Feurer, T.; Jäger, T.; Avancini, E.; Bissig, B.; Yoon, S.; Buecheler, S.; Tiwari, A.N. Low-Temperature-Processed Efficient Semi-Transparent Planar Perovskite Solar Cells for Bifacial and Tandem Applications. *Nat. Commun.* **2015**, *6*, 8932. [CrossRef] [PubMed]

121. McMeekin, D.P.; Sadoughi, G.; Rehman, W.; Eperon, G.E.; Saliba, M.; Hörantner, M.T.; Haghighirad, A.; Sakai, N.; Korte, L.; Rech, B.; et al. A Mixed-Cation Lead Mixed-Halide Perovskite Absorber for Tandem Solar Cells. *Science* **2016**, *351*, 151–155. [CrossRef] [PubMed]

122. Leguy, A.M.A.; Hu, Y.; Campoy-Quiles, M.; Alonso, M.I.; Weber, O.J.; Azarhoosh, P.; Van Schilfgaarde, M.; Weller, M.T.; Bein, T.; Nelson, J.; et al. Reversible Hydration of $CH_3NH_3PbI_3$ in Films, Single Crystals, and Solar Cells. *Chem. Mater.* **2015**, *27*, 3397–3407. [CrossRef]

123. Niu, G.; Guo, X.; Wang, L. Review of Recent Progress in Chemical Stability of Perovskite Solar Cells. *J. Mater. Chem. A* **2015**, *3*, 8970–8980. [CrossRef]

124. Frost, J.M.; Butler, K.T.; Brivio, F.; Hendon, C.H.; van Schilfgaarde, M.; Walsh, A. Atomistic Origins of High-Performance in Hybrid Halide Perovskite Solar Cells. *Nano Lett.* **2014**, *14*, 2584–2590. [CrossRef] [PubMed]

125. Yang, J.; Siempelkamp, B.D.; Liu, D.; Kelly, T.L. Investigation of $CH_3NH_3PbI_3$ Degradation Rates and Mechanisms in Controlled Humidity Environments Using in Situ Techniques. *ACS Nano* **2015**, *9*, 1955–1963. [CrossRef] [PubMed]

126. Leijtens, T.; Eperon, G.E.; Pathak, S.; Abate, A.; Lee, M.M.; Snaith, H.J. Overcoming Ultraviolet Light Instability of Sensitized TiO_2 with Meso-Superstructured Organometal Tri-Halide Perovskite Solar Cells. *Nat. Commun.* **2013**, *4*, 2885. [CrossRef] [PubMed]

127. Philippe, B.; Park, B.-W.; Lindblad, R.; Oscarsson, J.; Ahmadi, S.; Johansson, E.M.J.; Rensmo, H. Chemical and Electronic Structure Characterization of Lead Halide Perovskites and Stability Behavior under Different Exposures—A Photoelectron Spectroscopy Investigation. *Chem. Mater.* **2015**, *27*, 1720–1731. [CrossRef]

128. Conings, B.; Drijkoningen, J.; Gauquelin, N.; Babayigit, A.; D'Haen, J.; D'Olieslaeger, L.; Ethirajan, A.; Verbeeck, J.; Manca, J.; Mosconi, E.; et al. Intrinsic Thermal Instability of Methylammonium Lead Trihalide Perovskite. *Adv. Energy Mater.* **2015**, *5*, 1500477. [CrossRef]

129. Kim, N.-K.; Min, Y.H.; Noh, S.; Cho, E.; Jeong, G.; Joo, M.; Ahn, S.-W.; Lee, J.S.; Kim, S.; Ihm, K.; et al. Investigation of Thermally Induced Degradation in $CH_3NH_3PbI_3$ Perovskite Solar Cells Using In-Situ Synchrotron Radiation Analysis. *Sci. Rep.* **2017**, *7*, 4645. [CrossRef] [PubMed]

130. Grancini, G.; Roldán-Carmona, C.; Zimmermann, I.; Mosconi, E.; Lee, X.; Martineau, D.; Narbey, S.; Oswald, F.; De Angelis, F.; Graetzel, M.; et al. One-Year Stable Perovskite Solar Cells by 2D/3D Interface Engineering. *Nat. Commun.* **2017**, *8*, 15684. [CrossRef] [PubMed]

131. Noh, J.H.; Im, S.H.; Heo, J.H.; Mandal, T.N.; Seok, S.I. Chemical Management for Colorful, Efficient, and Stable Inorganic-Organic Hybrid Nanostructured Solar Cells. *Nano Lett.* **2013**, *13*, 1764–1769. [CrossRef] [PubMed]

132. Jeon, N.J.; Noh, J.H.; Kim, Y.C.; Yang, W.S.; Ryu, S.; Seok, S.I. Solvent Engineering for High-Performance Inorganic-Organic Hybrid Perovskite Solar Cells. *Nat. Mater.* **2014**, *13*, 897–903. [CrossRef] [PubMed]

133. Hoke, E.T.; Slotcavage, D.J.; Dohner, E.R.; Bowring, A.R.; Karunadasa, H.I.; McGehee, M.D. Reversible Photo-Induced Trap Formation in Mixed-Halide Hybrid Perovskites for Photovoltaics. *Chem. Sci.* **2015**, *6*, 613–617. [CrossRef] [PubMed]

134. Jiang, Q.; Rebollar, D.; Gong, J.; Piacentino, E.L.; Zheng, C.; Xu, T. Pseudohalide-Induced Moisture Tolerance in Perovskite $CH_3NH_3Pb(SCN)_2I$ Thin Films. *Angew. Chem.-Int. Ed.* **2015**, *54*, 7617–7620. [CrossRef] [PubMed]

135. Chen, Y.; Li, B.; Huang, W.; Gao, D.; Liang, Z. Efficient and Reproducible $CH_3NH_3PbI_{3-x}(SCN)_x$ Perovskite Based Planar Solar Cells. *Chem. Commun.* **2015**, *51*, 11997–11999. [CrossRef] [PubMed]

136. Ganose, A.M.; Savory, C.N.; Scanlon, D.O. (CH3NH3)2Pb(SCN)2I2: A More Stable Structural Motif for Hybrid Halide Photovoltaics? *J. Phys. Chem. Lett.* **2015**, *6*, 4594–4598. [CrossRef] [PubMed]

137. Tai, Q.; You, P.; Sang, H.; Liu, Z.; Hu, C.; Chan, H.L.W.; Yan, F. Efficient and Stable Perovskite Solar Cells Prepared in Ambient Air Irrespective of the Humidity. *Nat. Commun.* **2016**, *7*, 11105. [CrossRef] [PubMed]

138. Eperon, G.E.; Stranks, S.D.; Menelaou, C.; Johnston, M.B.; Herz, L.M.; Snaith, H.J. Formamidinium Lead Trihalide: A Broadly Tunable Perovskite for Efficient Planar Heterojunction Solar Cells. *Energy Environ. Sci.* **2014**, *7*, 982–988. [CrossRef]

139. Koh, T.M.; Fu, K.; Fang, Y.; Chen, S.; Sum, T.C.; Mathews, N. Formamidinium-Containing Metal-Halide: An Alternative Material for Near-IR Absorption Perovskite Solar Cells. *J. Phys. Chem. C* **2013**, *118*, 16458–16462. [CrossRef]

140. Qin, X.; Zhao, Z.; Wang, Y.; Wu, J.; Jiang, Q.; You, J. Recent Progress in Stability of Perovskite Solar Cells. *J. Semicond.* **2017**, *38*, 11002. [CrossRef]

141. Choi, H.; Jeong, J.; Kim, H.-B.; Kim, S.; Walker, B.; Kim, G.-H.; Kim, J.Y. Cesium-Doped Methylammonium Lead Iodide Perovskite Light Absorber for Hybrid Solar Cells. *Nano Energy* **2014**, *7*, 80–85. [CrossRef]

142. Jeon, N.J.; Noh, J.H.; Yang, W.S.; Kim, Y.C.; Ryu, S.; Seo, J.; Seok, S.I. Compositional Engineering of Perovskite Materials for High-Performance Solar Cells. *Nature* **2015**, *517*, 476–480. [CrossRef] [PubMed]

143. Yang, W.S.; Noh, J.H.; Jeon, N.J.; Kim, Y.C.; Ryu, S.; Seo, J.; Seok, S.I. High-Performance Photovoltaic Perovskite Layers Fabricated through Intramolecular Exchange. *Science* **2015**, *348*, 1234–1237. [CrossRef] [PubMed]

144. Wang, Q.; Zheng, X.; Deng, Y.; Zhao, J.; Chen, Z.; Huang, J. Stabilizing the α-Phase of $CsPbI_3$ Perovskite by Sulfobetaine Zwitterions in One-Step Spin-Coating Films. *Joule* **2017**, *1*, 371–382. [CrossRef]

145. Eperon, G.E.; Paterno, G.M.; Sutton, R.J.; Zampetti, A.; Haghighirad, A.A.; Cacialli, F.; Snaith, H.J. Inorganic Caesium Lead Iodide Perovskite Solar Cells. *J. Mater. Chem. A* **2015**, *3*, 19688–19695. [CrossRef]

146. Saliba, M.; Matsui, T.; Seo, J.-Y.; Domanski, K.; Correa-Baena, J.-P.; Nazeeruddin, M.K.; Zakeeruddin, S.M.; Tress, W.; Abate, A.; Hagfeldt, A.; et al. Cesium-Containing Triple Cation Perovskite Solar Cells: Improved Stability, Reproducibility and High Efficiency. *Energy Environ. Sci.* **2016**, *9*, 1989–1997. [CrossRef] [PubMed]

147. Sun, Y.; Peng, J.; Chen, Y.; Yao, Y.; Liang, Z. Triple-Cation Mixed-Halide Perovskites: Towards Efficient, Annealing-Free and Air-Stable Solar Cells Enabled by Pb(SCN)2 Additive. *Sci. Rep.* **2017**, *7*, 1–7. [CrossRef] [PubMed]

148. Bu, T.; Liu, X.; Zhou, Y.; Yi, J.; Huang, X.; Luo, L.; Xiao, J.; Ku, Z.; Peng, Y.; Huang, F.; et al. Novel Quadruple-Cation Absorber for Universal Hysteresis Elimination for High Efficiency and Stable Perovskite Solar Cells. *Energy Environ. Sci.* **2017**, *10*, 2509–2515. [CrossRef]

149. Hu, Y.; Bai, F.; Liu, X.; Ji, Q.; Miao, X.; Qiu, T.; Zhang, S. Bismuth Incorporation Stabilized α-CsPbI3 for Fully Inorganic Perovskite Solar Cells. *ACS Energy Lett.* **2017**, *2*, 2219–2227. [CrossRef]

150. Zhang, X.; Ren, X.; Liu, B.; Munir, R.; Zhu, X.; Yang, D.; Li, J.; Liu, Y.; Smilgies, D.-M.; Li, R.; et al. Stable High Efficiency Two-Dimensional Perovskite Solar Cells via Cesium Doping. *Energy Environ. Sci.* **2017**, *10*, 2095–2102. [CrossRef]

151. Wang, Z.; Lin, Q.; Chmiel, F.P.; Sakai, N.; Herz, L.M.; Snaith, H.J. Efficient Ambient-Air-Stable Solar Cells with 2D-3D Heterostructured Butylammonium-Caesium-Formamidinium Lead Halide Perovskites. *Nat. Energy* **2017**, *2*, 1–10. [CrossRef]

152. Batmunkh, M.; Shearer, C.J.; Bat-Erdene, M.; Biggs, M.J.; Shapter, J.G. Single-Walled Carbon Nanotubes Enhance the Efficiency and Stability of Mesoscopic Perovskite Solar Cells. *ACS Appl. Mater. Interfaces* **2017**, *9*, 19945–19954. [CrossRef] [PubMed]

153. Roose, B.; Pathak, S.K.; Steiner, U. Doping of TiO_2 for sensitised solar cells. *Chem. Soc. Rev.* **2015**, *44*, 8326–8349. [CrossRef] [PubMed]

154. Yang, J.; Siempelkamp, B.D.; Mosconi, E.; De Angelis, F.; Kelly, T.L. Origin of the Thermal Instability in $CH_3NH_3PbI_3$ Thin Films Deposited on ZnO. *Chem. Mater.* **2015**, *27*, 4229–4236. [CrossRef]

155. Arafat Mahmud, M.; Kumar Elumalai, N.; Baishakhi Upama, M.; Wang, D.; Gonçales, V.R.; Wright, M.; Justin Gooding, J.; Haque, F.; Xu, C.; Uddin, A. Cesium Compounds as Interface Modifiers for Stable and Efficient Perovskite Solar Cells. *Sol. Energy Mater. Sol. Cells* **2018**, *174*, 172–186. [CrossRef]

156. Cao, J.; Wu, B.; Chen, R.; Wu, Y.; Hui, Y.; Mao, B.W.; Zheng, N. Efficient, Hysteresis-Free, and Stable Perovskite Solar Cells with ZnO as Electron-Transport Layer: Effect of Surface Passivation. *Adv. Mater.* **2018**, *1705596*, 1–9. [CrossRef] [PubMed]

157. Mahmud, M.A.; Elumalai, N.K.; Upama, M.B.; Wang, D.; Zarei, L.; Gonçales, V.R.; Wright, M.; Xu, C.; Haque, F.; Uddin, A. Adsorbed Carbon Nanomaterials for Surface and Interface-Engineered Stable Rubidium Multi-Cation Perovskite Solar Cells. *Nanoscale* **2017**, 773–790. [CrossRef] [PubMed]

158. Pang, A.; Shen, D.; Wei, M.; Chen, Z.-N. Highly Efficient Perovskite Solar Cells Based on $Zn_2Ti_3O_8$ Nanoparticles as Electron Transport Material. *ChemSusChem* **2017**, *11*, 424–431. [CrossRef] [PubMed]

159. Wang, X.; Deng, L.-L.; Wang, L.-Y.; Dai, S.-M.; Xing, Z.; Zhan, X.-X.; Lu, X.-Z.; Xie, S.-Y.; Huang, R.-B.; Zheng, L.-S. Cerium Oxide Standing out as an Electron Transport Layer for Efficient and Stable Perovskite Solar Cells Processed at Low Temperature. *J. Mater. Chem. A* **2017**, *5*, 1706–1712. [CrossRef]

160. Shin, S.S.; Yeom, E.J.; Yang, W.S.; Hur, S.; Kim, M.G.; Im, J.; Seo, J.; Noh, J.H.; Seok, S.I. Colloidally Prepared La-Doped $BaSnO_3$ Electrodes for Efficient, Photostable Perovskite Solar Cells. *Science* **2017**, *356*, 167–171. [CrossRef] [PubMed]

161. Lee, K.; Ryu, J.; Yu, H.; Yun, J.; Lee, J.; Jang, J. Enhanced Efficiency and Air-Stability of NiO_X-Based Perovskite Solar Cells via PCBM Electron Transport Layer Modification with Triton X-100. *Nanoscale* **2017**, *9*, 16249–16255. [CrossRef] [PubMed]

162. Kim, G.-H.; Jang, H.; Yoon, Y.J.; Jeong, J.; Park, S.Y.; Walker, B.; Jeon, I.-Y.; Jo, Y.; Yoon, H.; Kim, M.; et al. Fluorine Functionalized Graphene Nano Platelets for Highly Stable Inverted Perovskite Solar Cells. *Nano Lett.* **2017**, *17*, 6385–6390. [CrossRef] [PubMed]

163. Kim, H.I.; Kim, M.-J.; Choi, K.; Lim, C.; Kim, Y.-H.; Kwon, S.-K.; Park, T. Improving the Performance and Stability of Inverted Planar Flexible Perovskite Solar Cells Employing a Novel NDI-Based Polymer as the Electron Transport Layer. *Adv. Energy Mater.* **2018**, *1702872*. [CrossRef]

164. Mahmud, M.A.; Elumalai, N.K.; Upama, M.B.; Wang, D.; Gonçales, V.R.; Wright, M.; Xu, C.; Haque, F.; Uddin, A. A High Performance and Low-Cost Hole Transporting Layer for Efficient and Stable Perovskite Solar Cells. *Phys. Chem. Chem. Phys.* **2017**, *19*, 21033–21045. [CrossRef] [PubMed]

165. Calió, L.; Momblona, C.; Gil-Escrig, L.; Kazim, S.; Sessolo, M.; Sastre-Santos, Á.; Bolink, H.J.; Ahmad, S. Vacuum Deposited Perovskite Solar Cells Employing Dopant-Free Triazatruxene as the Hole Transport Material. *Sol. Energy Mater. Sol. Cells* **2017**, *163*, 237–241. [CrossRef]

166. Matsui, T.; Petrikyte, I.; Malinauskas, T.; Domanski, K.; Daskeviciene, M.; Steponaitis, M.; Gratia, P.; Tress, W.; Correa-Baena, J.P.; Abate, A.; et al. Additive-free transparent triarylamine-based polymeric hole-transport materials for stable perovskite solar cells. *ChemSusChem* **2016**, *9*, 2567–2571. [CrossRef] [PubMed]

167. Li, Q.; Zhao, Y.; Fu, R.; Zhou, W.; Zhao, Y.; Lin, F.; Liu, S.; Yu, D.; Zhao, Q. Enhanced Long-Term Stability of Perovskite Solar Cells Using a Double-Layer Hole Transport Material. *J. Mater. Chem. A* **2017**, *5*, 14881–14886. [CrossRef]

168. Hawash, Z.; Ono, L.K.; Raga, S.R.; Lee, M.V.; Qi, Y. Air-Exposure Induced Dopant Redistribution and Energy Level Shifts in Spin-Coated Spiro-MeOTAD Films. *Chem. Mater.* **2015**, *27*, 562–569. [CrossRef]

169. Kim, G.-W.; Kang, G.; Malekshahi Byranvand, M.; Lee, G.-Y.; Park, T. Gradated Mixed Hole Transport Layer in a Perovskite Solar Cell: Improving Moisture Stability and Efficiency. *ACS Appl. Mater. Interfaces* **2017**, *9*, 27720–27726. [CrossRef] [PubMed]

170. Ginting, R.T.; Jeon, M.-K.; Lee, K.-J.; Jin, W.-Y.; Kim, T.-W.; Kang, J.-W. Degradation Mechanism of Planar-Perovskite Solar Cells: Correlating Evolution of Iodine Distribution and Photocurrent Hysteresis. *J. Mater. Chem. A* **2017**, *5*, 4527–4534. [CrossRef]

171. Habisreutinger, S.N.; Leijtens, T.; Eperon, G.E.; Stranks, S.D.; Nicholas, R.J.; Snaith, H.J. Carbon Nanotube/Polymer Composites as a Highly Stable Hole Collection Layer in Perovskite Solar Cells. *Nano Lett.* **2014**, *14*, 5561–5568. [CrossRef] [PubMed]

172. Luo, Q.; Ma, H.; Zhang, Y.; Yin, X.; Yao, Z.; Wang, N.; Li, J.; Fan, S.; Jiang, K.; Lin, H. Cross-Stacked Superaligned Carbon Nanotube Electrodes for Efficient Hole Conductor-Free Perovskite Solar Cells. *J. Mater. Chem. A* **2016**, *4*, 5569–5577. [CrossRef]

173. Mei, A.; Li, X.; Liu, L.; Ku, Z.; Liu, T.; Rong, Y.; Xu, M.; Hu, M.; Chen, J.; Yang, Y.; et al. A Hole-Conductor-Free, Fully Printable Mesoscopic Perovskite Solar Cell with High Stability. *Science* **2014**, *345*, 295–298. [CrossRef] [PubMed]

174. Koh, C.W.; Heo, J.H.; Uddin, M.A.; Kwon, Y.-W.; Choi, D.H.; Im, S.H.; Woo, H.Y. Enhanced Efficiency and Long-Term Stability of Perovskite Solar Cells by Synergistic Effect of Nonhygroscopic Doping in Conjugated Polymer-Based Hole-Transporting Layer. *ACS Appl. Mater. Interfaces* **2017**, *9*, 43846–43854. [CrossRef] [PubMed]

175. Wang, D.; Elumalai, N.K.; Mahmud, M.A.; Wright, M.; Upama, M.B.; Chan, K.H.; Xu, C.; Haque, F.; Conibeer, G.; Uddin, A. V2O5-PEDOT: PSS Bilayer as Hole Transport Layer for Highly Efficient and Stable Perovskite Solar Cells. *Org. Electron. Physics, Mater. Appl.* **2018**, *53*, 66–73.

176. Ma, S.; Qiao, W.; Cheng, T.; Zhang, B.; Yao, J.; Alsaedi, A.; Hayat, T.; Ding, Y.; Tan, Z.; Dai, S. Optical-Electrical-Chemical Engineering of PEDOT:PSS by Incorporation of Hydrophobic Nafion for Efficient and Stable Perovskite Solar Cells. *ACS Appl. Mater. Interfaces* **2018**, *10*, 3902–3911. [CrossRef] [PubMed]

177. Yang, L.; Cai, F.; Yan, Y.; Li, J.; Liu, D.; Pearson, A.J.; Wang, T. Conjugated Small Molecule for Efficient Hole Transport in High-Performance p-i-n Type Perovskite Solar Cells. *Adv. Funct. Mater.* **2017**, *27*, 1702613. [CrossRef]

178. Lee, J.; Kang, H.; Kim, G.; Back, H.; Kim, J.; Hong, S.; Park, B.; Lee, E.; Lee, K. Achieving Large-Area Planar Perovskite Solar Cells by Introducing an Interfacial Compatibilizer. *Adv. Mater.* **2017**, *29*, 1606363. [CrossRef] [PubMed]

179. Wang, Q.; Chueh, C.C.; Zhao, T.; Cheng, J.; Eslamian, M.; Choy, W.C.H.; Jen, A.K.Y. Effects of Self-Assembled Monolayer Modification of Nickel Oxide Nanoparticles Layer on the Performance and Application of Inverted Perovskite Solar Cells. *ChemSusChem* **2017**, *10*, 3794–3803. [CrossRef] [PubMed]

180. Zhao, X.; Kim, H.-S.; Seo, J.-Y.; Park, N.-G. Effect of Selective Contacts on the Thermal Stability of Perovskite Solar Cells. *ACS Appl. Mater. Interfaces* **2017**, *9*, 7148–7153. [CrossRef] [PubMed]

181. Guarnera, S.; Abate, A.; Zhang, W.; Foster, J.M.; Richardson, G.; Petrozza, A.; Snaith, H.J. Improving the Long-Term Stability of Perovskite Solar Cells with a Porous Al$_2$O$_3$ Buffer Layer. *J. Phys. Chem. Lett.* **2015**, *6*, 432–437. [CrossRef] [PubMed]

182. Koushik, D.; Verhees, W.J.H.; Kuang, Y.; Veenstra, S.; Zhang, D.; Verheijen, M.A.; Creatore, M.; Schropp, R.E.I. High-Efficiency Humidity-Stable Planar Perovskite Solar Cells Based on Atomic Layer Architecture. *Energy Environ. Sci.* **2017**, *10*, 91–100. [CrossRef]

183. Wei, Z.; Zheng, X.; Chen, H.; Long, X.; Wang, Z.; Yang, S. A Multifunctional C + Epoxy/Ag-Paint Cathode Enables Efficient and Stable Operation of Perovskite Solar Cells in Watery Environments. *J. Mater. Chem. A* **2015**, *3*, 16430–16434. [CrossRef]

184. Chang, C.-Y.; Lee, K.-T.; Huang, W.-K.; Siao, H.-Y.; Chang, Y.-C. High-Performance, Air-Stable, Low-Temperature Processed Semitransparent Perovskite Solar Cells Enabled by Atomic Layer Deposition. *Chem. Mater.* **2015**, *27*, 5122–5130. [CrossRef]

185. Kaltenbrunner, M.; Adam, G.; Głowacki, E.D.; Drack, M.; Schwödiauer, R.; Leonat, L.; Apaydin, D.H.; Groiss, H.; Scharber, M.C.; White, M.S.; et al. Flexible High Power-per-Weight Perovskite Solar Cells with Chromium Oxide–metal Contacts for Improved Stability in Air. *Nat. Mater.* **2015**, *14*, 1032. [CrossRef] [PubMed]

186. You, J.; Meng, L.; Song, T.-B.; Guo, T.-F.; Yang, Y.M.; Chang, W.-H.; Hong, Z.; Chen, H.; Zhou, H.; Chen, Q.; et al. Improved Air Stability of Perovskite Solar Cells via Solution-Processed Metal Oxide Transport Layers. *Nat. Nanotechnol.* **2015**, *11*, 75. [CrossRef] [PubMed]

187. Hwang, I.; Jeong, I.; Lee, J.; Ko, M.J.; Yong, K. Enhancing Stability of Perovskite Solar Cells to Moisture by the Facile Hydrophobic Passivation. *ACS Appl. Mater. Interfaces* **2015**, *7*, 17330–17336. [CrossRef] [PubMed]

188. Cheacharoen, R.; Rolston, N.J.; Harwood, D.; Bush, K.A.; Dauskardt, R.H.; McGehee, M.D. Design and Understanding of Encapsulated Perovskite Solar Cells to Withstand Temperature Cycling. *Energy Environ. Sci.* **2018**, *11*, 144–150. [CrossRef]

189. Liu, Z.; Sun, B.; Shi, T.; Tang, Z.; Liao, G. Enhanced Photovoltaic Performance and Stability of Carbon Counter Electrode Based Perovskite Solar Cells Encapsulated by PDMS. *J. Mater. Chem. A* **2016**, *4*, 10700–10709. [CrossRef]

190. Matteocci, F.; Cinà, L.; Lamanna, E.; Cacovich, S.; Divitini, G.; Midgley, P.A.; Ducati, C.; Di Carlo, A. Encapsulation for Long-Term Stability Enhancement of Perovskite Solar Cells. *Nano Energy* **2016**, *30*, 162–172. [CrossRef]

191. Han, Y.; Meyer, S.; Dkhissi, Y.; Weber, K.; Pringle, J.M.; Bach, U.; Spiccia, L.; Cheng, Y.-B. Degradation Observations of Encapsulated Planar CH$_3$NH$_3$PbI$_3$ Perovskite Solar Cells at High Temperatures and Humidity. *J. Mater. Chem. A* **2015**, *3*, 8139–8147. [CrossRef]

192. Dong, Q.; Liu, F.; Wong, M.K.; Tam, H.W.; Djurišić, A.B.; Ng, A.; Surya, C.; Chan, W.K.; Ng, W.K. Encapsulation of Perovskite Solar Cells for High Humidity Conditions. *ChemSusChem* **2016**, *9*, 2597–2603. [CrossRef] [PubMed]

193. Shi, L.; Young, T.L.; Kim, J.; Sheng, Y.; Wang, L.; Chen, Y.; Feng, Z.; Keevers, M.J.; Hao, X.; Verlinden, P.J.; et al. Accelerated Lifetime Testing of Organic-Inorganic Perovskite Solar Cells Encapsulated by Polyisobutylene. *ACS Appl. Mater. Interfaces* **2017**, *9*, 25073–25081. [CrossRef] [PubMed]

194. Sanehira, E.M.; Tremolet de Villers, B.J.; Schulz, P.; Reese, M.O.; Ferrere, S.; Zhu, K.; Lin, L.Y.; Berry, J.J.; Luther, J.M. Influence of Electrode Interfaces on the Stability of Perovskite Solar Cells: Reduced Degradation Using MoO$_x$/Al for Hole Collection. *ACS Energy Lett.* **2016**, *1*, 38–45. [CrossRef]

195. Bastos, J.P.; Manghooli, S.; Jaysankar, M.; Tait, J.G.; Qiu, W.; Gehlhaar, R.; De Volder, M.; Uytterhoeven, G.; Poortmans, J.; Paetzold, U.W. Low-Cost Electrodes for Stable Perovskite Solar Cells. *Appl. Phys. Lett.* **2017**, *110*, 233902. [CrossRef]

196. Zhao, J.; Zheng, X.; Deng, Y.; Li, T.; Shao, Y.; Gruverman, A.; Shield, J.; Huang, J. Is Cu a Stable Electrode Material in Hybrid Perovskite Solar Cells for a 30-Year Lifetime? *Energy Environ. Sci.* **2016**, *9*, 3650–3656. [CrossRef]

197. Gifford, J. Dyesol Claims Perovskite Breakthrough. Available online: https://www.pv-magazine.com/2015/09/09/dyesol-claims-perovskite-stability-breakthrough_100021002/#axzz3lF0szmdL (accessed on 14 December 2017).

198. Crystal Clear? Available online: https://www.economist.com/news/science-and-technology/21651166-perovskites-may-give-silicon-solar-cells-run-their-money-crystal-clear (accessed on 14 December 2017).

199. Beets, B. LCOE for Renewables Decreases, Fossil Fuels See Increase. Available online: https://www.pv-magazine.com/2015/10/06/lcoe-for-renewables-decreases-fossil-fuels-see-increase_100021404/ (accessed on 28 December 2017).

200. Global Wind and Solar Costs to Fall Even Faster, While Coal Fades Even in China and India. Available online: https://about.bnef.com/blog/global-wind-solar-costs-fall-even-faster-coal-fades-even-china-india/ (accessed on 28 December 2017).

201. Cai, M.; Wu, Y.; Chen, H.; Yang, X.; Qiang, Y.; Han, L. Cost-Performance Analysis of Perovskite Solar Modules. *Adv. Sci.* **2017**, *4*, 1600269. [CrossRef] [PubMed]

202. Branker, K.; Pathak, M.J.M.; Pearce, J.M. A Review of Solar Photovoltaic Levelized Cost of Electricity. *Renew. Sustain. Energy Rev.* **2011**, *15*, 4470–4482. [CrossRef]

203. PV EnergyTrend. Available online: http://pv.energytrend.com/pricequotes.html (accessed on 21 November 2017).

204. Maniarasu, S.; Korukonda, T.B.; Manjunath, V.; Ramasamy, E.; Ramesh, M.; Veerappan, G. Recent Advancement in Metal Cathode and Hole-Conductor-Free Perovskite Solar Cells for Low-Cost and High Stability: A Route towards Commercialization. *Renew. Sustain. Energy Rev.* **2018**, *82 Pt 1*, 845–857. [CrossRef]

205. Petrus, M.; Music, A.; Closs, A.C.; Bijleveld, J.C.; Sirtl, M.T.; Hu, Y.; Dingemans, T.J.; Bein, T.; Docampo, P. Design Rules for the Preparation of Low-Cost Hole Transporting Materials for Perovskite Solar Cells with Moisture Barrier Properties. *J. Mater. Chem. A* **2017**, *5*, 25200–25210. [CrossRef]

206. Petrus, M.L.; Bein, T.; Dingemans, T.J.; Docampo, P. A Low Cost Azomethine-Based Hole Transporting Material for Perovskite Photovoltaics. *J. Mater. Chem. A* **2015**, *3*, 12159–12162. [CrossRef]

207. Wu, J.; Liu, C.; Deng, X.; Zhang, L.; Hu, M.; Tang, J.; Tan, W.; Tian, Y.; Xu, B. Simple and Low-Cost Thiophene and Benzene-Conjugated Triaryamines as Hole-Transporting Materials for Perovskite Solar Cells. *RSC Adv.* **2017**, *7*, 45478–45483. [CrossRef]

208. Jeon, N.J.; Lee, J.; Noh, J.H.; Nazeeruddin, M.K.; Grätzel, M.; Seok, S.I. Efficient Inorganic–Organic Hybrid Perovskite Solar Cells Based on Pyrene Arylamine Derivatives as Hole-Transporting Materials. *J. Am. Chem. Soc.* **2013**, *135*, 19087–19090. [CrossRef] [PubMed]

209. Grisorio, R.; Roose, B.; Colella, S.; Listorti, A.; Suranna, G.P.; Abate, A. Molecular Tailoring of Phenothiazine-Based Hole-Transporting Materials for Highly Performing Perovskite Solar Cells. *ACS Energy Lett.* **2017**, *2*, 1029–1034. [CrossRef]

210. Pham, H.D.; Wu, Z.; Ono, L.K.; Manzhos, S.; Feron, K.; Motta, N.; Qi, Y.B.; Sonar, P. Low-Cost Alternative High-Performance Hole-Transport Material for Perovskite Solar Cells and Its Comparative Study with Conventional SPIRO-OMeTAD. *Adv. Electron. Mater.* **2017**, *3*, 1700139. [CrossRef]

211. Liu, X.; Kong, F.; Ghadari, R.; Jin, S.; Chen, W.; Yu, T.; Hayat, T.; Alsaedi, A.; Guo, F.; Tan, Z.; et al. Thiophene–Arylamine Hole-Transporting Materials in Perovskite Solar Cells: Substitution Position Effect. *Energy Technol.* **2017**, *5*, 1788. [CrossRef]

212. Do, K.; Choi, H.; Lim, K.; Jo, H.; Cho, J.W.; Nazeeruddin, M.K.; Ko, J. Star-Shaped Hole Transporting Materials with a Triazine Unit for Efficient Perovskite Solar Cells. *Chem. Commun.* **2014**, *50*, 10971–10974. [CrossRef] [PubMed]

213. Calió, L.; Kazim, S.; Grätzel, M.; Ahmad, S. Hole transport materials for perovskite solar cells. *Angew. Chem. Int. Ed.* **2016**, *55*, 14522. [CrossRef] [PubMed]

214. Tang, L.J.; Chen, X.; Wen, T.Y.; Yang, S.; Zhao, J.J.; Qiao, H.W.; Hou, Y.; Yang, H.G. A Solution-Processed Transparent NiO Hole-Extraction Layer for High-Performance Inverted Perovskite Solar Cells. *Chem. Eur. J.* **2018**, *24*, 2845. [CrossRef] [PubMed]

215. Qin, P.; Tanaka, S.; Ito, S.; Tetreault, N.; Manabe, K.; Nishino, H.; Nazeeruddin, M.K.; Grätzel, M. Inorganic Hole Conductor-Based Lead Halide Perovskite Solar Cells with 12.4% Conversion Efficiency. *Nat. Commun.* **2014**, *5*, 3834. [CrossRef] [PubMed]

216. Christians, J.A.; Fung, R.C.M.; Kamat, P.V. An Inorganic Hole Conductor for Organo-Lead Halide Perovskite Solar Cells. Improved Hole Conductivity with Copper Iodide. *J. Am. Chem. Soc.* **2014**, *136*, 758–764. [CrossRef] [PubMed]

217. Chu, L.; Liu, W.; Qin, Z.; Zhang, R.; Hu, R.; Yang, J.; Yang, J.; Li, X. Boosting Efficiency of Hole Conductor-Free Perovskite Solar Cells by Incorporating p-Type NiO Nanoparticles into Carbon Electrodes. *Sol. Energy Mater. Sol. Cells* **2018**, *178*, 164–169. [CrossRef]

218. Liu, S.; Cao, K.; Li, H.; Song, J.; Han, J.; Shen, Y.; Wang, M. Full Printable Perovskite Solar Cells Based on Mesoscopic TiO$_2$/Al$_2$O$_3$/NiO (Carbon Nanotubes) Architecture. *Sol. Energy* **2017**, *144*, 158–165. [CrossRef]

219. Bhandari, K.P.; Collier, J.M.; Ellingson, R.J.; Apul, D.S. Energy Payback Time (EPBT) and Energy Return on Energy Invested (EROI) of Solar Photovoltaic Systems: A Systematic Review and Meta-Analysis. *Renew. Sustain. Energy Rev.* **2015**, *47*, 133–141. [CrossRef]

220. Celik, I.; Mason, B.E.; Phillips, A.B.; Heben, M.J.; Apul, D. Environmental Impacts from Photovoltaic Solar Cells Made with Single Walled Carbon Nanotubes. *Environ. Sci. Technol.* **2017**, *51*, 4722–4732. [CrossRef] [PubMed]

221. Espinosa, N.; Serrano-Luján, L.; Urbina, A.; Krebs, F.C. Solution and Vapour Deposited Lead Perovskite Solar Cells: Ecotoxicity from a Life Cycle Assessment Perspective. *Sol. Energy Mater. Sol. Cells* **2015**, *137*, 303–310. [CrossRef]

222. Anctil, A.; Fthenakis, V. Life Cycle Assessment of Organic Photovoltaics. Third Generation Photovoltaics Vasilis Fthenakis. IntechOpen, 2006. Available online: https://www.intechopen.com/books/third-generation-photovoltaics/life-cycle-assessment-of-organic-photovoltaics (accessed on 17 September 2017).

223. Cleetus, R. Renewable Energy Surges Globally with China and India in the Lead. Available online: http://blog.ucsusa.org/rachel-cleetus/renewable-energy-china-india (accessed on 28 December 2017).

224. Arora, B. Rising Chinese Solar Prices May Put Indian Projects at Risk. Available online: https://www.bloombergquint.com/business/2017/08/20/rising-chinese-solar-panel-prices-may-put-projects-bid-at-record-low-tariff-at-risk (accessed on 28 December 2017).

225. India Solar Report. Bridge to India, 2017. Available online: http://www.bridgetoindia.com/reports/india-solar-handbook-2017/ (accessed on 5 February 2018).

226. Mahapatra, S. Renewable Energy Share Hits All-Time High in India-13.2% of Electricity. Available online: https://cleantechnica.com/2017/10/22/renewable-energy-share-hits-time-high-india-13-2-electricity/ (accessed on 5 February 2018).
227. Central Statistics Office, Government of India. Energy Statistics 2017. Available online: http://www.mospi.nic.in/sites/default/files/publication_reports/Energy_Statistics_2017r.pdf.pdf (accessed on 5 February 2018).
228. Central Electricity Authority, Ministry of Power, Government of India. Growth of Electricity Sector in India from 1947–2017. Available online: http://www.cea.nic.in/reports/others/planning/pdm/growth_2017.pdf (accessed on 5 February 2018).

Review

Emerging Characterizing Techniques in the Fine Structure Observation of Metal Halide Perovskite Crystal

Kongchao Shen [1,2], Jinping Hu [2], Zhaofeng Liang [2], Jinbang Hu [2], Haoliang Sun [2], Zheng Jiang [2] and Fei Song [2,*]

[1] Department of Physics, Zhejiang University, Hangzhou 310027, China; 11636007@zju.edu.cn
[2] Shanghai Institute of Applied Physics, Chinese Academy of Sciences, Shanghai 201204, China; hujinping@sinap.ac.cn (J.H.); liangzhaofeng@sinap.ac.cn (Z.L.); hujinbang@sinap.ac.cn (J.H.); sunhaoliang@sinap.ac.cn (H.S.); jiangzheng@sinap.ac.cn (Z.J.)
* Correspondence: songfei@sinap.ac.cn

Received: 30 March 2018; Accepted: 16 May 2018; Published: 23 May 2018

Abstract: Driven by its appealing application in the energy harvesting industry, metal halide perovskite solar cells are attracting increasing attention from various fields, such as chemistry, materials, physics, and energy-related industries. While the energy conversion efficiency of the perovskite solar cell is being investigated often by various research groups, the relationship between the surface structure and the property is still ambiguous and, therefore, becomes an urgent topic due to its wide application in the real environment. Recently, the fine structure characterization of perovskite crystals has been analysed by varying techniques, such as XRD, synchrotron-based grazing incidence XRD, XAFS, and STM, in addition to others. In this review article, we will summarize recent progresses in the monitoring of fine nanostructures of the surface and crystal structures of perovskite films, mainly by XAFS, XRD, and STM, focusing on the discussion of the relationship between the properties and the stability of perovskite solar cells. Furthermore, a prospective is given for the development of experimental approaches towards fine structure characterization.

Keywords: organic-inorganic crystal; surface structure; STM; calculation

1. Introduction

Metal halide perovskite structure has developed rapidly in the past decade since its first usage in solar cells. It is the champion of power conversion efficiency (PCE), increasing from 3.8% to more than 22% during recent years, while other types, such as Si-based, dye-sensitized, and organic solar cells, have had relatively less development in performance [1–4]. Perovskite films are commonly used in metal-halide solar cells with increasingly high PCE, however, the exact mechanism and the detailed structures of the perovskite film are still ambiguous. Consequently, much research has been performed on the stability [5–8], PCE [9,10], and toxicity [11,12] of the photosensitive film of the perovskite solar cell. In general, the perovskite solar cell is composed of the electrodes, electron transport layer, metal-halide perovskite film, and the hole transport layer. To reduce the influence of heterojunctions in solar cells, perovskite structures with single-crystal quality are a good candidate to detect related physic properties, such as giant photostriction [13], long range of electron-hole diffusion [14], structural and optoelectronic characteristics [15], harvesting of below-bandgap light absorption [16], optoelectronic properties [5], and ferroelectric and piezoelectric properties [17]. Moreover, perovskite crystals can also be utilized to detect related novel physical phenomenon, such as the quantification of re-absorption and re-emission processes, to determine photon recycling efficiency [18] and the detection of photons [19,20]. These investigations identifying such properties and physical phenomenon are performed in specific perovskite crystals. Furthermore, high-performance single-crystal and planar-type photodetector has also been fabricated using single-crystal metal halide perovskite [21], as well as the narrow-band perovskite single-crystal photodetectors [22].

It has been acknowledged that perovskite crystals play an important role in the further promotion of perovskite solar cells' performance and industrial usage, as perovskite crystals have the distinct advantage of high controllability. Therefore, using perovskite crystals to solve one of the main issues of instability in perovskite solar cells might be reasonable. The subsequent question, then, is how to determine the crystal quality and characterize the fine structure, accordingly. Presently, there are several state-of-the-art techniques to achieve the structure determination of perovskite crystals, which will, therefore, be discussed and summarized in this review paper. First, we briefly summarize the growth methods of perovskite crystals with considerable discussion on the quality of perovskite crystals, which are investigated mainly by XRD, GIXRD, and XAFS. This covers reports from the literature, in addition to our own work. Secondly, the surface structures of perovskite crystals are investigated by scanning tunneling microscopy (STM) and are analyzed and compared with reports from the literature. The advantage of STM is that it gives a direct morphology of the perovskite surface with angstrom resolution in real space.

2. Basic Structure of Perovskite Crystals

Before discussing the approaches of growth of single crystal perovskite, it is worth noting the unit cell of perovskite crystal structure (molecular formula: ABX_3, as depicted in Figure 1a). There are eight octahedral in one unit cell, where the cation, B, fits into an enclosed octahedral and the cation, A, is situated between the eight octahedral. The stability of the ABX_3 perovskite crystal structure can be determined by the tolerance factor (t), the range of which can be calculated by the formula (1) as shown below, where r_A, r_B, and r_X represent the corresponding ionic radii of the ABX_3 perovskite [23,24]. The site, A, usually represents the organic group (such as $CH_3NH_3^+$, $NH_2CH_2NH_3^+$, $C(NH_2)_3^+$, or Cs+). The cation, B, is typically Pb^{2+} or Sn^{2+}, while the anion, X, is the halide ion (Cl^-, Br^-, or I^-). Furthermore, the cation, B, can also be easily connected with the anion, X, to form the BX_6 octahedral ionic group [25]. In Figure 1b, tolerance factors of perovskite crystal structures are listed and, subsequently, we can conclude that the ideal tolerance factor is from 0.9 to 1.0. The t range, from 0.7 to 0.9, can also represent the crystal structure because of the smaller A or large B in the ABX_3 structure. If the factor t is larger than 1.0, this means that there may exist various layered crystal structures. In addition, the Goldschmidt tolerance factor (t) and octahedral factor (μ) are also necessary, and these two factors can also be used to predict the formability of perovskites [26–28].

$$t = \frac{r_A + r_X}{\sqrt{2}(r_B + r_X)} \tag{1}$$

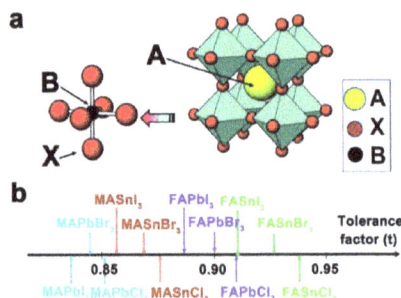

Figure 1. (a) Unit cell of ABX_3 with octahedral structure; (b) the tolerance factor of different crystal structures. Image is reproduced from Reference [24] with permission.

Apart from the phase (known as the α phase) of the perovskite structure shown in Figure 1, the crystal structure will also change in different temperature regions [23], such as the tetragonal phase (denoted as the β phase) and the orthorhombic phase (denoted as the γ phase) [29]. Accordingly, the phase transition in perovskite crystal structure can significantly alter its optical and electrical properties and, therefore, affect its applications significantly. In general, the most representative perovskite structure is MAPbI$_3$, and its phase change is summarized in Figure 2 [30]. Starting from the transitional point of phase change at around 330 K, the cubic perovskite structure changes into the tetragonal phase along with the point group from *Pm-3m* to *I4/mcm*. Contrastingly, the tetragonal changes into the orthorhombic phase, along with the structure phase evolution, from *I4/mcm* to *P4/mbm* when the temperature point drops below 160 K. Moreover, size-dependent orthorhombic-to-tetragonal phase transition has also been discovered by Li et al. [31]. Meanwhile, FA$_x$MA$_{1-x}$PbI$_3$ phase transition was discovered by Weber et al. [32]. It has been discovered that CsPbI$_3$ has two phases, α and δ, with the transition point at a temperature of approximately 600 K [33].

When single-crystal perovskite is exposed to ambient air dramatic changes occur, such as anisotropic moisture erosion of CH$_3$NH$_3$PbI$_3$ single crystals [34], and it can be estimated that the corresponding performance will be worsening over time [35]. The main explanation for this phenomenon is that the perovskite structure gets degraded and the crystal phase is destroyed by ambient molecules, which is confirmed by the observation of the color change in the appearance. Despite the structural degradation of perovskite crystals being well reported, little is known about what exactly happens at the nanoscale level.

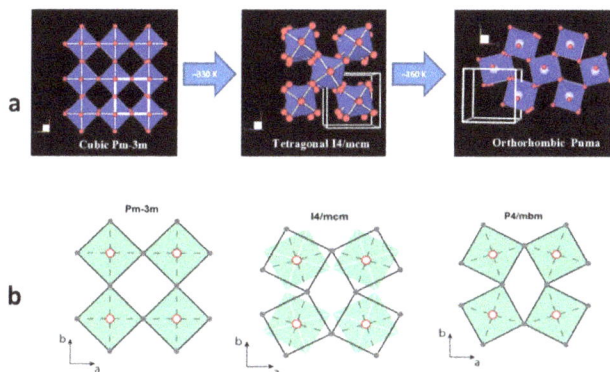

Figure 2. Crystal structures of MAPbI$_3$ at different temperatures. (**a**) The PbI$_6$ octahedral are blue and the iodine atoms are red, with MA cations not being shown to better highlight the distortions of the Pb-I network due to the structural phase transitions; (**b**) the relative rotations of neighboring layers of PbI$_6$ octahedral along the c axis are shown as filled green squares and unfilled black squares. Figures were taken from Reference [30] with permission.

Growth of ABX$_3$ Single Crystals

As discovered, the bottom seeded solution growth (BSSG) method is suitable to grow perovskite crystals from materials in particular solvents, which have low solubility at room temperature, but increasing solubility as temperature rises. The first MAPbI$_3$ bulk single crystal with centimeter-scale was grown using this approach (as shown in Figure 3a) [36]. It was discovered that preparation conditions, such as temperature and crystal seed, are important to the growth of MAPbI$_3$ crystals. Figure 3a,b show different sizes of MAPbI$_3$ single crystals using the same method, except different temperature ranges are utilized (a with dimensions of 10 mm × 10 mm × 8 mm, and b with the size of 12 mm × 12 mm × 7 mm, respectively) [21,36]. Meanwhile, another similar method, top seeded solution growth (TSSG), can produce MAPbI$_3$ single crystals with a size of 10 mm in length and 3.3 mm in height, as shown in Figure 3c. [14] Using the same approach, $CH_3NH_3SnI_3$ and $CH(NH_2)_2SnI_3$ can also be obtained with a size of approximately 10 mm (Figure 3d,e) [37].

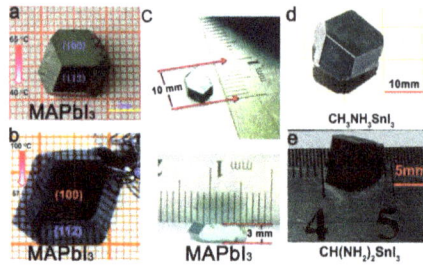

Figure 3. (**a,b**) Bottom seeded solution growth of MAPbI3 single crystal perovskites with different temperature ranges resulting in different sizes; (**c**) Top seeded solution growth of MAPbI3 single crystal; (**d**) MASnI3 single crystal and (**e**) $CH(NH_2)_2PbI_3$ single crystal. The figure was taken from References [14,21,36,37] with permission.

Moreover, the temperature-lowering method is simple, convenient, and applicable for the growth of high quality single crystal perovskites with even larger sizes, such as $MAPbX_3$ (X = Cl, Br, I), $MASnI_3$, and $FASnI_3$. However, it is not suitable for low solubility materials and precursors at high temperatures. Compared with the temperature-lowering method, the inverse temperature crystallization (ITC) method is suitable for those materials whose solubility in particular solvents is high at room temperature but decreases with increasing temperature. By employing the ITC method, millimeter-sized $MAPbX_3$ (X = Cl, Br, I) single crystals were obtained via different organic solvents as shown in Figures 4a [38] and 4b [39]. Similarly, millimeter-size $FAPbX_3$ (X = Br, I) single crystals were gained using the same ITC method (Figure 4c) [40]. To some extent, residual molecules from the solvent can remain inside crystals. [41,42]

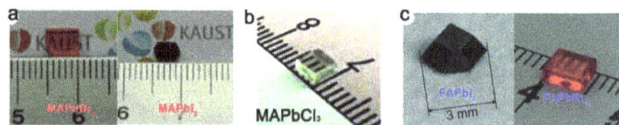

Figure 4. The inverse-temperature-crystallization method producing large-size crystals of (**a**) $MAPbX_3$ (**b**) MAPbCl3, and (**c**) FAPbX3. The figure was taken from refs [38–40] with permission.

The TSSP, BSSP, and ITC approaches are convenient to obtain single perovskite crystals and, therefore, has become the most widely used techniques in the laboratory. Despite this, other methods of growing perovskite crystals also exist. Firstly, millimeter-scale $MAPbBr_3$ crystals can be obtained by the anti-solvent vapor-assisted crystallization (AVC) method within one week [43], while single crystals (dimensions > 4 mm) were grown on the colorless $(C_6H_5C_2H_4NH_3)_2PbCl_4$ plate within several weeks [44]. Secondly, using the slow evaporation method, hybrid perovskite analogue (benzylammonium)$_2$PbX$_4$ (X = Cl, Br) crystals, with the dimensions of $5 \times 10 \times 2$ mm^3, were obtained (Figure 5a) [45]. Thirdly, by using the droplet-pinned crystallization (DPC) measures, micrometer-scale single crystals $CH_3NH_3PbI_3$ were formed under a conventional, but efficient, preparation process, which also provided the possibility to grow single crystalline thin film (Figure 5b) [46]. Fourthly, using the combined solution process and the vapor-phase conversion method, 2D MAPbI3 perovskite nanosheets were prepared via a twostep process by Liu et al. (Figure 5c) [47]. Last but not least, the hot casting method was also adopted for the growth of millimeter perovskite crystals [48].

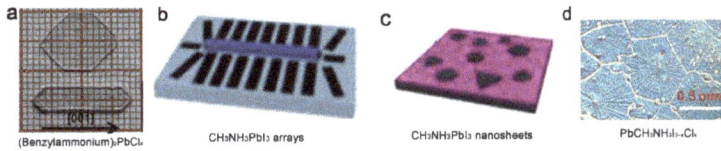

Figure 5. (**a**) Slow evaporation method produces the (benzylammonium)$_2$PbCl$_4$ perovskite single crystal; (**b**) Arrays of CH$_3$NH$_3$PbI$_3$ single crystal; (**c**) 2D MAPbI$_3$ perovskite nanosheets; (**d**) Milimeter-scale crystal grain [48]. The figure was taken from ref [45–48] with permission.

Presently, the most commonly used techniques for crystallization determination and fine structure exploitation are X-ray diffractions and other related measures, such as two-dimensional XRD (2DXRD) and grazing-incidence XRD. When the detecting area on the perovskite surface presents with single-crystal crystallization, patterns with short robs or dots will be seen in the 2D-XRD or GIXRD measurements, indicating the presence of a reciprocal lattice, as seen in Figure 6a [49]. In addition, we can also gather in-plane and out-of-plane information, as explained in Figure 6b,c. When polycrystalline orientation (Figure 6d) exists in the detecting area, a faint ring connecting short robs will be shown with a certain reciprocal space, q. Moreover, extra short robs also appear in the XRD pattern, which can be attributed to the formation of single crystals, but with different orientations. Compared to the pattern with short robs and faint rings, the complete ring presented in Figure 6e without short robs indicates that there are many single crystals with different orientations formed in the detected sample [50]. The difference between Figure 6d,e is the discrimination in growth methods, and it can be seen that the hot cast approach apparently improves the crystallinity of perovskite film and results in a better PCE.

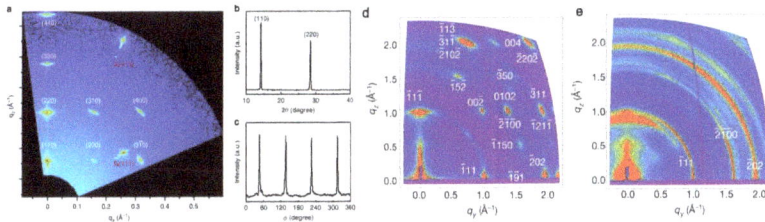

Figure 6. (**a**) 2D XRD of single-crystal patterned perovskite thin films prepared by the geometrically confined lateral crystal growth process; (**b**) Out-of-plane XRD scan. Diffraction peaks are consistent with the diffraction spots along the q$_z$ line obtained by 2D XRD; (**c**) In-plane scan obtained at the fixed angle of (002) plane. The four sharp peaks demonstrate the four-fold symmetry of tetragonal perovskite; (**d**) Hot-cast and (**e**) near-single-crystalline (BA)$_2$(MA)$_3$Pb$_4$I$_{13}$ perovskite films with Miller indices of the most prominent peaks are shown. Color scale is proportional to the X-ray scattering intensity. The figure was taken from References [49,50] with permission.

While XRD gives the direct crystallization information of the perovskite structure, the fine structure inside perovskite crystals can also be determined by X-ray absorption fine structure spectroscopy (XAFS). This is achieved by fitting the raw data measured at the Pb L$_3$-edge, in the CH$_3$NH$_3$PbI$_{3-x}$Cl$_x$ crystal, to get the information of the coordination number of Pb and the bonding length between Pb and I. Parameters obtained by fitting the Pb-L$_3$ EXAFS shows that there are vacancies and defects existing in this kind of perovskite crystal. Specifically, the multiple-shell fitting is better than the single-shell fitting, both in real space (R space) and reciprocal space (k space), as shown in Figure 7 [51]. Encouragingly, the XAFS approach can see more inside the perovskite crystal structure with complex unit cells, and get the atomic structure information, such as defects and so on.

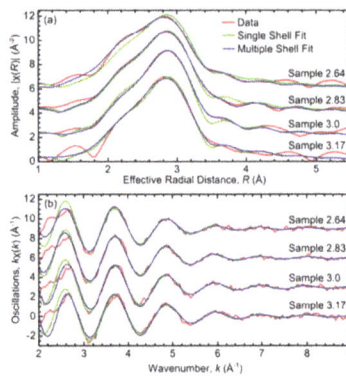

Figure 7. EXAFS spectra and fits in (**a**) R-space and (**b**) k-space for all samples. The figure was taken from Reference [51] with permission.

3. Surface Structure of ABX₃ Detecting by STM

Although ABX$_3$ perovskite crystals have been fabricated with a millimeter scale by using the different approaches discussed above, this is limited by the short lifetime of perovskite crystals in ambient atmosphere or the dirty crystal surface, and the surface morphology of perovskite crystals with atomic resolution was seldom mentioned. Initially, one method was proposed, with the codeposition of CH_3NH_3I and PbI_2, to form $CH_3NH_3PbI_3$ followed by in-situ imaging with STM [52]. The (001) surface of the orthorhombic MAPbI3 crystals were terminated either by MA-I or by Pb-I layers. As shown in Figure 8a–c, as the dimer and the zigzag structure are combined and potentially converted into each other at the (001) surface, the observed (001) surface by STM is, consequently, MA-I-terminated and bright protrusions in the STM images are assigned to iodine anions, due to the negligible contribution of electronic states near E_F from MA anions [53]. The detailed structure is shown in Figure 8d,e. Interestingly, all these results were supported by theoretical calculation. It is later discovered that the codeposition of PbI_2 and CH_3NH_3I is an ideal method for in-situ STM, and other measurements, in ultra-high vacuum (UHV). Koresh et al. suggested that the surface termination of the $CH_3NH_3PbBr_3$ (001) crystal surface is the methylammonium bromide (CH_3NH_3Br) layer, supported by angle resolved photoelectron spectroscopy (ARPES) and theoretical calculation [54]. The surface termination, together with the structural stability and electronic structures of MAPbI$_3$, that have been investigated so far are summarized in ref [55]. Haruyama et al. also reported the surface properties of $CH_3NH_3PbI_3$ perovskite crystals with STM and other UHV measurements [56], while the surface-related properties of $CH_3NH_3PbI_3$ perovskite thin film was also reported by Afzaal et al. [57].

Besides the growth of different kinds of ABX$_3$ perovskite crystals in UHV, cleaved surfaces from perovskite crystals have also been used in UHV. Interestingly, Márton Kollár presented a method which can guarantee the flat and clean surface of the $CH_3NH_3PbBr_3$ crystal and allows surface sensitive measurements., which is needed for the understanding and further engineering of this material family [58]. Furthermore, other in-situ cleave methods have also been used during STM experiments [59]. Ohmann et al. utilized in-situ cleaving of $CH_3NH_3PbBr_3$ crystal to exploit the molecular orientation on the perovskite surface. Exact orientation of the molecular dipole within the lattice, and especially the interplay of methylammonium (MA) groups with hosting anions detected by STM, shows that such perovskite crystals keeps the phenomenon of modified arrangements of atoms and molecules on the surface, and have structurally and electronically distinct domains with ferroelectric and antiferroelectric characteristics. All these behaviors can be explained by surface reconstruction and a substantial interplay of the orientations of the polar organic cations, $(CH_3NH_3)^+$, with the position of the hosting anions. The first observation of such surface reconstruction was very similar to the report produced by Limin She [52]. As an example, two surface structures (zigzag and dimer) are shown in Figure 9a,c. Figure 9b,d are simulated STM images of the (010) plane of the orthorhombic crystal with surface reconstruction. The ferroelectric property of this kind of perovskite is different in various domains, as shown in Figure 9e. Meanwhile, we can also discover, from Figure 9f,g, the same defect, which was discussed and shown in Figure 8. In principle, STM measurements can reveal surface morphologies in real space with atomic resolution, as converted from measured electronic states and tunneling current. However, the limited-scale measurement (usually within micrometer) prevents the investigation of large-size perovskite crystals. Therefore, a complementary method to STM, such as angle resolved photoelectron spectroscopy (ARPES), will help to obtain the electronic structure in reciprocal space and the averaged information for large samples [54].

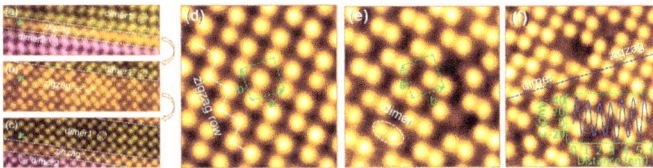

Figure 8. STM images of MAPbI₃ crystal films deposited on an Au(111) crystal surface. (**a–c**) Sequential images acquired at the same region showing the reversible transition between the dimer and zigzag structures (4.2 × 12.8 nm²; U = 2.0, 0.85, and −1.25 V; I = 30 pA). Phase boundaries are denoted by dotted lines. Iodine zigzag rows and dimers are denoted by dashed lines and ellipses, respectively; (**d,e**) Zoom-in view of the zigzag and dimer structures (4.3 × 4.3 nm²; 2.5 V; 50 pA). The unit cell is denoted by dashed rectangles; (**f**) STM image of the two phases coexisting at the same region (5.6 × 5.6 nm²; U = 2.5 V; I = 50 pA). The inset is the height profile along the dashed line. The figure was taken from Reference [52] with permission.

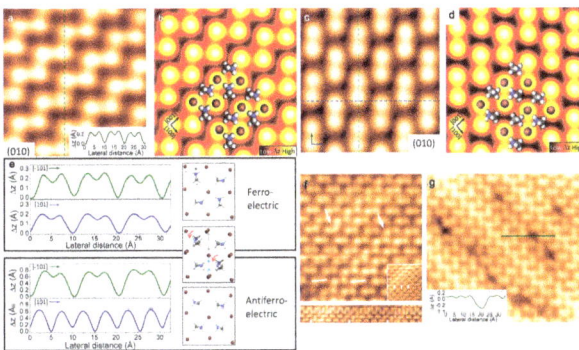

Figure 9. (**a**) Atomically resolved STM topography of the perovskite surface. Inset: Height profile along dashed line in the image. Image size: 31 × 31 Å² (U = −5 V, I = 0.1 nA); (**b**) Simulated STM image of the (010) plane of the orthorhombic crystal with surface reconstruction; (**c**) Image size: 31 × 31 Å² (U = −9 V, I = 0.1 nA); The arrows and dashed lines indicate the direction of the height profiles shown in part e; (**d**) Corresponding calculated image; (**e**) Comparison of height profiles and the corresponding model. The arrows indicate the rotational and positional change of the MA molecules and the Br ions, respectively, between the ferroelectric and antiferroelectric domain; (**f**) Start of the dislocation rows indicated by the white arrow. The angled lines indicate the modified Br atom positions adjacent to a dislocation row. Inset: Three dislocation rows beside each other. Bottom: Periodic arrangement of the dislocation rows; (**g**) STM image of defects on the surface. Image sizes: (**f**) 64 × 64 Å², inset 56 × 55 Å², bottom 164 × 23 Å²; (**g**) 92 × 103 Å². U = −9 V, I = 0.1 nA. The figure was taken from Reference [59] with permission.

As discovered, both approaches of codeposition of PbI_2 and CH_3NH_3I, and cleaving from the pristine perovskite crystals cannot avoid surface defects, which have been resolved by STM. However, this raises the question of: how these defects will affect the physical properties of perovskite crystals. Liu et al. revealed the influence of defects in the crystal, as discussed in Figure 10 [60]. Vacancy sites, V_{Br}, V_{MA}, V_{Br-Pb}, and V_{Br-MA}, can be formed easily when water molecules are adsorbed on the (010) surface of the $CH_3NH_3PbBr_3$ perovskite crystal because water reduces the formation energy of vacancy sites. On the other hand, the pristine $CH_3NH_3PbBr_3$ (010) surface is shown to be quite inert toward the adsorption of molecules or atoms. Contrastingly, in the presence of vacancy sites, the adsorption energies of water, oxygen, and acetonitrile molecules are significantly increased due to the formation of hydrogen bonds between the adsorbates and the defective surface. It is expected that, in contrast to the pristine surface, structural decomposition of perovskite crystals is more likely to occur on the surface vacancy sites [60]. The stability of perovskite crystals or films is, therefore, very sensitive to water or other adsorbed molecules. In addition, polar solvents can also degrade the surface of perovskite crystals [61]. Dramatic changes, such as crystal structure and band gap, will appear on the surface of perovskite crystals after exposure to water, as suggested by Kamat et al. [62]. The formula of the chemical reaction between perovskite and water has been discussed in detail in ref. [7], however, oxygen is not the main contaminant. If the solar cells are stored in dark and dry air conditions, there will not be any noticeable degradation [63,64].

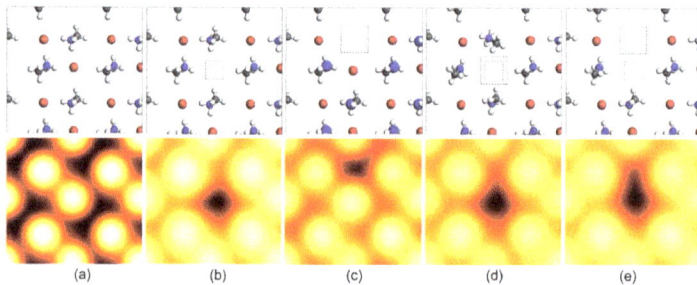

Figure 10. Simulated STM images of (**a**) the pristine $CH_3NH_3PbBr_3$ (010) surface and (**b**) the surface with a Br vacancy V_{Br}; (**c**) a MA vacancy V_{MA}; (**d**) a Br-Pb double vacancy V_{Br-Pb}; and (**e**) a Br-MA double vacancy V_{Br-MA}. All these STM images were calculated with a W(111) tip model. The squares show the position of missing Br, Pb atoms or MA cations. $V_{bias} = -3.0$ V. Color code: N (blue), C (gray), H (white), Br (brown). The figure was taken from Reference [60] with permission.

After the discussion of surface defects existing in perovskite crystal, which affects the stability of perovskite, the surface morphology of perovskite crystals exposed in ambient conditions will be discussed. The aged surfaces (from as-grown single crystals) of $CH_3NH_3PbBr_3$ are shown in Figure 11a,b, in which the period of the crystal can be seen clearly, while Figure 11c,d shows the surface structures of pristine crystals, which show 2D crystals or robs-like crystals. More detailed information is shown in Figure 11e,f with a magnified view. When the pristine surface was exposed in ambient conditions, the surface structure changed significantly, which originates from hydration-induced morphological changes. However, the reason that the water molecules cause surface structure restructuring is still unknown. Notably, surface reconstruction on aged surfaces (as visualized from atomic-scale scanning tunneling microscopy) might lead to changes in the composition and optical band gap, as well as the degradation of carrier dynamics, photocurrent, and solar cell device performance [65]. Additionally, ionic defects traveling through gaps between perovskite crystal grains to erode the perovskite's stability, as well as the band gap types (either direct or indirect band gap), will both influence the lifetime of charge carriers, the diffusion length, and the efficient absorption of perovskite semiconductors [66]. Therefore, a decreased grain boundary is favored for improving the perovskite film's stability and the corresponding solar cell's performance, and a similar relativistic effect has also been reported using theoretical calculation [67–69].

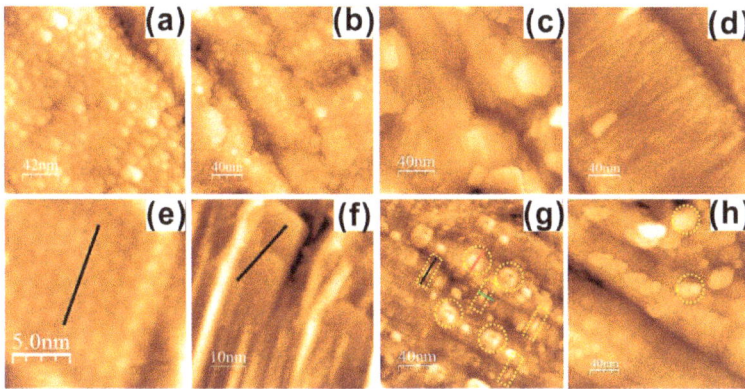

Figure 11. (**a**,**b**) STM image of the aged surface from the as-grown single crystal; (**c**,**d**) Magnified view at various positions from the pristine crystal; (**e**,**f**) Ordered structures having well-aligned stacked planes from the pristine perovskite crystal surface; (**g**,**h**) STM images showing the hydrate formation after overnight exposure to ambient air with ring-like protrusions (marked by circles) and the disordered 1D chain (marked by rectangles), indicating molecular water incorporation in the pristine surface. Scanning parameters: V_b = 2.5 V and It = 0.7 nA. The figure was taken from Reference [65] with permission.

The alignment of MA groups is theoretically predicted to play a decisive role in the structural and electronic properties of the perovskite, and the exact orientation of the MA ions on the surface has been precisely determined by STM. This is important for the development of stable perovskite materials. Promoting the development of perovskite solar cells toward their theoretical PCE requires delicate control over the charge carrier dynamics throughout the whole perovskite solar cell device. Based on a complementary understanding of the perovskite device's hybrid mechanism, the elaborate control over the carrier behavior in the perovskite film, especially across interfaces between different materials, is expected [70]. With better characterization of surface structures with XRD, STM, and so on, improved electronic properties of these materials could be gained by structure optimization from optimized growth methods, thus promoting device performance advantages. In pursuit of higher PCE, perovskite solar cells will definitely require an optimized carrier transport pathway, which is related to all device layers and corresponding interfaces. Presumably, flat perovskite crystal surfaces with less grain boundaries and stable surface structures will facilitate the charge extraction at both interfaces adjacent to the ETL, which is, essentially, helpful for obtaining a higher PCE. Moreover, atomic-scale defects can strongly influence the behavior of the solar cell absorber, and we have learned from the literature that the decomposition of perovskite crystal structures usually starts from surface defects. The identification of surface defects and their chemical properties should provide useful hints to understanding the chemical insights of the poor stability of $CH_3NH_3PbX_3$ perovskite crystals, which is the key origin of the device with high PCE but not long-term stability.

Based on these reports and investigations, a series of methods has, therefore, been proposed to improve the stability of perovskite films and crystals. For example, Long et al. report a high crystallized $MAPbI_3$ perovskite crystal with texture structure as prepared from $HPbI_3$ reacting with low partial pressure MA gas, which demonstrates substantially higher thermal and moisture stability compared to that of polycrystalline perovskite prepared from $MAI+PbI_2$ [8]. These observations suggest that the crystal-like perovskite sensitizer in solar cells possesses better performance than the three-dimensional (3D) perovskite structure, which can be utilized as an efficient approach to improve the stability of perovskite crystals. Moreover, external doping with other organics or inorganics can also improve the stability of perovskite [70,71] crystals, as well as the coordination between the perovskite film and the adjacent film [72,73]. Nevertheless, stability is still one of the crucial issues in the development of perovskite solar cells. The 3D organic-inorganic perovskites are one of the most appealing thin-film solar-cell materials as they can absorb light over a broad range of solar spectrum wavelengths. However, a photovoltaic material needs to be stable for long periods when exposed to sunlight, and this is unfortunately not the case for 3D perovskites. Ruddlesden–Popper phases (layered 2D perovskite films) [50], on the other hand, are photostable, but only have a poor PCE, which is because organic cations in the material act as insulating spacing layers between the conducting inorganic slabs present and so inhibit charge transport in out-of-plane layers. Nevertheless, PCE is obviously another important factor for perovskite devices. Consequently, an ideal photovoltaic material should have both high PCE and long-term stability. Based on these requirements, 2D halide perovskites, formed on the 3D perovskite structure to passivate interfacial defects and vacancies, and enhance moisture tolerance, seems to be a promising approach. Interestingly, hybrid 3D/2D perovskite films possess longer photoluminescence lifetimes, as well as lower trap state densities by the passivation of cationic and halide vacancies on the surface or grain boundaries, thereby reducing the non-radiative recombination pathways. More importantly, the hybrid 3D/2D perovskite exhibits higher ambient stability than a pure 3D perovskite where the hydrophobic nature of the long aliphatic carbon chains in the 2D perovskite provide an additional moisture repelling effect to the entire perovskite film.

4. Overlook and Prospective

Typically, great crystallinity is essential to the performance of the perovskite solar cell. In this review paper, the fabrication of single-crystal perovskite structure has been briefly discussed, followed by the fine structure determination with XRD and GIXRD, which are the most common and efficient characterization methods, currently. Secondly, the surface properties of fabricated perovskite crystals, including termination layer, ferroelectric and anti-ferroelectric effects induced, surface defects, and stability, have been elaborately discussed, to reveal the relationship between surface defects and structure stability via STM at a nanoscale level, as well as the surface structure effect on the performance of perovskite solar cell. However, limited by the instability of the perovskite single-crystal structure, there is still a paucity of research investigating the single-crystal surface of the metal halo perovskite structure by STM.

As generally agreed, fine structure engineering is a key point to further improve the power convention efficiency of perovskite solar cells. Consequently, substantial development of fine-structure characterization approaches will continually be a well-pursued topic in the field of organic-inorganic perovskite solar cells, especially in in-situ and operando conditions. X-ray fine structure absorption spectroscopy (XAFS) should be considerably employed in the future to characterize the detailed information behind crystal structure, such as the exact coordination information of metal ions in the hybridized perovskite structure (Pb, Cu, Sn, Cs, and so on), the bonding length, and the chemical state as-prepared and under operando situation. Moreover, atomic force microscopy (AFM) can also be widely utilized as a supplement to STM, which can measure the surface morphology in air or in situ. Scanning electron microscopy (SEM) has been widely used in the structure determination of perovskite films, however, its spatial resolution is still a limitation for fine structure characterization. Synchrotron-based scanning transmission electron microscopy can be a very helpful tool for both fine structure mapping and spatial element identification, which has not been mentioned in current literature. Besides XAFS, the promotion of other structure characterization methods under operando conditions is an appealing direction for research that should not be ignored. One possible way could be the combination of absorption spectroscopy and electrical measurements. Of course, X-ray photoelectron spectroscopy (XPS) together with ultraviolet photoelectron spectroscopy (UPS) are greatly helpful for the surface and interface electronic structure determination, such as the energy level alignment, valence band structure, and core level state, which are closely connected to the electrical properties of perovskite solar cells.

Acknowledgments: This work is supported by National Natural Science Foundation (91545101, U1732267), The National Key Research and Development Program of China (2016YFA0401302) and The Hundred Talents Program of the Chinese Academy of Sciences.

Conflicts of Interest: The authors declare no conflicts of interest.

References

1. Kojima, A.; Teshima, K.; Shirai, Y.; Miyasaka, T. Organometal Halide Perovskites as Visible-Light Sensitizers for Photovoltaic Cells. *J. Am. Chem. Soc.* **2009**, *131*, 6050–6051. [CrossRef] [PubMed]
2. Kromdijk, J.; Głowacka, K.; Leonelli, L.; Gabilly, S.T.; Iwai, M.; Niyogi, K.K.; Long, S.P. Improving Photosynthesis and Crop Productivity by Accelerating Recovery From Photoprotection. *Science* **2016**, *354*, 857–861. [CrossRef] [PubMed]
3. Xu, Q.; Yuan, D.; Mu, H.; Igbari, F.; Bao, Q.; Liao, L. Efficiency Enhancement of Perovskite Solar Cells by Pumping Away the Solvent of Precursor Film Before Annealing. *Nanoscale Res. Lett.* **2016**, *11*, 248. [CrossRef] [PubMed]
4. Yang, S.; Fu, W.; Zhang, Z.; Chen, H.; Li, C. Recent Advances in Perovskite Solar Cells: Efficiency, Stability and Lead-Free Perovskite. *J. Mater. Chem. A* **2017**, *5*, 11462–11482. [CrossRef]
5. Li, W.; Rao, H.; Chen, B.; Wang, X.; Kuang, D. A Formamidinium-Methylammonium Lead Iodide Perovskite Single Crystal Exhibiting Exceptional Optoelectronic Properties and Long-Term Stability. *J. Mater. Chem. A* **2017**, *5*, 19431–19438. [CrossRef]
6. Asghar, M.I.; Zhang, J.; Wang, H.; Lund, P.D. Device Stability of Perovskite Solar Cells—A Review. *Renew. Sustain. Energy Rev.* **2017**, *77*, 131–146. [CrossRef]
7. Leijtens, T.; Eperon, G.E.; Noel, N.K.; Habisreutinger, S.N.; Petrozza, A.; Snaith, H.J. Stability of Metal Halide Perovskite Solar Cells. *Adv. Energy Mater.* **2015**, *5*, 1500963. [CrossRef]
8. Long, M.; Zhang, T.; Zhu, H.; Li, G.; Wang, F.; Guo, W.; Chai, Y.; Chen, W.; Li, Q.; Xu, J.; et al. Textured CH$_3$NH$_3$PbI$_3$ Thin Film with Enhanced Stability for High Performance Perovskite Solar Cells. *Nano Energy* **2017**, *33*, 485–496. [CrossRef]
9. Zhang, F.; Wang, Z.; Zhu, H.; Pellet, N.; Luo, J.; Yi, C.; Liu, X.; Liu, H.; Wang, S.; Xiao, Y.; et al. Over 20% PCE Perovskite Solar Cells with Superior Stability Achieved by Novel and Low-Cost Hole-Transporting Materials. *Nano Energy* **2017**, *41*, 469–475. [CrossRef]
10. Seok, S.I.; Grätzel, M.; Park, N. Methodologies Toward Highly Efficient Perovskite Solar Cells. *Small* **2018**, 1704177. [CrossRef] [PubMed]
11. Babayigit, A.; Ethirajan, A.; Muller, M.; Conings, B. Toxicity of Organometal Halide Perovskite Solar Cells. *Nat. Mater.* **2016**, *15*, 247–251. [CrossRef] [PubMed]
12. Giustino, F.; Snaith, H.J. Toward Lead-Free Perovskite Solar Cells. *ACS Energy Lett.* **2016**, *1*, 1233–1240. [CrossRef]
13. Zhou, Y.; You, L.; Wang, S.; Ku, Z.; Fan, H.; Schmidt, D.; Rusydi, A.; Chang, L.; Wang, L.; Chen, L.; et al. Giant Photostriction in Organic-Inorganic Lead Halide Perovskites. *Nat. Commun.* **2016**, *7*, 11193. [CrossRef] [PubMed]
14. Dong, Q.F.; Fang, Y.; Shao, Y.; Mulligan, P.; Qiu, J.; Cao, L.; Huang, J. Electron-Hole Diffusion Lengths > 175 mm in Solution-Grown CH$_3$NH$_3$PbI$_3$ Single Crystals. *Science* **2015**, *347*, 967–970. [CrossRef] [PubMed]
15. Park, N. Crystal Growth Engineering for High Efficiency Perovskite Solar Cells. *Crystengcomm* **2016**, *18*, 5977–5985. [CrossRef]
16. Chen, Z.; Dong, Q.; Liu, Y.; Bao, C.; Fang, Y.; Lin, Y.; Tang, S.; Wang, Q.; Xiao, X.; Deng, Y.; et al. Thin Single Crystal Perovskite Solar Cells to Harvest Below-Bandgap Light Absorption. *Nat. Commun.* **2017**, *8*, 1890. [CrossRef] [PubMed]
17. Ding, R.; Zhang, X.; Sun, X.W. Organometal Trihalide Perovskites with Intriguing Ferroelectric and Piezoelectric Properties. *Adv. Funct. Mater.* **2017**, *27*, 21. [CrossRef]

18. Fang, Y.; Wei, H.; Dong, Q.; Huang, J. Quantification of Re-Absorption and Re-Emission Processes to Determine Photon Recycling Efficiency in Perovskite Single Crystals. *Nat. Commun.* **2017**, *8*, 14417. [CrossRef] [PubMed]

19. Yakunin, S.; Dirin, D.N.; Shynkarenko, Y.; Morad, V.; Cherniukh, I.; Nazarenko, O.; Kreil, D.; Nauser, T.; Kovalenko, M.V. Detection of Gamma Photons Using Solution-Grown Single Crystals of Hybrid Lead Halide Perovskites. *Nat. Photonics* **2016**, *10*, 585–589. [CrossRef]

20. Dirin, D.N.; Cherniukh, I.; Yakunin, S.; Shynkarenko, Y.; Kovalenko, M.V. Solution-Grown CsPbBr$_3$ Perovskite Single Crystals for Photon Detection. *Chem. Mater.* **2016**, *28*, 8470–8474. [CrossRef] [PubMed]

21. Lian, Z.; Yan, Q.; Lv, Q.; Wang, Y.; Liu, L.; Zhang, L.; Pan, S.; Li, Q.; Wang, L.; Sun, J. High-Performance Planar-Type Photodetector On (100) Facet of MAPbI$_3$ Single Crystal. *Sci. Rep.* **2015**, *5*, 16563. [CrossRef] [PubMed]

22. Fang, Y.; Dong, Q.; Shao, Y.; Yuan, Y.; Huang, J. Highly Narrowband Perovskite Single-Crystal Photodetectors Enabled by Surface-Charge Recombination. *Nat. Photonics* **2015**, *9*, 679–686. [CrossRef]

23. Li, C.; Lu, X.; Ding, W.; Feng, L.; Gao, Y.; Guo, Z. Formability of ABX$_3$ (X = F, Cl, Br, I) Halide Perovskites. *Acta Crystallogr. B* **2008**, *64*, 702–707. [CrossRef] [PubMed]

24. Fan, Z.; Sun, K.; Wang, J. Perovskites for Photovoltaics: A Combined Review of Organic-Inorganic Halide Perovskites and Ferroelectric Oxide Perovskites. *J. Mater. Chem. A* **2015**, *3*, 18809–18828. [CrossRef]

25. Saidaminov, M.I.; Mohammed, O.F.; Bakr, O.M. Low-Dimensional-Networked Metal Halide Perovskites: The Next Big Thing. *ACS Energy Lett.* **2017**, *2*, 889–896. [CrossRef]

26. Li, C.; Soh, K.C.K.; Wu, P. Formability of ABO$_3$ Perovskites. *J. Alloys Compd.* **2004**, *372*, 40–48. [CrossRef]

27. Sun, Q.; Yin, W. Thermodynamic Stability Trend of Cubic Perovskites. *J. Am. Chem. Soc.* **2017**, *139*, 14905–14908. [CrossRef] [PubMed]

28. Li, Z.; Yang, M.; Park, J.S.; Wei, S.H.; Berry, J.J.; Zhu, K. Stabilizing Perovskite Structures by Tuning Tolerance Factor: Formation of Formamidinium and Cesium Lead Iodide Solid-State Alloys. *Chem. Mater.* **2016**, *28*, 284–292. [CrossRef]

29. Feng, J.; Xiao, B. Crystal Structures, Optical Properties, and Effective Mass Tensors of CH3 NH$_3$PbX$_3$ (X = I and Br) Phases Predicted from HSE06. *J. Phys. Chem. Lett.* **2014**, *5*, 1278–1282. [CrossRef] [PubMed]

30. Whitfield, P.S.; Herron, N.; Guise, W.E.; Page, K.; Cheng, Y.Q.; Milas, I.; Crawford, M.K. Structures, Phase Transitions and Tricritical Behavior of the Hybrid Perovskite Methyl Ammonium Lead Iodide. *Sci. Rep.* **2016**, *6*, 35685. [CrossRef] [PubMed]

31. Li, D.; Wang, G.; Cheng, H.; Chen, C.; Wu, H.; Liu, Y.; Huang, Y.; Duan, X. Size-Dependent Phase Transition in Methylammonium Lead Iodide Perovskite Microplate Crystals. *Nat. Commun.* **2016**, *7*, 11330. [CrossRef] [PubMed]

32. Weber, O.J.; Charles, B.; Weller, M.T. Phase Behaviour and Composition in the Formamidinium-Methylammonium Hybrid Lead Iodide Perovskite Solid Solution. *J. Mater. Chem. A* **2016**, *4*, 15375–15382. [CrossRef]

33. Dastidar, S.; Hawley, C.J.; Dillon, A.D.; Gutierrez-Perez, A.D.; Spanier, J.E.; Fafarman, A.T. Quantitative Phase-Change Thermodynamics and Metastability of Perovskite-Phase Cesium Lead Iodide. *J. Phys. Chem. Lett.* **2017**, *8*, 1278–1282. [CrossRef] [PubMed]

34. Lv, Q.; He, W.; Lian, Z.; Ding, J.; Li, Q.; Yan, Q. Anisotropic Moisture Erosion of CH3 NH$_3$PbI$_3$ Single Crystals. *Crystengcomm* **2017**, *19*, 901–904. [CrossRef]

35. Grancini, G.; D'Innocenzo, V.; Dohner, E.R.; Martino, N.; Srimath Kandada, A.R.; Mosconi, E.; De Angelis, F.; Karunadasa, H.I.; Hoke, E.T.; Petrozza, A. CH$_3$NH$_3$PbI$_3$ Perovskite Single Crystals: Surface Photophysics and their Interaction with the Environment. *Chem. Sci.* **2015**, *6*, 7305–7310. [CrossRef] [PubMed]

36. Dang, Y.; Liu, Y.; Sun, Y.; Yuan, D.; Liu, X.; Lu, W.; Liu, G.; Xia, H.; Tao, X. Bulk Crystal Growth of Hybrid Perovskite Material CH$_3$NH$_3$PbI$_3$. *Crystengcomm* **2015**, *17*, 665–670. [CrossRef]

37. Dang, Y.; Zhou, Y.; Liu, X.; Ju, D.; Xia, S.; Xia, H.; Tao, X. Formation of Hybrid Perovskite Tin Iodide Single Crystals by Top-Seeded Solution Growth. *Angew. Chem. Int. Ed.* **2016**, *55*, 3447–3450. [CrossRef] [PubMed]

38. Saidaminov, M.I.; Abdelhady, A.L.; Murali, B.; Alarousu, E.; Burlakov, V.M.; Peng, W.; Dursun, I.; Wang, L.; He, Y.; Goriely, A.; et al. High-Quality Bulk Hybrid Perovskite Single Crystals within Minutes by Inverse Temperature Crystallization. *Nat. Commun.* **2015**, *6*, 7586. [CrossRef] [PubMed]

39. Maculan, G.; Sheikh, A.D.; Abdelhady, A.L.; Saidaminov, M.I.; Haque, M.A.; Murali, B.; Alarousu, E.; Mohammed, O.F.; Wu, T.; Bakr, O.M. $CH_3NH_3PbCl_3$ Single Crystals: Inverse Temperature Crystallization and Visible-Blind UV-Photodetector. *J. Phys. Chem. Lett.* **2015**, *6*, 3781–3786. [CrossRef] [PubMed]

40. Saidaminov, M.I.; Abdelhady, A.L.; Maculan, G.; Bakr, O.M. Retrograde Solubility of Formamidinium and Methylammonium Lead Halide Perovskites Enabling Rapid Single Crystal Growth. *Chem. Commun.* **2015**, *51*, 17658–17661. [CrossRef] [PubMed]

41. Hao, F.; Stoumpos, C.C.; Liu, Z.; Chang, R.P.H.; Kanatzidis, M.G. Controllable Perovskite Crystallization at a Gas-Solid Interface for Hole Conductor-Free Solar Cells with Steady Power Conversion Efficiency Over 10%. *J. Am. Chem. Soc.* **2014**, *136*, 16411–16419. [CrossRef] [PubMed]

42. Guo, Y.; Shoyama, K.; Sato, W.; Matsuo, Y.; Inoue, K.; Harano, K.; Liu, C.; Tanaka, H.; Nakamura, E. Chemical Pathways Connecting Lead(II) Iodide and Perovskite via Polymeric Plumbate(II) Fiber. *J. Am. Chem. Soc.* **2015**, *137*, 15907–15914. [CrossRef] [PubMed]

43. Shi, D.; Adinolfi, V.; Comin, R.; Yuan, M.; Alarousu, E.; Buin, A.; Chen, Y.; Losovyj, Y. Solar Cells. Low Trap-State Density and Long Carrier Diffusion in Organolead Trihalide Perovskite Single Crystals. *Science* **2015**, *347*, 519–522. [CrossRef] [PubMed]

44. Mitzi, D.B. A Layered Solution Crystal Growth Technique and the Crystal Structure of $(C_6H_5C_2NH_3)_2$ $PbCl_4$. *J. Solid State Chem.* **1999**, *145*, 694–704. [CrossRef]

45. Liao, W.; Zhang, Y.; Hu, C.; Mao, J.; Ye, H.; Li, P.; Huang, S.D.; Xiong, R. A Lead-Halide Perovskite Molecular Ferroelectric Semiconductor. *Nat. Commun.* **2015**, *6*, 7338. [CrossRef] [PubMed]

46. Ye, T.; Fu, W.; Wu, J.; Yu, Z.; Jin, X.; Chen, H.; Li, H. Single-Crystalline Lead Halide Perovskite Arrays for Solar Cells. *J. Mater. Chem. A* **2016**, *4*, 1214–1217. [CrossRef]

47. Liu, J.; Xue, Y.; Wang, Z.; Xu, Z.Q.; Zheng, C.; Weber, B.; Song, J.; Wang, Y.; Lu, Y.; Bao, Q.; et al. Two-Dimensional $CH_3NH_3PbI_3$ Perovskite: Synthesis and Optoelectronic Application. *ACS Nano* **2016**, *10*, 3536–3542. [CrossRef] [PubMed]

48. Nie, W.; Tsai, H.; Asadpour, R.; Blancon, J.C.; Neukirch, A.J.; Gupta, G.; Crochet, J.J.; Chhowalla, M.; Tretiak, S.; Wang, H.L.; et al. High-Efficiency Solution-Processed Perovskite Solar Cells with Millimeter-Scale Grains. *Science* **2015**, *347*, 522–525. [CrossRef] [PubMed]

49. Lee, L.; Baek, J.; Park, K.S.; Lee, Y.; Shrestha, N.K.; Sung, M.M. Wafer-Scale Single-Crystal Perovskite Patterned Thin Films Based On Geometrically-Confined Lateral Crystal Growth. *Nat. Commun.* **2017**, *8*, 15882. [CrossRef] [PubMed]

50. Tsai, H.; Nie, W.; Blancon, J.C.; Stoumpos, C.C.; Asadpour, R.; Harutyunyan, B.; Neukirch, A.J.; Verduzco, R.; Crochet, J.J.; Pedesseau, L.; et al. High-Efficiency Two-Dimensional Ruddlesden-Popper Perovskite Solar Cells. *Nature* **2016**, *536*, 312–316. [CrossRef] [PubMed]

51. McLeod, J.A.; Wu, Z.; Sun, B.; Liu, L. The Influence of the I/Cl Ratio on the Performance of $CH_3NH_3PbI_{3-x}Cl_x$-based Solar Cells: Why is $CH_3NH_3I:PbCl_{2=3}$: 1 the "Magic" Ratio? *Nanoscale* **2016**, *8*, 6361–6368. [CrossRef] [PubMed]

52. She, L.; Liu, M.; Zhong, D. Atomic Structures of $CH_3NH_3PbI_3$ (001) Surfaces. *ACS Nano* **2015**, *10*, 1126–1131. [CrossRef] [PubMed]

53. Wang, Y.; Gould, T.; Dobson, J.F.; Zhang, H.; Yang, H.; Yao, X.; Zhao, H. Density Functional Theory Analysis of Structural and Electronic Properties of Orthorhombic Perovskite $CH_3NH_3PbI_3$. *Phys. Chem. Chem. Phys.* **2013**, *16*, 1424–1429. [CrossRef] [PubMed]

54. Komesu, T.; Huang, X.; Paudel, T.R.; Losovyj, Y.B.; Zhang, X.; Schwier, E.F.; Zheng, M.; Iwasawa, H.; Shimada, K.; Saidaminov, M.I.; et al. Surface Electronic Structure of Hybrid Organo Lead Bromide Perovskite Single Crystals. *J. Phys. Chem. C* **2016**, *120*, 21710–21715. [CrossRef]

55. Haruyama, J.; Sodeyama, K.; Han, L.; Tateyama, Y. Termination Dependence of Tetragonal $CH_3NH_3PbI_3$ Surfaces for Perovskite Solar Cells. *J. Phys. Chem. Lett.* **2014**, *5*, 2903–2909. [CrossRef] [PubMed]

56. Haruyama, J.; Sodeyama, K.; Han, L.; Tateyama, Y. Surface Properties of $CH_3NH_3PbI_3$ for Perovskite Solar Cells. *Acc. Chem. Res.* **2016**, *49*, 554–561. [CrossRef] [PubMed]

57. Afzaal, M.; Salhi, B.; Al-Ahmed, A.; Yates, H.M.; Hakeem, A.S. Surface-Related Properties of Perovskite $CH_3NH_3PbI_3$ Thin Films by Aerosol-Assisted Chemical Vapour Deposition. *J. Mater. Chem. C* **2017**, *5*, 8366–8370. [CrossRef]

58. Kollár, M.; Ćirić, L.; Dil, J.H.; Weber, A.; Muff, S.; Ronnow, H.M.; Náfrádi, B.; Monnier, B.P.; Luterbacher, J.S.; Horváth, E.; et al. Clean, Cleaved Surfaces of the Photovoltaic Perovskite. *Sci. Rep.* **2017**, *7*, 695. [CrossRef] [PubMed]

59. Ohmann, R.; Ono, L.K.; Kim, H.; Lin, H.; Lee, M.V.; Li, Y.; Park, N.; Qi, Y. Real-Space Imaging of the Atomic Structure of Organic-Inorganic Perovskite. *J. Am. Chem. Soc.* **2015**, *137*, 16049–16054. [CrossRef] [PubMed]

60. Liu, Y.; Palotas, K.; Yuan, X.; Hou, T.; Lin, H.; Li, Y.; Lee, S. Atomistic Origins of Surface Defects in CH$_3$NH$_3$PbBr$_3$ Perovskite and their Electronic Structures. *ACS Nano* **2017**, *11*, 2060–2065. [CrossRef] [PubMed]

61. Suarez, B.; Gonzalez-Pedro, V.; Ripolles, T.S.; Sanchez, R.S.; Otero, L.; Mora-Sero, I. Recombination Study of Combined Halides (Cl, Br, I) Perovskite Solar Cells. *J. Phys. Chem. Lett.* **2014**, *5*, 1628–1635. [CrossRef] [PubMed]

62. Christians, J.A.; Miranda Herrera, P.A.; Kamat, P.V. Transformation of the Excited State and Photovoltaic Efficiency of CH$_3$NH$_3$PbI$_3$ Perovskite upon Controlled Exposure to Humidified Air. *J. Am. Chem. Soc.* **2015**, *137*, 1530–1538. [CrossRef] [PubMed]

63. Lee, M.M.; Teuscher, J.; Miyasaka, T.; Murakami, T.N.; Snaith, H.J. Efficient Hybrid Solar Cells Based On Meso-Superstructured Organometal Halide Perovskites. *Science* **2012**, *338*, 643–647. [CrossRef] [PubMed]

64. Kim, H.; Lee, C.R.; Im, J.H.; Lee, K.B.; Moehl, T.; Marchioro, A.; Moon, S.J.; Humphry-Baker, R.; Yum, J.H.; Grätzel, M.; et al. Lead Iodide Perovskite Sensitized All-Solid-State Submicron Thin Film Mesoscopic Solar Cell with Efficiency Exceeding 9%. *Sci. Rep.* **2012**, *2*, 591. [CrossRef] [PubMed]

65. Murali, B.; Dey, S.; Abdelhady, A.L.; Peng, W.; Alarousu, E.; Kirmani, A.R.; Cho, N.; Sarmah, S.P.; Parid, M.R.; Zhumekenov, A.A.; et al. Surface Restructuring of Hybrid Perovskite Crystals. *ACS Energy Lett.* **2016**, *1*, 1119–1126. [CrossRef]

66. Wang, T.; Daiber, B.; Frost, J.M.; Mann, S.A.; Garnett, E.C.; Walsh, A.; Ehrler, B. Indirect to Direct Bandgap Transition in Methylammonium Lead Halide Perovskite. *Energy Environ. Sci.* **2017**, *10*, 509–515. [CrossRef]

67. Zheng, F.; Tan, L.Z.; Liu, S.; Rappe, A.M. Rashba Spin-Orbit Coupling Enhanced Carrier Lifetime in CH$_3$NH$_3$PbI$_3$. *Nano Lett.* **2015**, *15*, 7794–7800. [CrossRef] [PubMed]

68. Kepenekian, M.; Robles, R.; Katan, C.; Sapori, D.; Pedesseau, L.; Even, J. Rashba and Dresselhaus Effects in Hybrid Organic-Inorganic Perovskites: From Basics to Devices. *ACS Nano* **2015**, *9*, 11557–11567. [CrossRef] [PubMed]

69. Etienne, T.; Mosconi, E.; De Angelis, F. Dynamical Origin of the Rashba Effect in Organohalide Lead Perovskites: A Key to Suppressed Carrier Recombination in Perovskite Solar Cells? *J. Phys. Chem. Lett.* **2016**, *7*, 1638–1645. [CrossRef] [PubMed]

70. Zhou, H.P.; Chen, Q.; Li, G.; Luo, S.; Song, T.B.; Duan, H.S.; Hong, Z.; You, J.; Liu, Y.; Yang, Y. Interface engineering of highly efficient perovskite solar cells. *Science* **2014**, *345*, 542–546. [CrossRef] [PubMed]

71. Li, X.; Guo, Y.; Luo, B. Improved Stability and Photoluminescence Yield of Mn^{2+}-Doped CH$_3$NH$_3$PbCl$_3$ Perovskite Nanocrystals. *Crystals* **2018**, *8*, 4. [CrossRef]

72. Koh, C.W.; Heo, J.H.; Uddin, M.A.; Kwon, Y.; Choi, D.H.; Im, S.H.; Woo, H.Y. Enhanced Efficiency and Long-Term Stability of Perovskite Solar Cells by Synergistic Effect of Nonhygroscopic Doping in Conjugated Polymer-Based Hole-Transporting Layer. *ACS Appl. Mater. Interfaces* **2017**, *9*, 43846–43854. [CrossRef] [PubMed]

73. Yin, G.; Ma, J.; Jiang, H.; Li, J.; Yang, D.; Gao, F.; Zeng, J.; Liu, Z.; Liu, S.F. Enhancing Efficiency and Stability of Perovskite Solar Cells through Nb-Doping of TiO$_2$ at Low Temperature. *ACS Appl. Mater. Interfaces* **2017**, *9*, 10752–10758. [CrossRef] [PubMed]

crystals

MDPI

Review

Twin Domains in Organometallic Halide Perovskite Thin-Films

Wei Liu [1], Yang Liu [1], Ju Wang [1], Cuncun Wu [1], Congyue Liu [1], Lixin Xiao [1], Zhijian Chen [1], Shufeng Wang [1,2,*] and Qihuang Gong [1,2]

[1]   State Key Laboratory for Artificial Microstructure and Mesoscopic Physics, Department of Physics, Peking University, Beijing 100871, China; wliu2016@pku.edu.cn (W.L.); liuyg@foxmail.com (Y.L.); 1701110204@pku.edu.cn (J.W.); cuncunwu@163.com (C.W.); liucongyue@pku.edu.cn (C.L.); lxxiao@pku.edu.cn (L.X.); zjchen@pku.edu.cn (Z.C.); qhgong@pku.edu.cn (Q.G.)

[2]   Collaborative Innovation Center of Extreme Optics, Shanxi University, Taiyuan 030006, China

*   Correspondence: wangsf@pku.edu.cn

Received: 30 March 2018; Accepted: 11 May 2018; Published: 16 May 2018

Abstract: The perovskite is a class of material with crystalline structure similar to $CaTiO_3$. In recent years, the organic-inorganic hybrid metallic halide perovskite has been widely investigated as a promising material for a new generation photovoltaic device, whose power conversion efficiency (PCE) record reaches 22.7%. One of its underlying morphological characteristics is the twin domain within those sub-micron sized crystal grains in perovskite thin films. This is important for discussion since it could be the key for understanding the fundamental mechanism of the device's high performance, such as long diffusion distance and low recombination rate. This review aims to summarize studies on twin domains in perovskite thin films, in order to figure out its importance, guide the current studies on mechanism, and design new devices. Firstly, we introduce the research history and characteristics of widely known twin domains in inorganic perovskite $BaTiO_3$. We then focus on the impact of the domain structure emerging in hybrid metallic halide perovskite thin films, including the observation and discussion on ferroelectricity/ferroelasity. The theoretical analysis is also presented in this review. Finally, we present a spectroscopic method, which can reveal the generality of twin domains within perovskite thin films. We anticipate that this summary on the structural and physical properties of organometallic halide perovskite will help to understand and improve the high-performance of photovoltaic devices.

Keywords: perovskite; twin domains; ferroelectricity; ferroelasity

1. Introduction

Perovskite, which is named after the Russian mineralogist Lev Perovski (1792–1856), is a widely known material whose discovery should be traced back to 1839 by Gustav Rose [1]. From the perspective of lattice structure, perovskite represents a group of material with ABX_3 stoichiometry, whose crystalline structure is similar to $CaTiO_3$ [2]. $CH_3NH_3PbI_3$ (MAPbI3) is an organic–inorganic hybrid metallic halide perovskite with the unit cell containing Pb^{2+} and I^-, forming a octahedral BX_6 3D network, and together with cuboctahedral AX_{12} where MA^+ occupies the center as seen in Figure 1 [3]. Since its first report with power conversion efficiency (PCE) 3.8% for a dye-sensitized solar cell in 2009 [4], its promising potentiality as the next generation photovoltaic material was soon discovered. A high PCE record of 22.7%, comparable to silicon-based solar cells, was achieved recently [5]. It has been widely applied to various forms of optoelectronic devices, such as lasers [6,7], light emitting diodes (LED) [8,9], photodetectors [10], and photovoltaic devices [11,12].

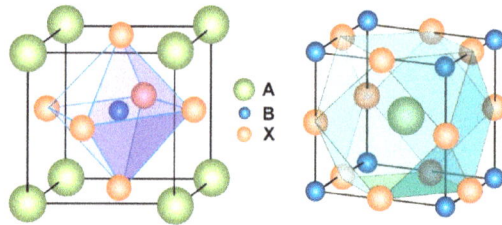

Figure 1. Typical crystal structure of perovskite ABX_3 with octahedral BX_6 (**left**) and cuboctahedral AX_{12} (**right**). As for $CH_3NH_3PbI_3$ circumstance, A = $CH_3NH_3^+$, B = Pb^+, X = I^+. Reprinted with permission from reference [3]. Copyright (2014) American Chemical Society.

The phase transition of organometallic halide perovskite within the heating process has been widely studied in recent years. In $MAPbI_3$, the octahedral (BX_6) tilting with halide elements, which arises from the disordered orientation of methylammonium cation depending on the temperature and represents the bending of Pb-I-Pb bonding angle, provides possibility of three crystalline configurations with increased temperature as seen in Figure 2 [13,14]. Cubic phase is an ideal perovskite structure without tilts ($\theta_{ac} = \theta_b = 180°$) which is maintained and stabilized above 327 K [13,15]. However, the tilts may present in the equatorial plane X_{eq} only ($\theta_{ac} < 180°$ and $\theta_b = 180°$, rotation leading to a tetragonal phase between 162 K and 327 K) or together with tilting along the axial direction X_{ax} ($\theta_{ac} < 180°$ and $\theta_b < 180°$, rotation leading to an orthorhombic phase below 162 K) [13,15]. The phase transition of $MAPbI_3$ from cubic crystalline structure to tetragonal together with lowered symmetry during the annealing process provides the possibility of intrinsic twin domains, similar to the ferroelectric ones within inorganic perovskite $BaTiO_3$ films [16]. Although they are hard to observe in $MAPbI_3$ thin films under SEM, the related possible ferroelectricity has been proposed to explain the *I–V* hysteresis experimentally [17,18] and theoretically [19], as well as the low recombination rate [20]. Recently, a TEM result revealed its presence, which provided the clue to disclose the unique optoelectronic performance of organometallic halide perovskite [21].

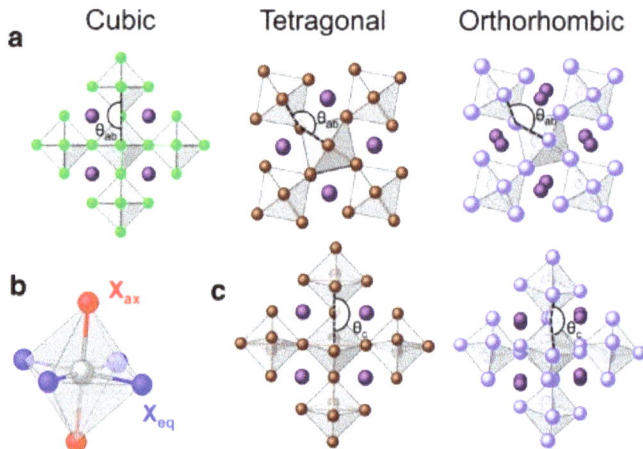

Figure 2. Possible crystalline phases of perovskite ABX_3. (**a**) Typical perovskite structure with octahedral BX_6 geometry (top view) of various phase cubic, tetragonal, orthorhombic and the relative side view pattern (**c**), where A = violet, B = gray, X = green, brown, cyan typically. (**b**) Illumination of octahedral BX_6 with equatorial plane halide (X_{eq}, red) and axial plane halide (X_{ax}, blue). Reprinted with permission of reference [13]. American Chemistry Society.

In this review, after briefly summarizing the twin domain in the typical inorganic perovskite $BaTiO_3$, we introduce the current studies on $MAPbI_3$ thin films. Though observations on twin domains are quite limited at the moment, they effectively proved the possibility of their existence, and the significance to discuss the charge carrier extraction based on the domain structure. E.g., the calculation based on first principle theory revealed that the structure could provide an additional channel for charge carrier transportation separately without recombination. At last, we introduce our newly developed optical method, which revealed the general existence of these subgrain microstructures in the organometallic perovskite thin films. At the moment, the connection between the twin domain and high performance is still missing and hard to fulfill. The direct observation on charge transportation along the structure is necessary to solve the connection. We summarize these studies in this review and expect the coming of this observation, in order to finally clarify the mechanism of these optoelectronic devices' high performance. This mechanism would be different to any other high performance solar cells, such as silicon solar cells, thin film solar cells like CIGS, dye sensitized solar cells, and organic solar cells. We hope that the understanding of a new mechanism would help to improve the perovskite solar cells for high performance and the investigation of new photovoltaic materials.

2. Twin Domains in Inorganic Perovskite $BaTiO_3$

Twin domain is an intrinsic morphological feature observed in non-central symmetric crystalline structures. Research on its origination and physical properties have a long history since the concept was first introduced with scientific description in 1783 [22]. However, the unambiguous definition of twinning is still missing in the literature owing to the structure complexity and various morphological phenomena [23]. One typical definition based on geometry of twinning crystal points out that "twin is a complex crystalline edifice built up of two or more homogeneous portions of the same crystal species in juxtaposition and oriented with respect to each other according to the well-defined twin laws" [23]. Twin domains could be classified into growth twinning, mechanical twinning, and transformation twinning under the consideration of formation mechanism, in spite of its various definitions [24].

Growth twinning can occur during the crystal growth, either at the nucleation stage [24] or by the oriented attachment [25]. For the former one, twinning is mainly the result of the growing mistake when some atoms or clusters of atoms arrive at the twin orientation and then continue to grow into a spread individual. A typical example is penetration twin domains observed in $FeBO_3$ [26], quartz–homeotypic $FePO_4$ [27], $GaPO_4$ [28], Al_2O_3 [29], and some metallic materials like Zr and Ti [30]. Alternatively, in supersaturated solution condition, another type of growth twinning may occur on the base of twinned nucleation according to the theory of Buerger [24]. Under the later circumstance, growth twinning that resulted from the oriented attachment has not been widely studied yet. Some direct and indirect evidences including alum and β-quartz were reviewed by Massimo Nespolo and Giovanni Ferraris [25].

A mechanical pressure can switch the orientation of some crystals to another state while keeping the identical crystalline structure [24]. This result is regarded as mechanical twinning [31], which can be found in metals and minerals with elasticity and plasticity. They are shown as the rolling and shear texture in brass [32], stress-induced ω phase transition in β titanium alloys [33], and stacked microtwins of a few tens of nanometers in high manganese content austenitic steel [34] etc. It is worth mentioning that the ferroelastic transformation twin can be classified into mechanical twinning since it can be created by the shear deformation, resulting from mechanical load [35], such as the domain reorientation of ZrO_2 under unidirectional compressive stress at high temperature [36], the prominent W domain walls observed in $BiVO_4$ [37], a completely reversible domain structure induced by the high stress applied at the crack tip in $BaTiO_3$ [38].

Twinning domains formed during crystalline phase transition in numerous optoelectronic materials are usually defined as transformation twinning. Generally, the transformation twinning is intimately related to the change of crystalline symmetry. Twin domains turn out during the cooling down process from high to low temperature, while the crystal phase transits from higher symmetry to lower symmetry, owing to the movements of cooperative atom in a nuclear region that is attributed to the loss of symmetry [39]. Well-known examples of the transformation twinning are revealed in perovskite, which obey the twin laws of high-low symmetry transition.

$BaTiO_3$ is a typical representative of inorganic perovskite, which is commonly used today for numerous applications due to its ferroelectricity and piezoelectricity. $BaTiO_3$ crystal is of cubic phase and centrosymmetric lattice structure above 120 °C [40]. The crystal would transform into a polar tetragonal phase with the emerging of the complex twin domains upon cooling. Basically, there are two kinds of twinnings according to the direction of the polar axis between adjacent twined domains: perpendicular (90° domain) or antiparallel (180° domain) [40]. One way to distinguish them is through polarized optical detection. Under the polarized light, responses are the same for the adjacent 180° domains. For 90° type, it could be justified by the observation of wedge-shaped lamella domains as seen the square-net pattern in Figure 3a, which were almost always identified under the polarized light in the tetragonal $BaTiO_3$ single crystal [41,42]. Actually, the twin domain could be more unambiguous by etching or external electric field when photographed by optical microscopy, TEM, and SEM [43,44]. A typical $BaTiO_3$ 90° domain with characteristic banded domain structure of herringbone is showed in Figure 3d. The corresponding TEM bright field image of the same sample as seen in Figure 3e, which illustrated twin domains with width varied in 300~500 nm. The insetting selected area diffraction pattern (SADP), which showed the spot splitting subtending an angle of ~35° along the [110] direction, indicated that they were the 90° a–a type domains [45]. The twinning photograph confirmed that the $BaTiO_3$ 90° domain was prominent in the twin domains [42,46]. On the other hand, the Figure 3b,c illustrated the classical $BaTiO_3$ 180° twin domains, which were presented with image contrast resembling watermark patterns characterized by irregular lines, were embedded in the lamellar-shaped 90° domains. One more thing to be mentioned is that the growth twinning of $BaTiO_3$ may emerge due to the mistake of nucleation during the crystal growing process. Figure 3f shows the famous growth twinning butterfly of $BaTiO_3$ which relates to the cooling rate and the presence of certain impurities [47,48].

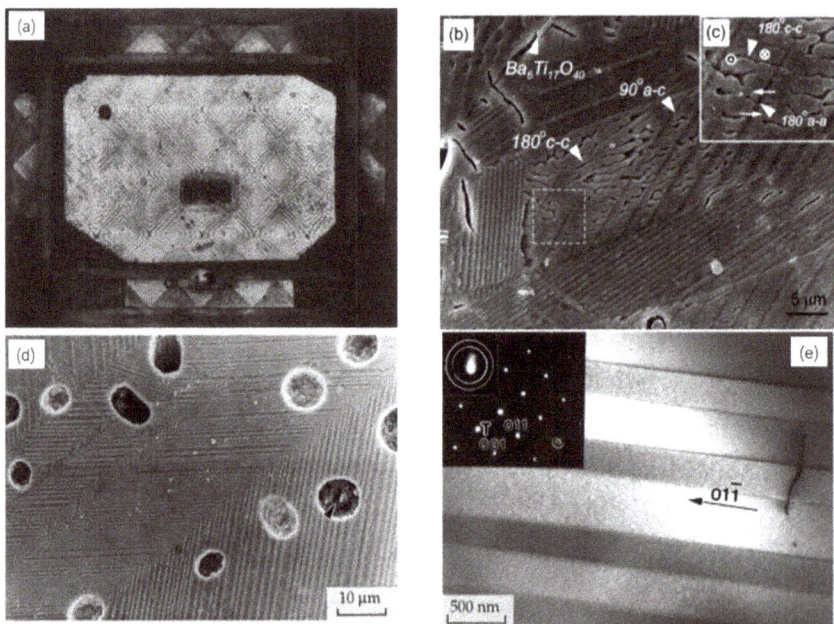

Figure 3. *Cont.*

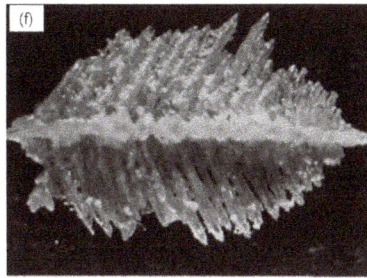

Figure 3. (a) Square-net pattern and edge viewed single crystal. (b,c) The 180° type domains characterized by irregular lines. (d) The ferroelectric domains of the herringbone-type BaTiO$_3$. (e) The TEM image with SAED of the same sample in (d). (f) The butterfly growth twinning of BaTiO$_3$. Part (a) is adapted from reference [41] with permission from American Physical Society. Part (b,c) are adapted from reference [42] with permission from John Wiley and Sons. Part (d,e) are adapted from reference [45] with permission from Elsevier. Part (f) is adapted from reference [48] with permission from John Wiley and Sons.

3. Subgrain Twin Domains Observed in MAPbI$_3$

Photovoltaic devices based on methylammonium lead iodide perovskite have astonished breakthroughs in solar energy conversion efficiency in these years. Light absorber layers of perovskite solar cells are usually polycrystalline thin films, in which subgrain microstructures are rarely observed. Therefore, early research on the impact of microstructure in perovskite films was mainly focused only on the grain, including grain size and grain boundary. Choi et al. and Grancini et al. reported that the absorption spectrum of MAPbI$_3$ in the mesoporous TiO$_2$ with smaller grain size (few tens of nanometers) performed a blue shift compared to perovskite thin films deposited the flat substrate with larger grain size (hundreds of nanometers) [49,50]. Bastiani et al. and D'Innocenzo et al. claimed that the smaller grain size of the perovskite material, the shorter lifetime of the photoluminescence from the perspective of charge carrier dynamics [51,52], which was contrary to the result of Nie et al. [53]. The evidence showed that the fluorescence quenching in single crystal perovskite was even stronger [54,55]. The effect of perovskite grain boundary on the optoelectronic prosperities also received a large amount of attention. Dane W. de Quilette et al. reported that the fluorescence intensity of the MAPbI$_{3-x}$Cl$_x$ grain boundary decreased to 65% compared to the inter grain by using confocal fluorescence microscopy [56]. The result meant that the grain boundary of perovskite material provided much excess non-radiation recombination pathway [56]. Interestingly, Yun et al. utilized KPFM and AFM to explore the impact of perovskite grain boundary with sunlight illumination. Their result showed that there were higher surface photovoltage and short circuit current in the grain boundary rather than the inter grain [57]. Their later work revealed that the grain boundary may benefit the charge separation and charge transportation [58].

There is no consensus about the impact of grain size and grain boundary on charge carrier dynamics and photovoltaic performance according to these contradictory studies. It results in challenge to the explanations about the remarkable photovoltaic efficiency, the mechanism of current-voltage hysteresis (*I–V* hysteresis) [17], and the impact of ion migration [59]. One fundamental question is about the origin of low recombination rate, which is contradicted to the Langvin recombination model [60]. The ambiguous understanding of these questions indicates the urgency of exploring the underlying mechanisms.

Recently, subgrain twin domains were found in MAPbI$_3$ thin films [21,61–64]. These intra microstructures within grains may be the intrinsic potential factors in charging the high performance of perovskite-based photovoltaic devices [65]. Prior to the discovery of twin domain, the organolead perovskite had been proposed as ferroelectric [66–68]. However, whether the organometallic perovskite such as MAPbI$_3$ possesses ferroelectricity remains ambiguous. We summarize some of the debates in Table 1.

Table 1. Summary of controversial results about ferroelectricity in MAPbI$_3$.

Sample	Main Method	Phenomenon	Result	Reference
β CH$_3$NH$_3$PbI$_3$ film	PFM	180° phase change after DC poling	ferroelectricity	[66]
CH$_3$NH$_3$PbI$_3$(Cl) film	PFM	Striped twin domain without ferroelectric switching	ferroelasticity	[61]
CH$_3$NH$_3$PbI$_3$ device	PFM	*P-E* hysteresis loops without switching	Non-ferroelectricity	[69]
Powder and single crystal of CH$_3$NH$_3$PbI$_3$	SHG XRD P-E loop	a. SHG efficiency below any detectable limit b. XRD result showed the centrosymmetric space group c. Approximately linear P-E loop	Non-ferroelectricity	[70]
CH$_3$NH$_3$PbI$_3$ film	P-E loop/PFM	Ferroelectric properties and switchable polarization	ferroelectricity	[71]
CH$_3$NH$_3$PbI$_3$ single crystal	P-E loop/SHG	*P-E* hysteresis loops in tetragonal phase and SHG response	ferroelectricity	[72]
CH$_3$NH$_3$PbI$_3$ film	P-E loop	No clear P-E switching	Non-ferroelectricity	[73]

SHG: Second-harmonic generation; XRD: X-ray diffraction; P-E loop: Polarization Intensity-Electric field loop curve.

Whether the twinning structure represent ferroelectric or not keeps coming up as a question. Hermes et al. employed piezoresponse force microscopy (PFM) to characterize the a relative large and flat surfaces of MAPbI$_3$ films [61]. Their results revealed the existence of subgrain twin domains in the tetragonal phase. The nanoscale periodic striped domain within single grain is shown in Figure 4a,b. These striped domains have their identical width within 100–300 nm in one grain, but vary between different grains. The 90° angle of stripes within grains shown in the magnified areas was a fingerprint-like structure for ferroelastic twin domains. It was supported by an additional PFM experiment on 90° rotated sample, for which the inverted phase contrast was found, with almost unchanged amplitude contrast (Figure 4d,e) [61]. Hermes et al. also proposed that the orientation of the polarization in the observed twin domains was a$_1$–a$_2$ phase, as seen in Figure 4c [61].

Furthermore, Strelcov et al. observed the 70° and 109° domain structure in MAPbI$_3$ thin films with PFM, as seen in Figure 5 [62]. It was proven that the nanoscale domain behaved as a function of the applied stress, which would not return to the pristine state when the stress was relieved. It was in good agreement with the ferroelasticity [62]. In addition, the unchanged domain pattern under different electric field strength, observed by PTIR (Photothermal Induced Resonance), excluded the possibility of the ferroelectricity. Instead, it supported the ferroelastic nature of the structure [62].

Recently, the alternating polarization striped domain with an average width of 90 nm was reported in solution-processed MAPbI$_3$ film, as seen in Figure 6 [63]. Contrary to the previous report, these twin domains observed by PFM were considered as ferroelectric domains [63]. The 90° continuation of the domain within grains was observed clearly in Figure 6c,d, similar to the domains revealed by Hermes et al. [61]. Further investigation showed that the structure could provide 25% enhancement of the charge carriers extraction ability measured by the photo-conductive AFM (PC-AFM), as showed in Figure 6b [63].

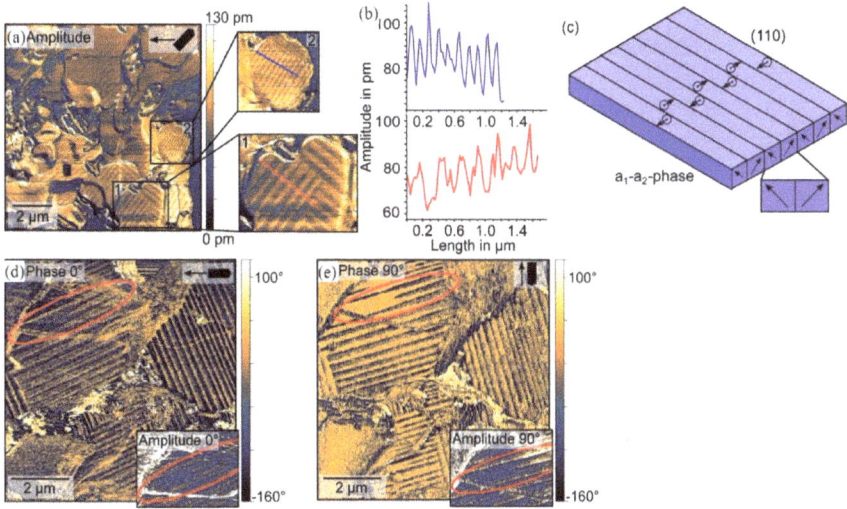

Figure 4. PFM measurement on MAPbI$_3$ films: (**a**) PFM amplitude image with magnified areas 1 and 2; (**b**) amplitude profiles extracted along the red line in area 1 and the blue line in area 2; (**c**) proposed a$_1$-a$_2$ phase in a 45° angle out of plane and 90° with respect to each other; lateral PFM image for PFM phase at 0° (**d**) and 90° (**e**) sample rotation, insets in the right corner show the PFM amplitude both with the same data scale. Reproduced from reference [61] with permission from American Chemical Society.

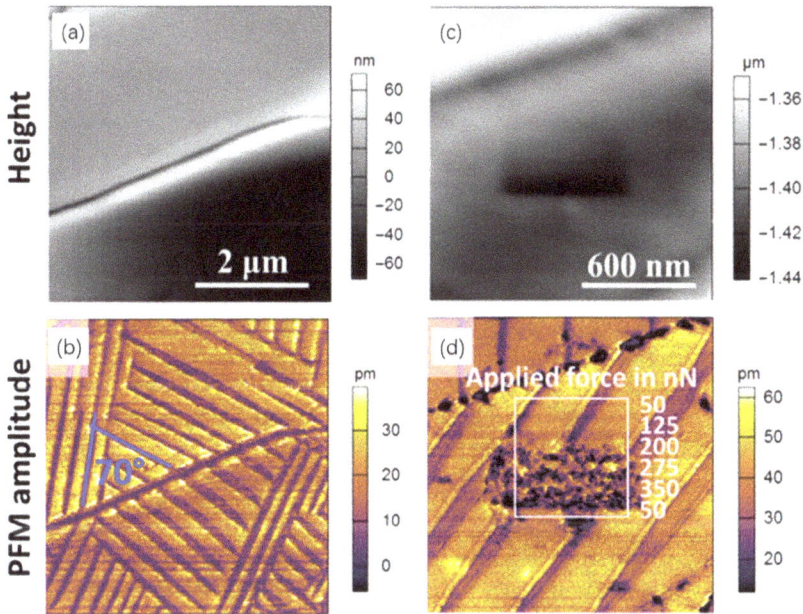

Figure 5. Pattern of typical ferroelastic domains in perovskite thin films. (**a**) The topography of perovskite thin films with processed with doctor blade–coating and corresponding (**b**) PFM amplitude images. (**c**) High-resolution topography image and (**d**) PFM amplitude image with an AFM tip loading. Reprinted with permission from reference [62], AAAS.

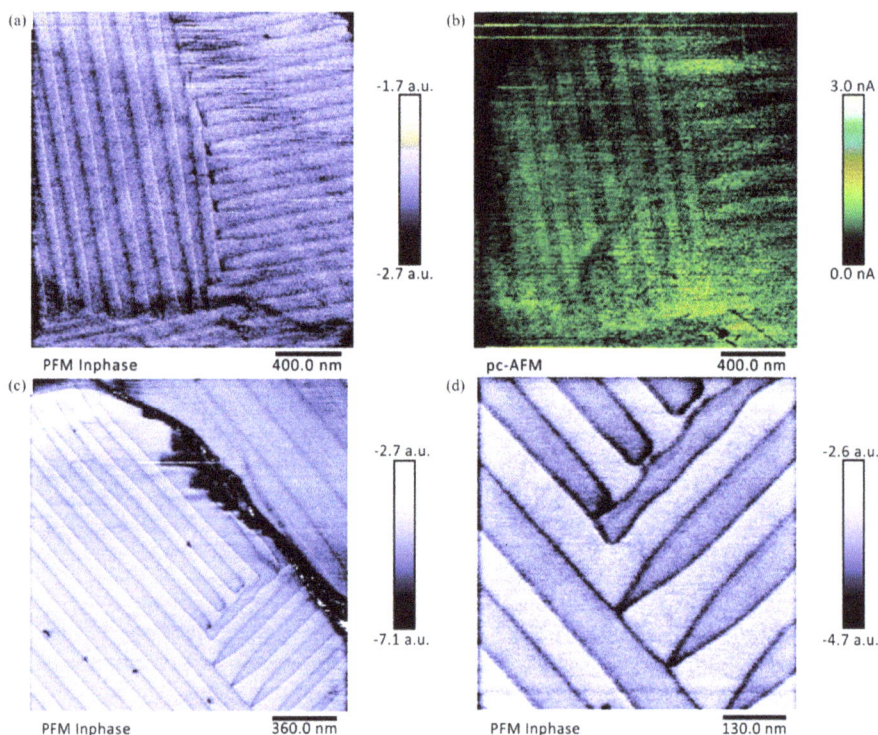

Figure 6. Domain structure with alternating polarization in perovskite thin films. (**a**) PFM phase image revealed patterns of polarized domain; (**b**) the same pattern in pc-AFM for the same area in (**a**); (**c**) PFM phase image of a ferroelectric MAPbI$_3$(Cl) domain pattern with the dark line represents the grain boundary; (**d**) Magnified PFM image of a smaller area in (**c**) illustrating a 90° continuation of the polarized domain. Reprinted with permission from reference [63]. Royal Society of Chemistry.

Employing PFM to detect subgrain twin domains is available for surface detection. However, the flat grain surface is needed to exclude the cross-talk of the surface structure with twinning morphology [63]. Hence, TEM, as a classical method to image twin domains in inorganic perovskite, is applied for investigating organometallic perovskite. Rothmann et al. reported intrinsic twin domains performed as ~100–300-nm-wide stripe in tetragonal phase MAPbI$_3$ thin films as seen in Figure 7 [21]. These twin domains were observed with TEM under the condition of low electron dose and rapid acquisition to carefully avoid surface damage [21]. The selected area electron diffraction (SAED) result implied that the diffraction spot from adjacent domains were mirrored to each other across the twin axis and the twin axis was parallel [112] in direction. They claimed that the separation angle θ, which would be attributed to the difference of the lattice space distance of {110}$_t$ and {002}$_t$ planes at the tetragonal phase in MAPbI$_3$, underpinned the twinning and formation of the {112}$_t$ twin plane [21]. They also found that the twin domain owed memorized characteristics. The striped contrast which existed at room temperature (tetragonal phase) disappeared in 70 °C (cubic phase). It turned out again when the sample cooled down to room temperature. This phase transformation-related phenomenon is consistent with the twin domain observed in inorganic perovskite BaTiO$_3$ [16].

Another interesting study showed that the tetragonal and cubic phase coexisted at room temperature, observed by HRTEM in MAPbI$_3$ thin films, as seen in Figure 8 [64]. The refinement of the HRTEM image, which was based on the Fourier filtering and reconstruction technique, showed that there were two types of spontaneously formed superlattices without composition change: triple layer superlattices made up of tetragonal/cubic/tetragonal stacking sequences and double layer superlattices made up of tetragonal/cubic stacking sequences. These spontaneous-formed superlattices were attributed to the structural transition under mixed tetragonal and cubic phase [64]. In addition, the atomically organized double layer superlattice was observed in the MAPbI$_3$–TiO$_2$ hetero-interface acting as a buffer layer, which provided possibility to benefit the photovoltaic performance due to enhanced charge transportation [64].

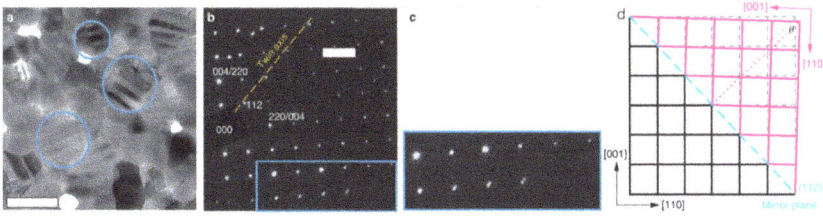

Figure 7. Twin domains revealed by TEM and the corresponding SAED pattern. (**a**) Bright-field TEM image of pristine CH$_3$NH$_3$PbI$_3$ thin film at room temperature. (**b**) The SAED pattern taken from a grain owing stripe contrast showed two single-crystal patterns with a mirrored relationship and magnified in (**c**); (**d**) Schematic of the proposed twinning geometry. Reprinted with permission from reference [21]. Springer Nature.

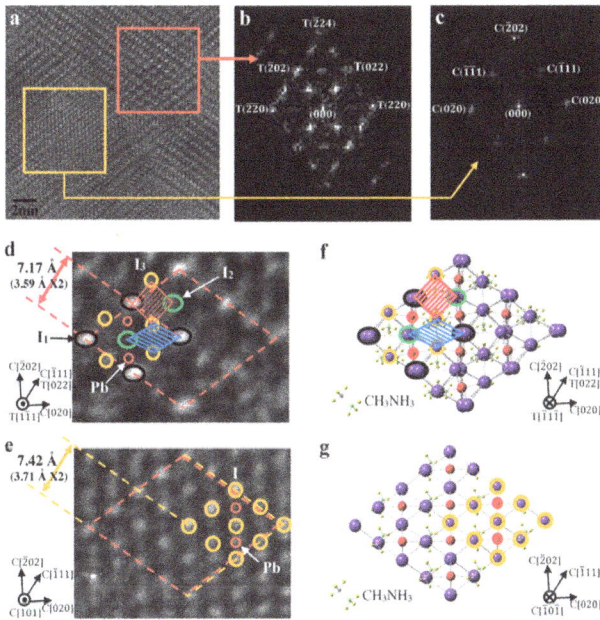

Figure 8. The coexistence of tetragonal and cubic phase in MAPbI$_3$ thin films. (**a**) HRTEM ($\times$1000 K) image showing the coexistence of the tetragonal and cubic domains in perovskite. (**b**) The FFT image of $[1\bar{1}1]_t$ (red rectangle, tetragonal phase) zone axis FFT. (**c**) The FFT image of $[101]_c$ (yellow rectangle, cubic phase). HRTEM ($\times$5500 K) images of tetragonal phase (**d**) at $[1\bar{1}1]_t$ zone axis and cubic phase (**c**) at $[101]_c$ zone axis. (**f**), (**g**) illustrate the atomic configurations simulated along $[1\bar{1}1]_t$ and $[\bar{1}0\bar{1}]_c$. Reprinted with permission from reference [64]. John Wiley and Sons.

Rare observations about twin domains in MAPbI$_3$ had been presented, although MAPbI$_3$ films have been widely adopted as active layers in solar cells. The morphological characteristics of the twin domain in MAPbI$_3$ films is hard to be summarized and some related growth conditions have benn summarized briefly in Table 2. Fortunately, there are abundant studies on inorganic perovskites. The studies showed that the morphology could be accounted to some factors such as grain size (*g*) and film thickness (*t*).

i. According to the classical understanding, twin domains in ferroelastic or ferroelectric materials originate from the minimum energy evolution of the grain, with the expense of domain wall energy [16]. Therefore, the occurrence of twin domain related to the grain size *g*. A semi-quantitative model based on typical BaTiO$_3$ twin domains revealed that the total elastic energy due to homogenous strain of phase transformation increased with g^3 while the domain wall energy increased with g^2. Research on BaTiO$_3$ ceramics concluded that the domain width obeyed the $g^{1/2}$ rule while $g > 1$ μm and the domain structure would no longer be unique with $g < 1$ μm [16]. Small twin structure of ~30 nm are observed in small grains of 300 nm [16].

ii. In ferroic material rhombohedral perovskite LSMO, it was revealed that the domain period followed the $t^{1/2}$ rule in ultrathin films which implied high density of twin patterns at small film thickness and was, in turn, more apparent [74].

The normal grain size in MAPbI$_3$ films is of a few hundred nanometers and a small twinning structure can be expected. Rough film surface is another obstacle for observation. High flatness of MAPbI$_3$ films is needed for observing twin domain patterns during the PFM characterization process as suggested by Holger Röhm et al. [63]. However, the requirement for a real working layer has no strict requirement for surface flatness. Therefore, the twinning structures in MAPbI$_3$ films are still hard to be studied.

Table 2. Summary of material growth condition for MAPbI$_3$ with twin domains.

Material	Preparation Method	Thickness	Grain Size	Annealing Process	Reference
MAPbI$_3$(Cl) polycrystalline films	One-step solution-processed	/	Up to 10 μm	140 °C, 20 min	[61]
MAPbI$_3$ polycrystalline films	Doctor blade-coating	600 nm	0.5~2 μm	100 °C, 30 min	[62]
MAPbI$_3$ polycrystalline films	Two-step spin-coating	/	0.5~2 μm	100 °C, 10 min	[62]
MAPbI$_3$(Cl) polycrystalline films	Two-step solution-processed	300 nm	Several micrometers	100 °C, 20 min	[63]
MAPbI$_3$ polycrystalline films	One-step solution-processed	300 nm	~1 μm	100 °C, 10 min	[21]
MAPbI$_3$ polycrystalline films	One-step solution-processed	/	/	100 °C, 30 min	[64]

4. Effect of the Subgrain Microstructure on Perovskite Solar Cells

The subgrain twin domain in MAPbI$_3$ could be related to ferroelasticity or ferroelectricity, which could explain the low recombination rate and long carrier diffusion distance. In inorganic perovskite ABX$_3$ like BaTiO$_3$, the B could move away from the center of the BX$_6$ cage in the tetragonal phase leading to the broken crystal centro-symmetry and a spontaneous electric field in turn [75,76]. A similar mechanism was proposed by Frost et al., who suggested that the organic cation rotating in the inorganic cage relative to MAPbI$_3$ could distort the lattice symmetry [20]. A corresponding electronic polarization field of 38 μC/cm^2, comparable to the inorganic ferroelectric perovskite, was calculated [20]. Liu et al. presented the impact of ferroelectricity in the charge separation process based on the DFT calculation in MAPbI$_3$ [77]. Their results illustrated that two possible mechanisms could benefit charge separation, for which the ferroelectric domain wall would be charged or uncharged due to the flexible rotation of MA$^+$ in the crystal structure, as seen in Figure 9a,b. One was the electron-hole pair separation channel between the charged domain wall existing in 90° domain and 180° domain [77]. The other was the small polarization field normal to ferroelectric domain walls in the uncharged 90° domain wall [77]. Frost et al. explained the enhancement of charge carrier transport induced by the ferroelectric domain in detail with the ferroelectric highway configuration [20]. According to the proposed ferroelectric highway, the electron-hole pair could be separated efficiently within one ferroelectric domain due to the polarization field and the charge carrier diffused along the domain boundary freely toward the electrodes without recombination, as seen in Figure 9c,d [20]. Furthermore, the calculation based on Landau–Ginzburg–Devonshire theory revealed that the carrier diffused to ferroelectric domain wall and accumulated there, as seen in Figure 9e. The accumulation of charge carriers increased the static conductivity of the domain wall by 3–4 orders compared to the bulk state [78]. Recently, the density function theory based model revealed a similar segregation transport of electrons and holes with increased conductance and reduced bandgap in MAPbI$_3$ [79].

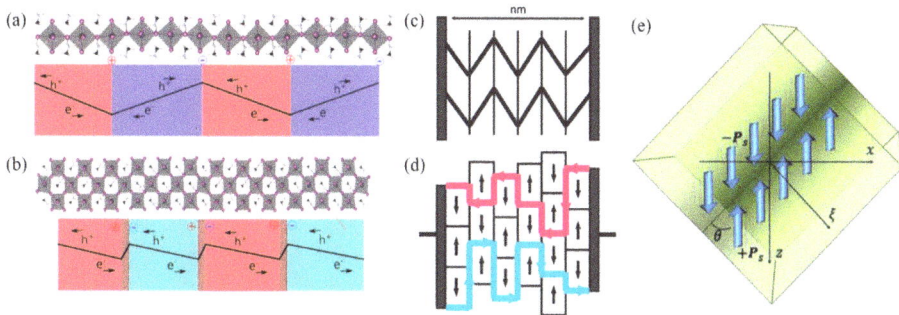

Figure 9. (**a,b**) Illustration of atomistic structure and electrostatic potential distribution corresponded for the CDWs (Charged domain walls) and UCDWs (Uncharged domain walls). (**c,d**) Schematic of the ferroelectric highway with 1D built-in potential (**c**) and associated 2D electron and hole separation pathway (**d**). (**e**) Charged domain walls in the uniaxial ferroelectric n-type semiconductors for the case polarization field inclined head-to-head. Part (**a,b**) are adapted from reference [77] with permission from American Chemical Society. Part (**c,d**) are adapted from reference [20] with permission from American Chemical Society. Part (**e**) is adapted from reference [78] with permission from Springer Nature.

Ambient experimental evidence to evaluate the positive or negative effect of subgrain microstructure observed in MAPbI$_3$ thin films on perovskite solar cells is still missing although theoretical results stand by the positive side. However, some indirect evidence has been reported recently. Daniele Rossi et al. developed a drift-diffusion model to simulate the actual benefit of polarization field to the perovskite solar cells based on the revealed herring-bone like ferroelectric domains in their earlier work [63]. Their results illustrated that the ferroelectric domain suppressed the defect-mediated recombination and direct recombination owing to the discontinued polarization field which acted as an electron-hole separate pathway [80]. Hsinhan Tsai et al. investigated the structural dynamic evolution under continued light illumination, whose results illustrated the light-induced uniform lattice expansion leading to enhanced power conversion efficiency (from 18.5% to 20.5%) and stability (more than 1500 h) of perovskite solar cells [81]. Actually, the light-soaking process performed as annealing procedure, during which the lattice strain originated from phase transition was nonnegligible. Its rational to propose that the occurrence of twin domains resulted from lattice strain during the light illumination given by the formation mechanism of transformation twin domains. More indirect evidence of twin domains' positive effect on photovoltaic devices has been suggested by the high performance and stability perovskite solar cells with nanoscale stripe patterns, especially the mixed-cation or mixed-halide perovskite such as (FAPbI$_3$)$_{0.85}$(MAPbI$_3$)$_{0.15}$ [82], MAPbI$_{3-x}$Cl$_x$ [83], Cs$_x$FA$_{1-x}$PbI$_3$ [84].

5. Optical Method to Observe the Broad Existence of Subgrain Twin Domains

The subgrain twin domains, which are widely found in inorganic perovskite BaTiO$_3$, remain mysterious in organometallic perovskite MAPbI$_3$ working layers, though they have been observed with PFM and TEM in specified samples. As we mentioned in former parts, the surface roughness, the small scale of the domain, and the a_1–a_2 phase characteristics are the obstacles for direct observation in working layers [21,63]. Fortunately, we developed a convenient optical method to identify the broad existence of subgrain twin domains hidden in MAPbI$_3$ active layers [85].

It is revealed that the photo-generated products, such as carriers, excitons, trions, and bi-exciton etc. strongly correlated to material morphology. This relationship requires some conditions. In principle, the morphology variation may not lead to change of photoproducts. E.g., larger and smaller crystal grains may have similar photophysical behaviors. However, on the other hand, if we reduce the grain size to the nano scale or lower its dimension to nanowire and two-dimensional layers, the photophysical behavior will be significantly different [86,87]. Therefore, if the variation on photoproducts is found, it would represent significant change in morphology. Based on this idea, the relationship between photoexcited exciton and free carriers was analyzed by our recently developed density-resolved spectroscopic method [88]. The method focuses on the excitation density-dependent photoproduct system, since the correlation of exciton and free carriers in semiconductors is density-dependent. According to the Saha–Langmuir equation [89] and fixed fluorescence intensity $I(n)$ [88]:

$$\frac{x^2}{1-x} = \frac{1}{n}\left(\frac{2\pi\mu k_B T}{h^2}\right)^{3/2} e^{\frac{E_B}{k_B T}} \tag{1}$$

$$I(n) \propto A_1(1-x)n + A_2(xn)^2 + A_3(1-x)n \cdot xn \tag{2}$$

where n is the total carrier intensity under light illumination, x is the ratio of free carriers, μ, k_B, h, E_B are the reduced effective mass, the Boltzmann constant, the Plank's constant, the exciton binding energy. A_1 and A_2 represent the decay rate of monomolecular and bimolecular recombination, A_3 is another decay rate that originated from exciton–carrier collision (ECC). At high excitation density, the fixed fluorescence intensity $I(n)$ performs as power index 1 representing coexistence of exciton and free carriers without ECC, while power index 3/2 which implies the dominance of ECC behavior at perovskite thin films [85].

For freshly made perovskite films without heat annealing (or slightly heated for removing solvent), the exciton and free carriers co-existed and interconverted to each other dynamically as shown in Figure 10a with obvious power index 1 at high excitation density. For thin films experiencing adequate heat annealing, the collision-induced quenching of exciton and free carriers (ECC) happened as illustrated in Figure 10b with power index 3/2 at high excitation density [85]. Since the non-annealed and annealed film were all with tetragonal phase, there must be some intrinsic morphological change appearing. In addition, when the fully annealed films (3/2 index) were heated to 340 K (above the phase transition temperature from tetragonal to cubic phase), the ECC disappeared (the index become 1), as seen in Figure 10c. However, the 3/2 index came back when the films were cooled down to room temperature. This initial annealing and phase transition dependent behavior was akin to the twin domain formation in BaTiO$_3$ film and MAPbI$_3$ films revealed by TEM [16,21]. Therefore, we confirmed that the subgrain morphology generated in MAPbI$_3$ films and such subgrain morphology should be twin domains proposed in previous studies [85]. The photoproduct behavior and corresponding proposed mechanism are summarized in Figure 10d [85]. Our full analysis of the photoproduct system revealed the existence of twin domains in sufficiently annealed MAPbI$_3$ thin films. By knowing the generality of the annealing process in perovskite thin film preparation process, it could be deduced that the broad existence of twin domains in perovskite solar cells which is in agreement with the previous surface morphology characterization [82–84], we are continuing further research to directly observe the charge transportation under the twin domain morphology.

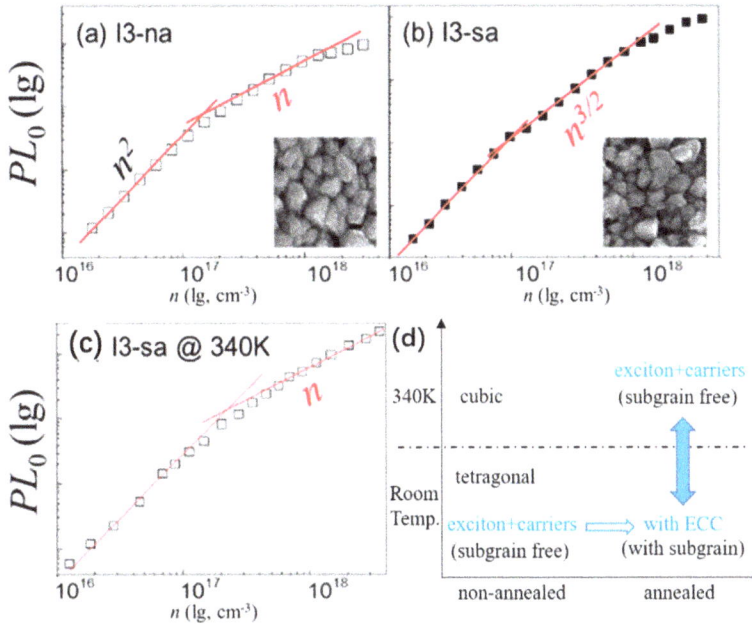

Figure 10. (**a–c**) Relation of photoluminescence intensity (PL_0) and excitation density (n) for MAPbI$_3$ thin films with non-annealed (I$_3$-na), sufficient annealed (I$_3$-sa), sufficient annealed while investigated at 340 K (I$_3$-sa @340 K). (**d**) Schematic of photoproduct system and the conversion relation for MAPbI$_3$ which illustrated that the typical ECC observed in annealed perovskite thin films at tetragonal phase. Reprinted with permission from reference [85]. Springer Nature.

6. Conclusions and Outlook

In this review, we give a detailed description of twin domains observed in inorganic and organometallic perovskite. We summarize the origin of twin domains emerged in metals and ceramics, together with explicit characterization of twin domain structure in typical inorganic perovskite BaTiO$_3$. Some typical corresponding subgrain twin domains emerged in MAPbI$_3$ thin films were revealed by PFM and TEM, given that the organometallic perovskite MAPbI$_3$ possesses the lattice structure analogous to BaTiO$_3$. Theoretical calculations about domain structure in MAPbI$_3$ indicating its benefit to charge transport is summarized. A new spectroscopic method which could reveal the generality of subgrain twin domain in perovskite films is also introduced.

Despite the relatively ambiguous understanding on the subgrain twin domain of organometallic perovskite, the experimental characterization and theoretical calculation presented above illustrate its importance in effective charge transportation and high photovoltaic performance. To further understand and improve perovskite solar cells' photovoltaic performance, tremendous attention on the origin and impact of the subgrain twin domain should be paid. One thing to be mentioned is that the subgrain twin domain of organometallic perovskite seems to correlate with the phase transformation from cubic phase to tetragonal phase akin to transformation twin of BaTiO$_3$. It is reasonable to infer the normal existence of the subgrain twin domain in organometallic films, providing that the indispensable annealing during the thin-film preparation process, which emphasizes the significance to explore the subgrain twin domain. To our understanding, the study on twinning domain in MAPbI$_3$ will finally reveal the fundamental mechanism of the material and the devices' high performance.

Acknowledgments: This work supported by the National Basic Research Program of China 2016YFB0401003; National Natural Science Foundation of China under grant Nos. 11527901, 61775004, 61575005, 11574009, U1605244, and 91750203.

Conflicts of Interest: The authors declare no conflicts of interest.

References

1. Perovskite Mineral Data. Available online: http://webmineral.com/data/Perovskite.shtml#.Wqd1auhubIU (accessed on 13 March 2018).
2. Bhalla, A.S.; Guo, R.; Roy, R. The perovskite structure—A review of its role in ceramic science and technology. *Mater. Res. Innov.* **2000**, *4*, 3–26. [CrossRef]
3. Kim, H.-S.; Im, S.H.; Park, N.-G. Organolead Halide Perovskite: New Horizons in Solar Cell Research. *J. Phys. Chem. C* **2014**, *118*, 5615–5625. [CrossRef]
4. Kojima, A.; Teshima, K.; Shirai, Y.; Miyasaka, T. Organometal Halide Perovskites as Visible-Light Sensitizers for Photovoltaic Cells. *J. Am. Chem. Soc.* **2009**, *131*, 6050–6051. [CrossRef] [PubMed]
5. Efficiency-Chart. Available online: Https://www.nrel.gov/pv/assets/images/efficiency-chart.png (accessed on 10 March 2018).
6. Liu, P.; He, X.; Ren, J.; Liao, Q.; Yao, J.; Fu, H. Organic–Inorganic Hybrid Perovskite Nanowire Laser Arrays. *ACS Nano* **2017**, *11*, 5766–5773. [CrossRef] [PubMed]
7. Zhu, H.; Fu, Y.; Meng, F.; Wu, X.; Gong, Z.; Ding, Q.; Gustafsson, M.V.; Trinh, M.T.; Jin, S.; Zhu, X.-Y. Lead halide perovskite nanowire lasers with low lasing thresholds and high quality factors. *Nat. Mater.* **2015**, *14*, 636–642. [CrossRef] [PubMed]
8. Qin, X.; Dong, H.; Hu, W. Green light-emitting diode from bromine based organic-inorganic halide perovskite. *Sci. China Mater.* **2015**, *58*, 186–191. [CrossRef]
9. Wang, N.; Cheng, L.; Ge, R.; Zhang, S.; Miao, Y.; Zou, W.; Yi, C.; Sun, Y.; Cao, Y.; Yang, R.; et al. Perovskite light-emitting diodes based on solution-processed self-organized multiple quantum wells. *Nat. Photonics* **2016**, *10*, 699–704. [CrossRef]
10. Dou, L.; Yang, Y.; You, J.; Hong, Z.; Chang, W.-H.; Li, G.; Yang, Y. Solution-processed hybrid perovskite photodetectors with high detectivity. *Nat. Commun.* **2014**, *5*, 5404. [CrossRef] [PubMed]
11. Liu, M.; Johnston, M.B.; Snaith, H.J. Efficient planar heterojunction perovskite solar cells by vapour deposition. *Nature* **2013**, *501*, 395–398. [CrossRef] [PubMed]
12. Green, M.A.; Ho-Baillie, A.; Snaith, H.J. The emergence of perovskite solar cells. *Nat. Photonics* **2014**, *8*, 506–514. [CrossRef]
13. Bertolotti, F.; Protesescu, L.; Kovalenko, M.V.; Yakunin, S.; Cervellino, A.; Billinge, S.J.L.; Terban, M.W.; Pedersen, J.S.; Masciocchi, N.; Guagliardi, A. Coherent Nanotwins and Dynamic Disorder in Cesium Lead Halide Perovskite Nanocrystals. *ACS Nano* **2017**, *11*, 3819–3831. [CrossRef] [PubMed]
14. Weller, M.T.; Weber, O.J.; Henry, P.F.; Pumpo, A.M.D.; Hansen, T.C. Complete structure and cation orientation in the perovskite photovoltaic methylammonium lead iodide between 100 and 352 K. *Chem. Commun.* **2015**, *51*, 4180–4183. [CrossRef] [PubMed]
15. Poglitsch, A.; Weber, D. Dynamic Disorder in Methylammoniumtrihalogenoplumbates(II) Observed by Millimeter-wave Spectroscopy. *J. Chem. Phys.* **1987**, *87*, 6373–6378. [CrossRef]
16. Arlt, G. Twinning in ferroelectric and ferroelastic ceramics: Stress relief. *J. Mater. Sci.* **1990**, *25*, 2655–2666. [CrossRef]
17. Snaith, H.J.; Abate, A.; Ball, J.M.; Eperon, G.E.; Leijtens, T.; Noel, N.K.; Stranks, S.D.; Wang, J.T.-W.; Wojciechowski, K.; Zhang, W. Anomalous Hysteresis in Perovskite Solar Cells. *J. Phys. Chem. Lett.* **2014**, *5*, 1511–1515. [CrossRef] [PubMed]
18. Chen, H.-W.; Sakai, N.; Ikegami, M.; Miyasaka, T. Emergence of Hysteresis and Transient Ferroelectric Response in Organo-Lead Halide Perovskite Solar Cells. *J. Phys. Chem. Lett.* **2015**, *6*, 164–169. [CrossRef] [PubMed]
19. Frost, J.M.; Butler, K.T.; Walsh, A. Molecular ferroelectric contributions to anomalous hysteresis in hybrid perovskite solar cells. *APL Mater.* **2014**, *2*, 081506. [CrossRef]

20. Frost, J.M.; Butler, K.T.; Brivio, F.; Hendon, C.H.; van Schilfgaarde, M.; Walsh, A. Atomistic Origins of High-Performance in Hybrid Halide Perovskite Solar Cells. *Nano Lett.* **2014**, *14*, 2584–2590. [CrossRef] [PubMed]

21. Rothmann, M.U.; Li, W.; Zhu, Y.; Bach, U.; Spiccia, L.; Etheridge, J.; Cheng, Y.-B. Direct observation of intrinsic twin domains in tetragonal CH3NH3PbI3. *Nat. Commun.* **2017**, *8*. [CrossRef] [PubMed]

22. Janovec, V.; Hahn, T.; Klapper, H. Twinning and domain structures. In *International Tables for Crystallography Volume D: Physical Properties of Crystals*; International Tables for Crystallography; Springer: Dordrecht, The Netherlands, 2006; pp. 377–392, ISBN 978-1-4020-0714-9.

23. Hahn, T.; Klapper, H. Twinning of crystals. In *International Tables for Crystallography Volume D: Physical Properties of Crystals*; Springer: Dordrecht, The Netherlands, 2006; pp. 393–448.

24. Buerger, M.J. The genesis of twin crystals. *Am. Mineral.* **1945**, *30*, 469–482.

25. Nespolo, M.; Ferraris, G. The oriented attachment mechanism in the formation of twins-a survey. *Eur. J. Mineral.* **2004**, *16*, 401–406. [CrossRef]

26. Kotrbová, M.; Kadečková, S.; Novák, J.; Brádler, J.; Smirnov, G.V.; Shvydko, Y.V. Growth and perfection of flux grown FeBO3 and 57FeBO3 crystals. *J. Cryst. Growth* **1985**, *71*, 607–614. [CrossRef]

27. Ng, H.N.; Calvo, C. Refinement of the Crystal Structure of the Low-quartz Modification of Ferric Phosphate. *Can. J. Chem.* **1975**, *53*, 2064–2067. [CrossRef]

28. Engel, G.; Klapper, H.; Krempl, P.; Mang, H. Growth twinning in quartz-homeotypic gallium orthophosphate crystals. *J. Cryst. Growth* **1989**, *94*, 597–606. [CrossRef]

29. Fang, X.-S.; Ye, C.-H.; Zhang, L.-D.; Xie, T. Twinning-Mediated Growth of Al2O3 Nanobelts and Their Enhanced Dielectric Responses. *Adv. Mater.* **2005**, *17*, 1661–1665. [CrossRef]

30. Song, S.G.; Gray, G.T. Structural interpretation of the nucleation and growth of deformation twins in Zr and Ti—II. Tem study of twin morphology and defect reactions during twinning. *Acta Metall. Mater.* **1995**, *43*, 2339–2350. [CrossRef]

31. Clayton, J.D. Mechanical Twinning in Crystal Plasticity. In *Nonlinear Mechanics of Crystals*; Solid Mechanics and Its Applications; Springer: Dordrecht, The Netherlands, 2011; pp. 379–421, ISBN 978-94-007-0349-0.

32. Van Houtte, P. Simulation of the rolling and shear texture of brass by the Taylor theory adapted for mechanical twinning. *Acta Metall.* **1978**, *26*, 591–604. [CrossRef]

33. Hanada, S.; Izumi, O. Transmission electron microscopic observations of mechanical twinning in metastable beta titanium alloys. *Metall. Trans. A* **1986**, *17*, 1409–1420. [CrossRef]

34. Allain, S.; Chateau, J.-P.; Dahmoun, D.; Bouaziz, O. Modeling of mechanical twinning in a high manganese content austenitic steel. *Mater. Sci. Eng. A* **2004**, *387–389*, 272–276. [CrossRef]

35. Březina, B.; Fousek, J. Twinning in Crystals. In *Crystal Growth in Science and Technology*; NATO ASI Series; Springer: Boston, MA, USA, 1989; pp. 185–195. ISBN 978-1-4612-7861-0.

36. Chan, C.-J.; Lange, F.F.; Rühle, M.; Jue, J.-F.; Virkar, A.V. Ferroelastic Domain Switching in Tetragonal Zirconia Single Crystals—Microstructural Aspects. *J. Am. Ceram. Soc.* **1991**, *74*, 807–813. [CrossRef]

37. Lim, A.R.; Choh, S.H.; Jang, M.S. Prominent ferroelastic domain walls in BiVO4 crystal. *J. Phys. Condens. Matter* **1995**, *7*, 7309. [CrossRef]

38. Meschke, F.; Kolleck, A.; Schneider, G.A. R-curve behaviour of BaTiO3 due to stress-induced ferroelastic domain switching. *J. Eur. Ceram. Soc.* **1997**, *17*, 1143–1149. [CrossRef]

39. Cahn, R.W. Twinned crystals. *Adv. Phys.* **1954**, *3*, 363–445. [CrossRef]

40. Cook, W.R. Domain Twinning in Barium Titanate Ceramics. *J. Am. Ceram. Soc.* **1956**, *39*, 17–19. [CrossRef]

41. Forsbergh, P.W. Domain Structures and Phase Transitions in Barium Titanate. *Phys. Rev.* **1949**, *76*, 1187–1201. [CrossRef]

42. Cheng, S.-Y.; Ho, N.-J.; Lu, H.-Y. Transformation-Induced Twinning: The 90° and 180° Ferroelectric Domains in Tetragonal Barium Titanate. *J. Am. Ceram. Soc.* **2006**, *89*, 2177–2187. [CrossRef]

43. Hooton, J.A.; Merz, W.J. Etch Patterns and Ferroelectric Domains in BaTiO3 Single Crystals. *Phys. Rev.* **1955**, *98*, 409–413. [CrossRef]

44. Merz, W.J. Domain Formation and Domain Wall Motions in Ferroelectric BaTiO3 Single Crystals. *Phys. Rev.* **1954**, *95*, 690–698. [CrossRef]

45. Chou, J.-F.; Lin, M.-H.; Lu, H.-Y. Ferroelectric domains in pressureless-sintered barium titanate. *Acta Mater.* **2000**, *48*, 3569–3579. [CrossRef]

46. Klassen-Neklyudova, M.V. *Mechanical Twinning of Crystals*; Springer: Boston, MA, USA, 1964; ISBN 978-1-4684-1541-4.

47. Nielsen, J.W.; Linares, R.C.; Koonce, S.E. Genesis of the Barium Titanate Butterfly Twin. *J. Am. Ceram. Soc.* **1962**, *45*, 12–17. [CrossRef]

48. De Vries, R.C. Observations on Growth of BaTiO₃, Crystals from KF Solutions. *J. Am. Ceram. Soc.* **1959**, *42*, 547–558. [CrossRef]

49. Choi, J.J.; Yang, X.; Norman, Z.M.; Billinge, S.J.L.; Owen, J.S. Structure of Methylammonium Lead Iodide Within Mesoporous Titanium Dioxide: Active Material in High-Performance Perovskite Solar Cells. *Nano Lett.* **2014**, *14*, 127–133. [CrossRef] [PubMed]

50. Grancini, G.; Marras, S.; Prato, M.; Giannini, C.; Quarti, C.; De Angelis, F.; De Bastiani, M.; Eperon, G.E.; Snaith, H.J.; Manna, L.; et al. The Impact of the Crystallization Processes on the Structural and Optical Properties of Hybrid Perovskite Films for Photovoltaics. *J. Phys. Chem. Lett.* **2014**, *5*, 3836–3842. [CrossRef] [PubMed]

51. De Bastiani, M.; D'Innocenzo, V.; Stranks, S.D.; Snaith, H.J.; Petrozza, A. Role of the crystallization substrate on the photoluminescence properties of organo-lead mixed halides perovskites. *APL Mater.* **2014**, *2*, 081509. [CrossRef]

52. D'Innocenzo, V.; Srimath Kandada, A.R.; De Bastiani, M.; Gandini, M.; Petrozza, A. Tuning the Light Emission Properties by Band Gap Engineering in Hybrid Lead Halide Perovskite. *J. Am. Chem. Soc.* **2014**, *136*, 17730–17733. [CrossRef] [PubMed]

53. Nie, W.; Tsai, H.; Asadpour, R.; Blancon, J.-C.; Neukirch, A.J.; Gupta, G.; Crochet, J.J.; Chhowalla, M.; Tretiak, S.; Alam, M.A.; et al. High-efficiency solution-processed perovskite solar cells with millimeter-scale grains. *Science* **2015**, *347*, 522–525. [CrossRef] [PubMed]

54. Yamada, Y.; Yamada, T.; Phuong, L.Q.; Maruyama, N.; Nishimura, H.; Wakamiya, A.; Murata, Y.; Kanemitsu, Y. Dynamic Optical Properties of CH₃NH₃PbI₃ Single Crystals as Revealed by One- and Two-Photon Excited Photoluminescence Measurements. *J. Am. Chem. Soc.* **2015**, *137*, 10456–10459. [CrossRef] [PubMed]

55. Bi, Y.; Hutter, E.M.; Fang, Y.; Dong, Q.; Huang, J.; Savenije, T.J. Charge Carrier Lifetimes Exceeding 15 μs in Methylammonium Lead Iodide Single Crystals. *J. Phys. Chem. Lett.* **2016**, *7*, 923–928. [CrossRef] [PubMed]

56. De Quilettes, D.W.; Vorpahl, S.M.; Stranks, S.D.; Nagaoka, H.; Eperon, G.E.; Ziffer, M.E.; Snaith, H.J.; Ginger, D.S. Impact of microstructure on local carrier lifetime in perovskite solar cells. *Science* **2015**, *348*, 683–686. [CrossRef] [PubMed]

57. Yun, J.S.; Ho-Baillie, A.; Huang, S.; Woo, S.H.; Heo, Y.; Seidel, J.; Huang, F.; Cheng, Y.-B.; Green, M.A. Benefit of Grain Boundaries in Organic–Inorganic Halide Planar Perovskite Solar Cells. *J. Phys. Chem. Lett.* **2015**, *6*, 875–880. [CrossRef] [PubMed]

58. Yun, J.S.; Seidel, J.; Kim, J.; Soufiani, A.M.; Huang, S.; Lau, J.; Jeon, N.J.; Seok, S.I.; Green, M.A.; Ho-Baillie, A. Critical Role of Grain Boundaries for Ion Migration in Formamidinium and Methylammonium Lead Halide Perovskite Solar Cells. *Adv. Energy Mater.* **2016**, *6*. [CrossRef]

59. Yuan, Y.; Huang, J. Ion Migration in Organometal Trihalide Perovskite and Its Impact on Photovoltaic Efficiency and Stability. *Acc. Chem. Res.* **2016**, *49*, 286–293. [CrossRef] [PubMed]

60. Ponseca, C.S.; Savenije, T.J.; Abdellah, M.; Zheng, K.; Yartsev, A.; Pascher, T.; Harlang, T.; Chabera, P.; Pullerits, T.; Stepanov, A.; et al. Organometal Halide Perovskite Solar Cell Materials Rationalized: Ultrafast Charge Generation, High and Microsecond-Long Balanced Mobilities, and Slow Recombination. *J. Am. Chem. Soc.* **2014**, *136*, 5189–5192. [CrossRef] [PubMed]

61. Hermes, I.M.; Bretschneider, S.A.; Bergmann, V.W.; Li, D.; Klasen, A.; Mars, J.; Tremel, W.; Laquai, F.; Butt, H.-J.; Mezger, M.; et al. Ferroelastic Fingerprints in Methylammonium Lead Iodide Perovskite. *J. Phys. Chem. C* **2016**, *120*, 5724–5731. [CrossRef]

62. Strelcov, E.; Dong, Q.; Li, T.; Chae, J.; Shao, Y.; Deng, Y.; Gruverman, A.; Huang, J.; Centrone, A. CH3NH3PbI3 perovskites: Ferroelasticity revealed. *Sci. Adv.* **2017**, *3*, e1602165. [CrossRef] [PubMed]

63. Roehm, H.; Leonhard, T.; Hoffmannbc, M.J.; Colsmann, A. Ferroelectric domains in methylammonium lead iodide perovskite thin-films. *Energy Environ. Sci.* **2017**, *10*, 950–955. [CrossRef]

64. Kim, T.W.; Uchida, S.; Matsushita, T.; Cojocaru, L.; Jono, R.; Kimura, K.; Matsubara, D.; Shirai, M.; Ito, K.; Matsumoto, H.; et al. Self-Organized Superlattice and Phase Coexistence inside Thin Film Organometal Halide Perovskite. *Adv. Mater.* **2018**, *30*. [CrossRef] [PubMed]

65. Rothmann, M.U.; Li, W.; Etheridge, J.; Cheng, Y.-B. Microstructural Characterisations of Perovskite Solar Cells—From Grains to Interfaces: Techniques, Features, and Challenges. *Adv. Energy Mater.* **2017**, *7*, 1700912. [CrossRef]
66. Kutes, Y.; Ye, L.; Zhou, Y.; Pang, S.; Huey, B.D.; Padture, N.P. Direct Observation of Ferroelectric Domains in Solution-Processed CH3NH3PbI3 Perovskite Thin Films. *J. Phys. Chem. Lett.* **2014**, *5*, 3335–3339. [CrossRef] [PubMed]
67. Wei, J.; Zhao, Y.; Li, H.; Li, G.; Pan, J.; Xu, D.; Zhao, Q.; Yu, D. Hysteresis Analysis Based on the Ferroelectric Effect in Hybrid Perovskite Solar Cells. *J. Phys. Chem. Lett.* **2014**, *5*, 3937–3945. [CrossRef] [PubMed]
68. Chen, B.; Yang, M.; Priya, S.; Zhu, K. Origin of J–V Hysteresis in Perovskite Solar Cells. *J. Phys. Chem. Lett.* **2016**, *7*, 905–917. [CrossRef] [PubMed]
69. Fan, Z.; Xiao, J.; Sun, K.; Chen, L.; Hu, Y.; Ouyang, J.; Ong, K.P.; Zeng, K.; Wang, J. Ferroelectricity of CH3NH3PbI3 Perovskite. *J. Phys. Chem. Lett.* **2015**, *6*, 1155–1161. [CrossRef] [PubMed]
70. Pratibha Mahale, S.G.; Mahale, P.; Kore, B.P.; Mukherjee, S.; Pavan, M.S.; De, C.; Ghara, S.; Sundaresan, A.; Pandey, A.; Guru Row, T.N.; et al. Is CH3NH3PbI3 Polar? *J. Phys. Chem. Lett.* **2016**, *7*, 2412–2419. [CrossRef]
71. Kim, Y.-J.; Dang, T.-V.; Choi, H.-J.; Park, B.-J.; Eom, J.-H.; Song, H.-A.; Seol, D.; Kim, Y.; Shin, S.-H.; Nah, J.; et al. Piezoelectric properties of CH3NH3PbI3 perovskite thin films and their applications in piezoelectric generators. *J. Mater. Chem. A* **2016**, *4*, 756–763. [CrossRef]
72. Rakita, Y.; Bar-Elli, O.; Meirzadeh, E.; Kaslasi, H.; Peleg, Y.; Hodes, G.; Lubomirsky, I.; Oron, D.; Ehre, D.; Cahen, D. Tetragonal CH3NH3PbI3 is ferroelectric. *Proc. Natl. Acad. Sci. USA* **2017**, *114*, E5504–E5512. [CrossRef] [PubMed]
73. Hoque, M.N.F.; Yang, M.; Li, Z.; Islam, N.; Pan, X.; Zhu, K.; Fan, Z. Polarization and Dielectric Study of Methylammonium Lead Iodide Thin Film to Reveal its Nonferroelectric Nature under Solar Cell Operating Conditions. *ACS Energy Lett.* **2016**, *1*, 142–149. [CrossRef]
74. Santiso, J.; Balcells, L.; Konstantinovic, Z.; Roqueta, J.; Ferrer, P.; Pomar, A.; Martínez, B.; Sandiumenge, F. Thickness evolution of the twin structure and shear strain in LSMO films. *CrystEngComm* **2013**, *15*, 3908–3918. [CrossRef]
75. Zhao, P.; Xu, J.; Ma, C.; Ren, W.; Wang, L.; Bian, L.; Chang, A. Spontaneous polarization behaviors in hybrid halide perovskite film. *Scr. Mater.* **2015**, *102*, 51–54. [CrossRef]
76. Yin, W.-J.; Yang, J.-H.; Kang, J.; Yan, Y.; Wei, S.-H. Halide perovskite materials for solar cells: A theoretical review. *J. Mater. Chem. A* **2015**, *3*, 8926–8942. [CrossRef]
77. Liu, S.; Zheng, F.; Koocher, N.Z.; Takenaka, H.; Wang, F.; Rappe, A.M. Ferroelectric Domain Wall Induced Band Gap Reduction and Charge Separation in Organometal Halide Perovskites. *J. Phys. Chem. Lett.* **2015**, *6*, 693–699. [CrossRef] [PubMed]
78. Rashkeev, S.N.; El-Mellouhi, F.; Kais, S.; Alharbi, F.H. Domain Walls Conductivity in Hybrid Organometallic Perovskites and Their Essential Role in CH3NH3PbI3 Solar Cell High Performance. *Sci. Rep.* **2015**, *5*. [CrossRef] [PubMed]
79. Bi, F.; Markov, S.; Wang, R.; Kwok, Y.; Zhou, W.; Liu, L.; Zheng, X.; Chen, G.; Yam, C. Enhanced Photovoltaic Properties Induced by Ferroelectric Domain Structures in Organometallic Halide Perovskites. *J. Phys. Chem. C* **2017**, *121*, 11151–11158. [CrossRef]
80. Rossi, D.; Pecchia, A.; der Maur, M.A.; Leonhard, T.; Röhm, H.; Hoffmann, M.J.; Colsmann, A.; Carlo, A.D. On the importance of ferroelectric domains for the performance of perovskite solar cells. *Nano Energy* **2018**, *48*, 20–26. [CrossRef]
81. Tsai, H.; Asadpour, R.; Blancon, J.-C.; Stoumpos, C.C.; Durand, O.; Strzalka, J.W.; Chen, B.; Verduzco, R.; Ajayan, P.M.; Tretiak, S.; et al. Light-induced lattice expansion leads to high-efficiency perovskite solar cells. *Science* **2018**, *360*, 67–70. [CrossRef] [PubMed]
82. Gratia, P.; Grancini, G.; Audinot, J.-N.; Jeanbourquin, X.; Mosconi, E.; Zimmermann, I.; Dowsett, D.; Lee, Y.; Grätzel, M.; De Angelis, F.; et al. Intrinsic Halide Segregation at Nanometer Scale Determines the High Efficiency of Mixed Cation/Mixed Halide Perovskite Solar Cells. *J. Am. Chem. Soc.* **2016**, *138*, 15821–15824. [CrossRef] [PubMed]
83. Tan, H.; Jain, A.; Voznyy, O.; Lan, X.; de Arquer, F.P.G.; Fan, J.Z.; Quintero-Bermudez, R.; Yuan, M.; Zhang, B.; Zhao, Y.; et al. Efficient and stable solution-processed planar perovskite solar cells via contact passivation. *Science* **2017**, eaai9081. [CrossRef] [PubMed]

84. Jiang, Y.; Leyden, M.R.; Qiu, L.; Wang, S.; Ono, L.K.; Wu, Z.; Juarez-Perez, E.J.; Qi, Y. Combination of Hybrid CVD and Cation Exchange for Upscaling Cs-Substituted Mixed Cation Perovskite Solar Cells with High Efficiency and Stability. *Adv. Funct. Mater.* **2018**, *28*. [CrossRef]

85. Wang, W.; Li, Y.; Wang, X.; Liu, Y.; Lv, Y.; Wang, S.; Wang, K.; Shi, Y.; Xiao, L.; Chen, Z.; et al. Interplay between Exciton and Free Carriers in Organolead Perovskite Films. *Sci. Rep.* **2017**, *7*, 14760. [CrossRef] [PubMed]

86. Telfah, H.; Jamhawi, A.; Teunis, M.B.; Sardar, R.; Liu, J. Ultrafast Exciton Dynamics in Shape-Controlled Methylammonium Lead Bromide Perovskite Nanostructures: Effect of Quantum Confinement on Charge Carrier Recombination. *J. Phys. Chem. C* **2017**. [CrossRef]

87. Makarov, N.S.; Guo, S.; Isaienko, O.; Liu, W.; Robel, I.; Klimov, V.I. Spectral and Dynamical Properties of Single Excitons, Biexcitons, and Trions in Cesium–Lead-Halide Perovskite Quantum Dots. *Nano Lett.* **2016**, *16*, 2349–2362. [CrossRef] [PubMed]

88. Wang, W.; Li, Y.; Wang, X.; Lv, Y.; Wang, S.; Wang, K.; Shi, Y.; Xiao, L.; Chen, Z.; Gong, Q. Density-dependent dynamical coexistence of excitons and free carriers in the organolead perovskite $CH_3NH_3PbI_3$. *Phys. Rev. B* **2016**, *94*, 140302. [CrossRef]

89. D'Innocenzo, V.; Grancini, G.; Alcocer, M.J.P.; Kandada, A.R.S.; Stranks, S.D.; Lee, M.M.; Lanzani, G.; Snaith, H.J.; Petrozza, A. Excitons versus free charges in organo-lead tri-halide perovskites. *Nat. Commun.* **2014**, *5*. [CrossRef] [PubMed]

Review

Growth of Metal Halide Perovskite, from Nanocrystal to Micron-Scale Crystal: A Review

Haijiao Harsan Ma [1],*, Muhammad Imran [2], Zhiya Dang [2],* and Zhaosheng Hu [3],*

1   Low Dimensional Quantum Physics & Device Group, State Key Discipline Laboratory of Wide Band Gap Semiconductor Technology, School of Microelectronics, Xidian University, 2 South Taibai Road, Xi'an 710071, China
2   Department of Nanochemistry, Istituto Italiano di Tecnologia, via Morego 30, 16163 Genova, Italy; muhammad.imran@iit.it
3   Graduate School of Life Science and Systems Engineering, Kyushu Institute of Technology, 2-4 Hibikino, Wakamatsu-ku, Kitakyushu 808-0196, Japan
*   Correspondence: mahj07@xidian.edu.cn (H.H.M.); Zhiya.Dang@iit.it (Z.D.); huzhaosheng07@gmail.com (Z.H.)

Received: 27 March 2018; Accepted: 17 April 2018; Published: 24 April 2018

Abstract: Metal halide perovskite both in the form of nanocrystal and thin films recently emerged as the most promising semiconductor material covering a huge range of potential applications from display technologies to photovoltaics. Colloidal inorganic and organic–inorganic hybrid metal halide perovskite nanocrystals (NCs) have received tremendous attention due to their high photoluminescence quantum yields, while large grain perovskite films possess fewer defects, and a long diffusion length providing high-power conversion efficiency in planar devices. In this review, we summarize the different synthesis routes of metal halide perovskite nanocrystals and the recent methodologies to fabricate high-quality micron scale crystals in the form of films for planar photovoltaics. For the colloidal synthesis of halide perovskite NCs, two methods including ligand-assisted reprecipitation and hot injection are mainly applied, and the doping of metal ions in NCs as well as anion exchange reactions are widely used to tune their optical properties. In addition, recent growth methods and underlying mechanism for high-quality micron size crystals are also investigated, which are summarized as solution-process methods (including the anti-solvent method, solvent vapor annealing technology, Ostwald ripening, additive engineering and geometrically-confined lateral crystal growth) and the physical method (vapor-assisted crystal growth).

Keywords: halide perovskite; nanocrystals; thin film; ligand-assisted reprecipitation method; hot injection method; anion exchange; anti-solvent method; solvent vapor annealing technology; vapor-assisted crystal growth

---

1. Introduction

Metal halide perovskite is emerging as a promising semiconductor and efficient light-harvester in high-performance, low-cost and large-coverage photovoltaics [1–3]. The high optical coefficient makes it a good candidate as the thin absorber layer, and the long diffusion length of both electrons and holes leads to high short-circuit current density. On the one hand, halide perovskite nanocrystals (NCs) are used for photovoltaics, which have received tremendous attention due to their high photoluminescence quantum yields, reaching almost 100%. Moreover, nanocrystals have the advantage of optical tunability compared to their bulk. On the other hand, for planar perovskite devices, a verity of preparation methods for micron-scale crystals have been promoted including the one-step methylammonium lead iodide (MAPbI$_3$) solution, two-step sequence spin-coating and vapor-coevaporation. However, the quality of the crystals and devices are not as good as one would expect as these methods usually leads to polynanocrystal films. Although these methods lead to more than 20% power-conversion efficiency in photovoltaics, recent studies have shown that impurities and defects lead to high carrier recombination, limiting further device performance. In addition, the small size of crystal grains results in obvious hysteresis phenomena, and also leads to a diffusion length of around 1 μm, which is not comparable with single crystals with a typical diffusion length of 100 μm [4–6]. In terms of environmentally friendly devices and materials, due to the containment of the toxic element lead in the metal halide perovskite, thin films are preferred in the real applications. Furthermore, because of the rapid degradation of light absorption with the decrease in thickness, the quality of grown perovskite crystals needs to be improved. Recent advances in the technique and methods have made it possible to grow high-quality metal halide perovskite cyrstals both at nano-scale and micron-scale. In this review, we focus on the synthesis of halide perovskite nanocrystals and their post-synthesis transformations as well as recent technology to fabricate high-quality large crystal films for planar photovoltaics.

2. The Fabrication of Lead Halide Perovskite Colloidal Nanocrystals

Lead halide perovskites (LHP) in the form of colloidal nanocrystals (NCs), such as organic–inorganic $CH_3NH_3PbX_3$, (MAPbX$_3$), $CH(NH_2)_2PbX_3$ (FAPbX$_3$) and all-inorganic CsPbX$_3$ LHPs (X = Cl, Br, I) have been intensively investigated for various applications including light-emitting devices (LEDs) and photodetectors, due to their color-tunable and narrow-band emissions as well as easy synthesis, convenient solution-based processing, and low fabrication cost [7–13]. Various approaches have been proposed for the direct synthesis of metal and organometal halide perovskite colloidal NCs (e.g., CsPbX$_3$, MAPbX$_3$, FAPbX$_3$, $X^- = Cl^-$, Br^-, I^-), among which the most common are the hot-injection and the ligand-assisted reprecipitation (LARP) approaches.

2.1. Colloidal Synthesis of MAPbX$_3$ Nanocrystals (NCs)

Motivated by the rapid development of the lead halide perovskite (LHP) thin films as a light-harvesting material for solar-cell applications, first colloidal synthesis of MAPbBr$_3$ NCs was carried out by the Pérez-Prieto et al. by sthe olvent-induced reprecipitation approach [15]. Nanocrystals were stabilized by using octylammonium bromide and octadecylammonium bromide as surfactants and were colloidally stable for up to 3 months. The absorption and photoluminescence (PL) peaks of these highly crystalline of MAPbBr$_3$ NCs were at 527 and 530 nm, respectively with a photoluminescence quantum yield (PLQY) of ~20%. Later, Zhang et al. modified the aforementioned procedure and introduced the LARP approach for the synthesis of MAPbX$_3$ NCs by replacing octylammonium bromide and octadecylammonium bromide with n-octylamine and oleic acid as a co-ligands system into the reprecipitation process [14]. As shown in Figure 1, this method is based on the reprecipitation of lead halide (PbX$_2$, X = Cl, Br, I) and organic halide (CH$_3$NH$_3$X, X = Cl, Br, I) salts in the presence of ligands; for instance, lead halide and organic halide salts are dissolved in strongly polar solvents like dimethylformamide (DMF) and are subsequently added dropwise to a solution of a non-polar medium like toluene in the presence of ligands. The miscibility gap between polar and non-polar solvents solubility consequently triggers the recrystallization of lead halide perovskite NCs. The NCs obtained were brightly luminescent with absolute PLQYs up to 70% in the case of CH$_3$NH$_3$PbBr$_3$ quantum dots (QDs), for which the transmission electron microscope (TEM) image and X-ray diffraction (XRD) results are shown in Figure 2a,b. The ligands' role was further investigated and it was observed that in absence of the octylamine, precursors undergo fast crystallization that leads to larger NCs with very low PLQY and subsequently precipitated out of the solution.

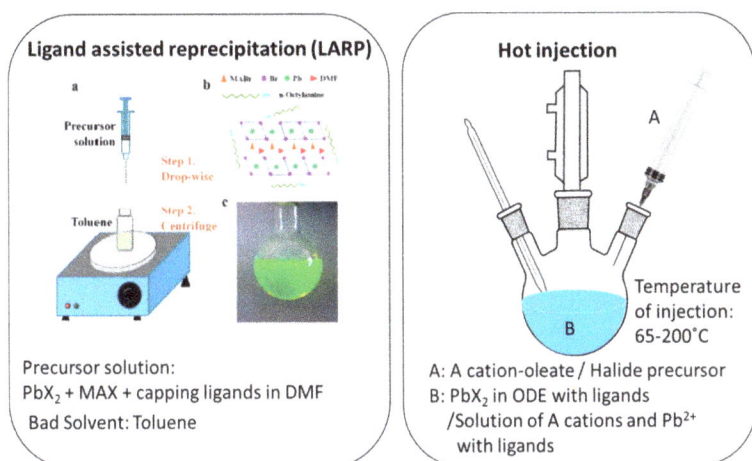

Figure 1. Schematics showing two methods of colloidal synthesis of lead-based halide perovskite (LHP) nanocrystals (NCs), in which the schematics for ligand-assisted reprecipitation (LARP) is reproduced with permission from Ref. [14], Copyright (2015) American Chemical Society.

The same approach was further extended to synthesize $CH_3NH_3PbX_3$ QDs (X = Cl, I) through halide substitutions. Interestingly, they successfully demonstrated a series of colloidal $CH_3NH_3PbX_3$ QDs with tunable compositions by simply adjusting the ratios of PbX_2 salts in the precursor solution, see Figure 2c shows that the PL spectra can be finely tuned from 407 to 734 nm by varying the halide composition (X = Br, Cl, and I).

Later, hot injection-based synthesis of $MAPbX_3$ (X = Br, I) was reported by Vybornyi et al. [16] As shown in Figure 1, this alternative synthesis route is basically an ionic metathesis approach that does not involve any polar solvent. Methylamine solution in THF was injected into a solution of PbX_2 (X = Br, I) in Octadecene (ODE)-containing long-chain capping ligands (an octylamine (OAm)/oleic acid (OA) mixture). The proton needed to form CH3NH3+ is provided by OA whereas PbX_2 serves as both the Pb^{2+} and X^- source, releasing Pb-oleate as a byproduct. $MAPbBr_3$ NPLs, NWs and nearly cubic shape $MAPbI_3$ NCs were successfully obtained by varying the amounts of surfactants (OAm/OA). Resultant NCs have poor optical properties compared to the NCs synthesized by LARP. Basically, the main limitation of the LARP method and two-precursor hot-injection method (precursor 1: Cs-oleate, precursor 2: Lead halide salt complex) is that both methods employ PbX_2 (X = Cl, Br, or I) salts as both lead and halide precursors. Therefore, in both of these cases, one is not allowed to precisely tune the amount of reaction species. In order to overcome the restrictions associated with the aforementioned synthetic procedures, recently Imran et al. introduced a new colloidal synthesis approach that can lead to either all-inorganic or organic–inorganic lead-based halide perovskite NCs [17]. The synthesis relies on the use of acyl halides as halide precursors that can be easily injected, at any temperature, into a solution of metal cations to trigger the nucleation and the growth of the halide NCs. Acyl halides, commonly used as versatile building blocks in organic chemistry reactions, are well known for their strong reactivity toward nucleophilic compounds (e.g., amines, alcohols, carboxylic acids) to form carboxylic acid derivatives (e.g., amides, esters, anhydrides) and releasing at the same time halide anions [18]. By simply adjusting the relative amount of cation precursors, ligands, solvents, benzoyl halides and the injection temperature, it was possible to synthesize either all-inorganic or organic-inorganic $APbX_3$ (A = Cs, MA or FA and X = Cl, Br or I) NCs with excellent control over the size distribution, very high phase purity and excellent optical properties [17]. $MAPbX_3$ NCs synthesized by this method have nearly cubic morphology and very high phase purity in all cases (see Figure 2d–i).

Resultant NCs were characterized with narrow emission line width from 15 nm to 43 nm along with very high PLQY up to 92% in case of MAPbBr3 (see Figure 2j–l).

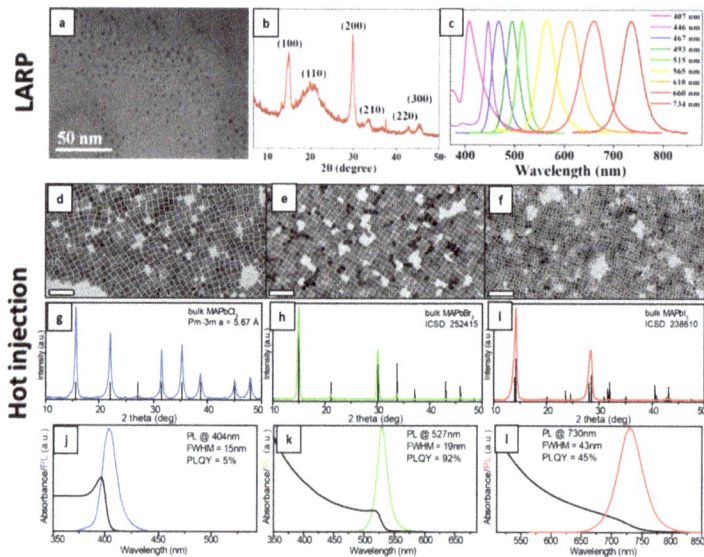

Figure 2. (**a–c**) MAPbX3 NCs synthesized by LARP reproduced with permission from Ref. [14], Copyright (2015) American Chemical Society: (**a**) Transmission electron microscope (TEM) image and (**b**) X-ray diffraction (XRD) pattern of MAPbBr3 NCs, (**c**) photoluminescence (PL) spectra of MAPbX3 NCs. (**d–l**) MAPbX3 NCs synthesized by the hot-injection method reproduced with permission from Ref. [17], Copyright (2018) American Chemical Society: Bright field TEM images (Scale bars are 100 nm in all images), XRD patterns, along with absorption and PL spectra of MAPbCl3, MAPbBr3, and MAPbI3 NCs respectively.

2.2. Colloidal Synthesis of CsPbX3 NCs

The formerly reported LARP strategy was further extended by Sun et al. for the preparation of fully inorganic LHP NCs [19]. The synthesis of CsPbX3 (X$^-$ = Cl$^-$, Br$^-$, I$^-$) spherical quantum dots was performed at room temperature (i.e., 25 °C) by mixing a solution of precursors in good solvent (such as *N,N*-dimethylformamide, DMF; tetrahydrofuran, THF; and dimethyl sulfoxide, DMSO) into a poor solvent (such as toluene and hexane). The shape control of CsPbX3 NCs was also demonstrated such as nanocubes, one-dimensional nanorods, and two-dimensional nanoplatelets a few unit cells in thickness by choosing different organic acid and amine ligands (see Figure 3a,b). Another liquid phase method is to trigger the nucleation and growth of platelets at room temperature (RT) by the injection of acetone in a mixture of precursors [20].

Protesescu et al. first developed a polar solvent-free two step ionic metathesis approach for the colloidal synthesis of brightly luminescent CsPbX3 NCs [21]. As shown in Figure 1, typically the CsPbX3 NCs were obtained by reacting the Cs-oleate with lead halide in boiling ODE solvent at 140–200 °C in the presence of a binary ligand system composed of aliphatic carboxylic acids and primary amines as surfactants to stabilize the NCs. Owing to the ionic nature of the ternary compound, very fast nucleation and growth kinetics of the NCs (1–3 s) were witnessed by in situ PL measurement with a charged coupled device (CCD) array detector, and the overall reaction mechanism can be summarized in the following equation:

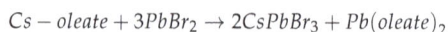

$$Cs - oleate + 3PbBr_2 \rightarrow 2CsPbBr_3 + Pb(oleate)_2$$

These CsPbX$_3$ NCs possess bright luminescence with narrow emission-line width typically from 11–42 nm. The PL peak position can be engineered across the entire visible spectrum by simply adjusting the halide composition (Cl:Br or Br:I) or by altering the size of the NCs. Later, a three precursor-based modified hot-injection method was developed by the Manna group which overcame the limitation of using lead halide salts as a source of both lead and halide ions. Very briefly, CsPbX$_3$ NCs were synthesized by dissolving metal cations (cesium carbonate and lead acetate) in octadecene by using oleylamine and oleic acid as a surfactants. Subsequently, the solution was heated up to the desired temperature (170–200 °C) and the benzoyl halide precursor was swiftly injected into the reaction flask, triggering the immediate nucleation and growth of the NCs. This simple synthesis method results in very monodisperse, strongly fluorescent NCs with high-phase purity and with narrow emission linewidth along with PLQY as high as 92% (see Figure 3c–k) [17]. In general, cesium lead halide-based perovskite NCs exhibit excellent optical properties, while CsPbCl$_3$ NCs are typically characterized by a significant non-radiative decay [22,23]. Interestingly, CsPbCl$_3$ NCs were of particular interest from the recently developed synthesis protocol by Manna group where they reported PLQY for CsPbCl$_3$ NCs as high as 65%, which is a record value, as shown in Figure 3c,f,i [17]. It is worth mentioning that such a high PLQY was observed only when employing a large excess of the Cl precursor, i.e., 1.8 mmol of benzoyl chloride and 0.2 mmol of the Pb precursor while with weak PL emission when the same NCs were prepared using a lower amount of benzoyl chloride [17].

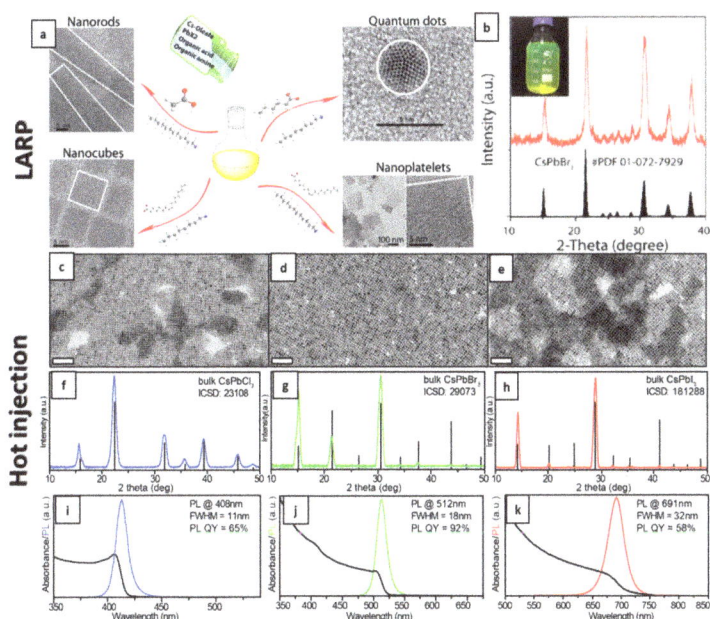

Figure 3. (**a,b**) CsPbX$_3$ NCs synthesized by LARP reproduced with permission from Ref. [19], Copyright (2016) American Chemical Society: (**a**) Bright field TEM image and (**b**) XRD pattern of CsPbX$_3$ NCs. (**c–k**) CsPbX$_3$ NCs synthesized by hot injection method reproduced with permission from Ref. [17], Copyright (2018) American Chemical Society. Bright field TEM images (scale bars are 100 nm in all images), XRD patterns of highly oriented NCs, along with absorption and PL spectra of CsPbCl$_3$, CsPbBr$_3$, and CsPbI$_3$ NCs respectively.

The recent studies on the crystal structure of halide perovskite show that in small NCs, the reduced sizes result in broad Bragg peaks, causing the structure analysis to be more difficult than the bulk analysis. Recently, a consensus has been reached that the average structure of $CsPbBr_3$ NCs is assigned to be the orthorhombic phase at RT [24,25]. More recently, Bertolotti et al. used advanced atomistic modeling of synchrotron wide-angle X-ray total scattering data to demonstrate that orthorhombic subdomains form which are hinged through a 2D or 3D network of twin boundaries into a pseudocubic phase [26]. By contrast with the previous reports, the as-synthesized highly luminescent in red $CsPbI_3$ NCs, which were considered to be ideally cubic [21], were also found to have the same structure as the well-known orthorhombic $CsPbBr_3$ phase [26]. Structural analysis for most of the lead halide perovskite nanocrystals reported in the literature was carried out by the XRD technique which is probably the most convenient option but not precise enough to conclude the formation of different phases. For instance, the cubic phase was initially reported for $CsPbBr_3$ NCs in a majority of the reports but later it turns out to be orthorhombic by in depth investigation. Therefore, particular attention is required for the structural analysis and, most importantly, the use of reliable technique such as synchrotron which is sensitive enough to give precise information about different phases.

In the $CsPbX_3$ NCs family, $CsPbI_3$ with a band gap of 1.73–1.80 eV is the most interesting one for photovoltaics application but it suffers from undesired phase transition from the 3D phase (perovskite) to the 1D (non-perovskite) yellow phase of $CsPbI_3$. Several strategies have been proposed to overcome the instability of the 3D (perovskite) phase of $CsPbI_3$ both in films and solution. Protesecu et al. proposed that the 3D phase of $CsPbI_3$ NCs can be stabilized by incorporating large organic cation (10% of FA^+) [27] while Liu et al. recently demonstrated the increased stability of $CsPbI_3$ NCs by introducing trioctylphosphine (TOP) as a surfactant [28]. A later method of preparation of TOP-PbI_2 precursor takes at least one week, while the same approach does not work for other lead halide counterparts that limits its potential [28]. Akkerman et al. also reported the increase in the stability of $CsPbI_3$ perovskite phase over up to a month by replacing Pb^{2+} with Mn^{2+} without altering the optical properties of the host material [29]. Imran et al. introduced benzoyl halide as an efficient halide precursor for the synthesis of lead halide perovskite NCs which allows them to work under halide excess conditions. They observed that synthesized $CsPbI_3$ NCs were stable in both films and colloidal solution for several weeks under ambient conditions. The improved stability of 3D-phase $CsPbI_3$ NCs was ascribed to the formation of lead halide terminated surfaces, in which Cs cations were partially replaced by alkylammonium ions [17]. Lead halide perovskite NCs with this type of surface have been reported to have improved stability and enhanced optical properties. Apart from the stability issues of $CsPbI_3$ NCs, the quantum-confined blue emitting $CsPbBr_3$ nano platelets/nanosheets with thickness of a few unit cells aggregate and turn green when dry solid films were prepared [12,30]. This shifting of emission band to the bulk band gap of $CsPbBr_3$ is preventing several potential applications including blue light-emitting diodes (LEDs).

As the most used procedure for $CsPbX_3$ NCs synthesis, the hot-injection method is a very promising methodology to achieve high-quality perovskite NCs. Xianghong He et al. [31] reviewed the effect of various parameters (ligand, reaction temperature) in the solution-fabrication strategies of all-inorganic trihalide perovskite NCs. In this approach, OA and OLA can break the cubic symmetry and lead to anisotropic growth of $CsPbBr_3$. Shorter chain carboxylic acids give rise to larger size nanocubes from a high-temperature reaction, while shorter chain amines result in thinner nanoplatelets [32]. The Manna group has successfully synthesized the lead-based perovskite NCs with different morphology by adjusting the ligands [12,33]. Adjusting the ratio of short (octanoic acid (OA) and octylamine (OAm)) to long (OA and OAm) ligands leads to 2–3 unit cell thick $CsPbBr_3$ nanosheets with lateral size from 300 nm up to 5 μm [12]. While adjusting the ratio of Amine with hex acid leads to $CsPbBr_3$ nanowires with widths from 5.1 nm to 2.8 nm at a reaction temperature of 65 °C [33].

Reaction temperature also plays a role in the shape and the size of the resultant NCs. The Kovalenko group [21] showed that the size of $CsPbX_3$ NCs is tuned in the range of 4–15 nm by varying the reaction temperature (140–200 °C). The $CsPbX_3$ NCs were changed from nanocubes to nanoplates at lower temperatures (between 90 °C and 130 °C) [34].

2.3. Colloidal Synthesis of FAPbX₃ NCs

Although the synthesis of CsPbX$_3$ and MAPbX$_3$ NCs has been optimized over the last few years, FAPbX$_3$ NCs with optimal optical properties as well as a narrow size distribution and phase purity have not yet been prepared by either hot-injection techniques or by the LARP approach. Recently, FAPbX$_3$ in the form of NCs have received considerable interest due to several potential advantages over their cesium and methylammonium counterparts, for instance higher stability due to a more symmetrical and tightly packed crystal structure, and impressive optical properties [35–37].

Protesescu et al. reported a polar solvent-free three-step hot-injection method for the synthesis of FAPbX$_3$ (X = Br, I) NCs. In a typical hot-injection synthesis, FA-Pb precursor solution was prepared by reacting Pb and FA acetates with oleic acid in octadecene. Subsequently, temperature was increased to 130 °C and oleylammonium bromide dissolved in toluene was rapidly injected. The NCs obtained have nearly cubic morphology and very high PLQY up to 85%. But its potential versatility was limited by the poor reactivity of the alkylammonium halide salts that leads to the formation of undesired secondary phases and the resultant NC contains 10% of phase impurity (NH$_4$Pb$_2$Br$_5$) [27,39]. Later on Levchuk et al. reported colloidal synthesis of FAPbX$_3$ (X = Cl, Br, I, or mixed Cl/Br and Br/I) NCs of 15−25 nm sizes by LARP, see Figure 4a–c [38]. The emission wavelength can be tuned from 415−740 nm by tailoring halide composition as well as by their thickness with narrow full width at half-maximum (FWHM) of 20−44 nm (see Figure 4d). Resulting NCs are characterized by radiative lifetimes of 5−166 ns and very high PLQY (up to 85%) but the shape control was poor in all the cases [38,40]. To overcome the limitation imposed by traditional synthesis methods, the Manna group recently reported a modified three-precursor hot-injection method for the synthesis of FAPbX$_3$ NCs by using benzoyl halide as a halide precursor. Interestingly, this approach allows one to work with the desired stoichiometry of the ions, since the halide ions and the metal cation sources are not delivered together, i.e., they are not delivered with the same chemical precursor. In a typical synthesis, formamidinium acetate and lead acetate were dissolved and degassed in oleylamine, oleic acid and octadecene at 125 °C in a three-neck flask. Subsequently, the solution was cooled down to the desired temperature (70–95 °C) and the benzoyl halide precursor was swiftly injected into the reaction flask, triggering the immediate nucleation and growth of the NCs. Typical TEM images of FAPbCl$_3$ and FAPbBr$_3$ NCs evidenced a narrow size distribution, which became slightly broader in the case of FAPbI$_3$ NCs (see Figure 4e–g). XRD analysis of FAPbX$_3$ NCs shows excellent phase purity and matches nicely with the corresponding bulk crystals (see Figure 4h–j) [41]. On the other hand, FAPbCl$_3$ NCs were synthesized for the first time by this approach and no cubic bulk structure has been reported so far. The refinement of the XRD pattern of FAPbCl$_3$ NCs led to a cubic structure (space group Pm-3m) with a = 5.67 Å. Furthermore, as shown in Figure 4k–m, FAPbBr$_3$ and FAPbI$_3$ NCs exhibited excellent optical properties, and had a high PLQY (90% for FAPbBr$_3$ and 65% for FAPbI$_3$) and narrow PL emission (20 nm for FAPbBr$_3$ and 48 nm for FAPbI$_3$). The FAPbCl$_3$ NCs were characterized by having a narrow PL (FWHM ▬ 16nm), but a low PLQY (about 2%) [17].

In general, both LARP and hot-injection approaches are applicable to the synthesis of the entire family of lead halide perovskite nanocrystals. LARP is performed at low temperature in air whereas the hot-injection method needs air-free conditions and relatively higher temperature. However, one major setback about the LARP approach is the presence of polar solvent such as DMF originating from the chemical synthesis, which results in the dissolution/decomposition of the perovskite NCs formed to convert back to precursors. Furthermore, the shape control of NCs and doped lead halide perovskite nanocrystals is mainly achieved by the hot-injection method.

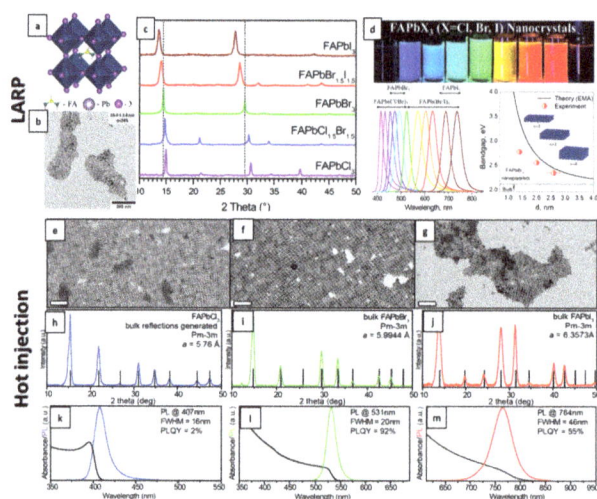

Figure 4. (**a–d**) FAPbX$_3$ NCs synthesized by LARP reproduced with permission from Ref. [38], Copyright (2017) American Chemical Society. (**a**) Cubic (FAPbBr$_3$, Pm-3m space group), (**b**) bright field TEM image of FAPbBr$_3$ NCs and (**c**) XRD pattern of FAPbX$_3$ NCs. (**d**) Photographs of FAPbX$_3$ NCs and their optical properties along with size dependent band gap variation for FAPbBr$_3$ NCs. (**e–m**) FAPbX$_3$ NCs synthesized by hot injection method reproduced with permission from Ref. [17], Copyright (2018) American Chemical Society: Bright field TEM images (Scale bars are 100 nm in all images), XRD patterns, along with absorption and PL spectra of FAPbCl$_3$, FAPbBr$_3$, and FAPbI$_3$ NCs respectively.

2.4. Tailoring the Properties of Halide Perovskite NCs

One way to tailor the properties of halide perovskite is the doping of metal ions (Mn^{2+}, Zn^{2+}, Cd^{2+}, Sn^{2+}, and Bi^{3+}) into the lattice of CsPbX$_3$ NCs [42,43]. In general, these metal ion-doped CsPbX$_3$ NCs were prepared by modifying the hot-injection method described above by adding MnX$_2$ together with PbX$_2$ solubilized in the mixture of OA and OAm in ODE. The doped halide perovskite NCs successfully exhibit tailored properties. For example, Parobek et al. reported that Mn-doped CsPbCl$_3$ NCs have a strong sensitized Mn luminescence, arising from the exchange coupling between the exciton and Mn [44]. Begum et al. doped colloidal CsPbBr$_3$ perovskite NCs with heterovalent Bi^{3+} ions by hot injection and showed that interfacial charge transfer can be tuned and facilitated by metal doping [45]. Another direct synthesis approach for doped halide perovskite NC is as follows. CsAc and PbAc$_2$ with MnAc$_2$ are mixed in toluene at room temperature with oleic acid (OA) and oleylamine (OLAM) as ligands. Subsequently, the HCl acid in water was added under vigorous stirring. After centrifuging, the supernatant which contains the NCs was collected and further treated to get a stable colloidal dispersion of Mn^{2+}-doped CsPbCl$_3$ NCs [46]. The doped halide perovskite NCs are significantly more stable because of the enhanced formation energy due to the doping [29,47].

Apart from doping, anion exchange is the most simple and commonly used post-synthesis method to tune the properties of perovskite NCs. Usually the anion exchange is achieved by using a range of different halide precursors to mix with the halide perovskite NC solutions, to tune the chemical composition and the optical properties of colloidal CsPbX$_3$ NCs in all the visible spectrums. Besides, by mixing solutions containing perovskite NCs emitting in different spectral ranges can produce NCs emitting in an intermediate range [48,49]. The anion exchange can also be achieved in a dihalomethane solution of halide perovskite NCs without any reacting anion source using photoexcitation to trigger the halide ion exchange [50].

Anion-exchange reactions provide an alternative path for compositional fine tuning while maintaining the parental structure and luminescent emission (tunable across the entire visible spectrum) of the final products especially in case of cesium lead halide perovskites ($CsPbX_3$, X = Cl, Br, I). So far most commonly used precursors for anion exchange reactions are lead halides salt, oleylammonium halides, tetrabutylammonium halides or solid metal halides MX_2 (M = Zn, Mg, Cu, Ca; X = Cl, Br, I). All the aforementioned cases either need complicated preparation steps or suffer from limited reactivity at room temperature for halide replacement; for instance, oleylammonium halides could be synthesized by reacting oleylamine with HX (X = Cl, Br, I) in ethanol overnight under a flow of nitrogen gas followed by multiple purification steps and the drying of products in vacuum oven. Lead halide salts are limited by their poor solubility under ambient conditions in a relatively non-polar solvent environment and show slow reactivity towards anion exchange reactions at room temperature. While in the case of TBA-X the exchange reaction worked only from $Br^- \rightarrow Cl^-$ and $I^- \rightarrow Br^-$ reverse exchange does not work (Cl^- to Br^- and Br^- to I^-) [48–52]. Considering the strong reactivity of the benzoyl halide precursor even at room temperature, we tested them for the post-synthesis transformations of $CsPbX_3$ NCs.

The addition of benzoyl chloride or benzoyl iodide into pre-synthesized $CsPbBr_3$ NCs at room temperature leads to a blue shift or a red shift, respectively, of both the PL and the absorption spectra of the product NCs (see Figure 5a). In both cases the XRD patterns of the resulting NCs confirmed the retention of the parent perovskite structure, with a systematic shift of the peaks (see Figure 5d). Interestingly, the back exchange reactions from $CsPbCl_3 \rightarrow CsPbBr_3$ and from $CsPbI_3 \rightarrow CsPbBr_3$ also worked efficiently when benzoyl bromide was added to $CsPbCl_3$ and $CsPbI_3$ NCs, respectively. Also, it is worth mentioning here no mixed-halide of $CsPb(Cl/I)_3$ could be obtained by the addition of benzoyl chloride to the $CsPbI_3$ NCs but rather a very slow and complete exchange occurred over time. Indeed this could be due to the larger difference in ionic radii between Cl^- and I^- causing the instability of the $CsPb(Cl/I)_3$ system.

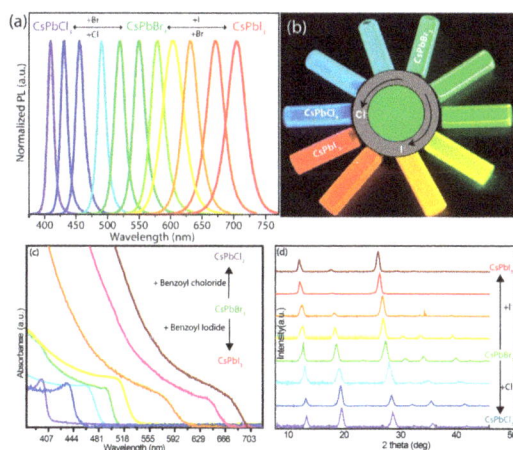

Figure 5. (a) Evolution of the PL spectra of $CsPbBr_3$ NCs by the addition of benzoyl chloride or benzoyl iodide. (b) Photograph of the different $CsPbX_3$ NC solutions obtained by anion exchange under an ultraviolet (UV) lamp. (c) Evolution of the absorbance spectra of representative anion-exchanged NCs. (d) XRD patterns of the pristine $CsPbBr_3$ NCs and the anion-exchanged samples, reproduced with permission from Ref. [17], Copyright (2018) American Chemical Society.

2.5. Lead-Free Halide Perovskite NCs

Two critical problems are still unresolved and hindering commercial applications for lead halide perovskite: the toxicity of lead and poor stability. Tremendous efforts have also been put into partially replacing lead with non-toxic metals such as Mn(II) or completely replacing lead with Sn(II), Sn(IV), Bi(III), Sb(III), Cu(II) etc. $CsPb_xMn_{1-x}Cl_3$ QDs in colloidal solution were synthesized through phosphine-free hot-injection via partial replacement of Pb with Mn, and the Mn substitution ratio is up to 46% [53]. Besides, synthesis of lead-free metal halide ($CsSnX_3$, Cs_2SnX_6, and $Cs_3Bi_2X_9$, $MA_3Bi_2X_9$, $Cs_3Sb_2X_9$, Cs_2AgBiX_6) NCs have been intensively explored, for which the unit cells for several types of these materials are illustrated in Figure 6 and the reported results are summarized in Table 1.

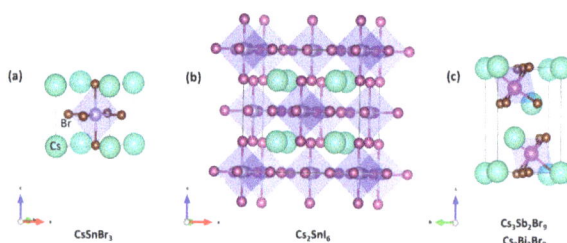

Figure 6. Unit cells of several types of lead-free halide perovskites: (**a**) $CsSnBr_3$ (inorganic crystal structure database (ICSD): 4071, cubic), (**b**) Cs_2SnI_6 (ICSD: 250743, cubic), (**c**) $Cs_3Sb_2Br_9$ (ICSD: 39824, hexagonal) or $Cs_3Bi_2Br_9$ (ICSD: 1142, hexagonal) (The atomic models are built by VESTA [54]).

Table 1. Summary of lead-free halide perovskite NCs from literature.

Type of Material	Synthesis Method	Morphology	Reference
$CsSnX_3$ (X = Cl, Br, I)	Hot injection	Nanocubes (about 15 nm)	[55]
Cs_2SnI_6	Hot injection	quasi-spherical diameters between 35 and 80 nm	[56]
Cs_2SnI_6	Hot injection	spherical QDs, nanorods, nanowires, and nanobelts to nanoplatelets	[57]
$MA_3Bi_2Br_9$	LARP	QDs with 3.05 nm ± 0.9 nm	[58]
$Cs_3Bi_2X_9$ (X = Cl, Br, I)	LARP	QDs with 3.88 nm ± 0.67 nm	[59]
$Cs_3Sb_2Br_9$	LARP	QDs with 3.07 ± 0.6 nm	[60]
Cs_2AgBiX_6 (X = Cl, Br, I)	Hot injection	Nanocubes (edge length about 8 nm)	[61]
$Cs_2AgBiBr_6$	Hot injection	Nanocubes (edge length about 9 nm)	[62]

Jellicoe et al. reported the synthesis of $CsSnX_3$ NCs by the hot-injection method [55]. In this attempt, the tin precursor was prepared by dissolving SnX_2 in the mildly reducing and coordinating solvent tri-n-octylphosphine. Then this solution was injected into a Cs_2CO_3 precursor solution containing oleic acid and oleylamine at 170 °C to obtain colloidally stable $CsSnX_3$ NCs, as shown in Figure 7 [55] Besides, $CsSnX_3$ nanocages were also synthesized by the hot-injection colloidal approach [56] Substitution of Pb with Sn reduces toxicity but compromises the air stability partly due to the oxidation process from Sn(II) to Sn(IV). Therefore, Cs_2SnI_6 NCs which consist of air-stable tetravalent Sn^{4+} are explored by the hot-injection method by Wang et al [57] and Dolzhnikov et al. [58]. However, Sn-based perovskites still contain abundant intrinsic defect sites leading to low PLQYs.

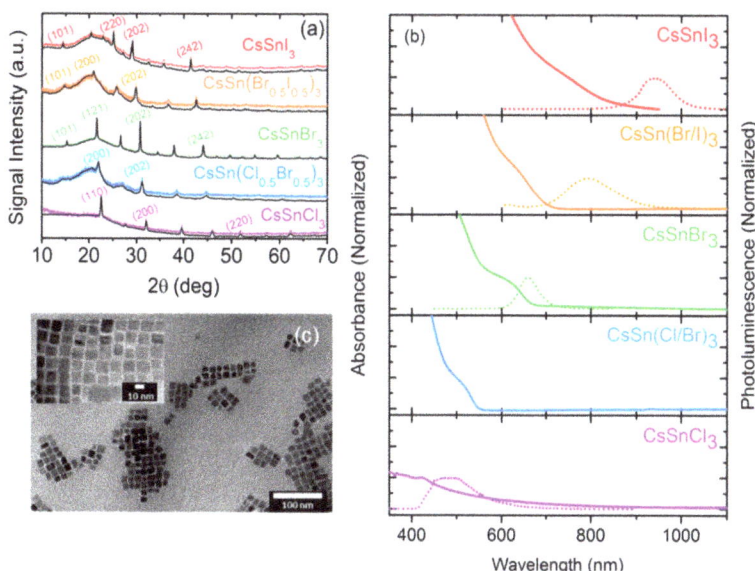

Figure 7. (a) Powder X-ray diffraction spectra of $CsSnX_3$ (X = Cl, $Cl_{0.5}Br_{0.5}$, Br, $Br_{0.5}I_{0.5}$, I) perovskite NCs. (b) Absorbance and steady-state PL of NCs containing pure and mixed halides. The PL spectrum of $CsSn(Cl_{0.5}Br_{0.5})_3$ particles was identical to the pure chloride-containing NCs. (c) TEM image of $CsSnI_3$ NCs. Reproduced with permission from Ref. [55], Copyright (2016) American Chemical Society.

Bismuth has much lower toxicity than lead and recently has been found to be promising for perovskite because Bi^{3+} is isoelectronic with Pb^{2+} and, meanwhile, more stable than Sn^{2+}. Recently, Leng et al. reported the synthesis of $MA_3Bi_2Br_9$ [59] and $Cs_3Bi_2X_9$ NCs [60] by LARP. For $MA_3Bi_2Br_9$ NCs, the dimethylformamide (DMF) and ethyl acetate are used as the "good" solvents to dissolve MABr and $BiBr_3$, and Octane acted as the "poor" solvent to precipitate QDs when the precursor solution was injected into octane. While $Cs_3Bi_2X_9$ NCs were synthesized by first dissolving CsBr and $BiBr_3$ in dimethyl sulfoxide (DMSO), and the precursor solution was added to the antisolvent–ethanol to crystallize $Cs_3Bi_2X_9$, for which the results are shown in Figure 8. Also, OLAm and OA were added to control the crystallization rate and stabilize the colloidal solution. Yang et al. also reported lead-free all-inorganic perovskite $Cs_3Bi_2X_9$ (X = Cl, Br, I) NCs synthesized by a similar approach, which used DMSO as the solvents to dissolve CsBr and $BiBr_3$ while isopropanol was used as the antisolvent to precipitate NCs [61]. Other lead-free layered $Cs_3Sb_2Br_9$ inorganic perovskite quantum dots were obtained by a modified LARP at RT within few seconds reaction [62]. In this approach, the precursor solution is prepared as a mixture of $SbBr_3$, CsBr, and oleylamine was dissolved in *N,N*-DMF or DMSO. Then the precursor is dropped into a mixed solution of octane and OA, inducing rapid recrystallization of $Cs_3Sb_2Br_9$ NCs.

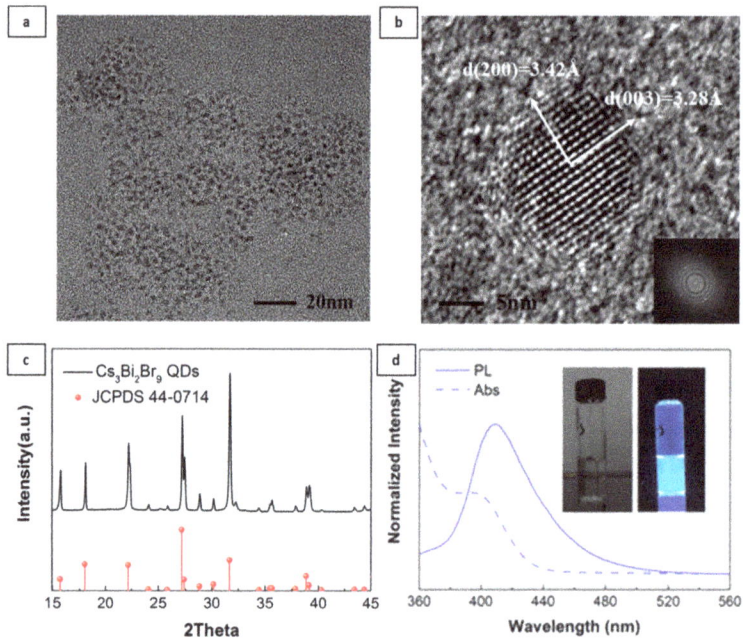

Figure 8. Characterizations of $Cs_3Bi_2Br_9$ QDs. (**a**) TEM image. (**b**) High-resolution TEM (HRTEM) image of a typical QD. (**c**) XRD patterns of QD powder. (**d**) Absorption and PL spectra. Insets: Typical optical images of the QD solution in ambient under a 325 nm UV lamp illumination. Reproduced with permission from Ref. [60], Copyright (2017) John Wiley and Sons.

Derived from the idea of oxide double perovskites ($A_2BB'O_6$), double perovskites appear as a new type of lead-free materials, where two divalent Pb^{2+} can be replaced by one monovalent ion M^+ and one trivalent ion M^{3+}. Cs_2AgBiX_6 NCs are obtained by the hot-injection approach as reported by Creutz et al. [63] and Zhou et al. [64]. In the approach by Creutz et al., neat TMSBr or TMSCl (2.7 mmol, TMS = trimethylsilyl) was swiftly injected into a boiling solvent containing Cs(OAc), Ag(OAc), and Bi(OAc)₃ dissolved in a combination of octadecene, oleic acid, and oleylamine at 140 °C [63]. In the approach by Zhou et al., $Cs_2AgBiBr_6$ NCs with pure cubic shape and high crystallinity were synthesized via a hot-injection method as shown in Figure 9, in which the Cs-oleate was injected into a high boiling organic solvent containing $BiBr_3$ and $AgNO_3$ at 200 °C. In this approach, a small amount of hydrobromic acid (HBr) additive ensures the full ionization of Ag^+. These NCs were applied into photochemical conversion of CO_2 into solar fuels, displaying their great potential as environment-friendly halide perovskite photocatalysts [64].

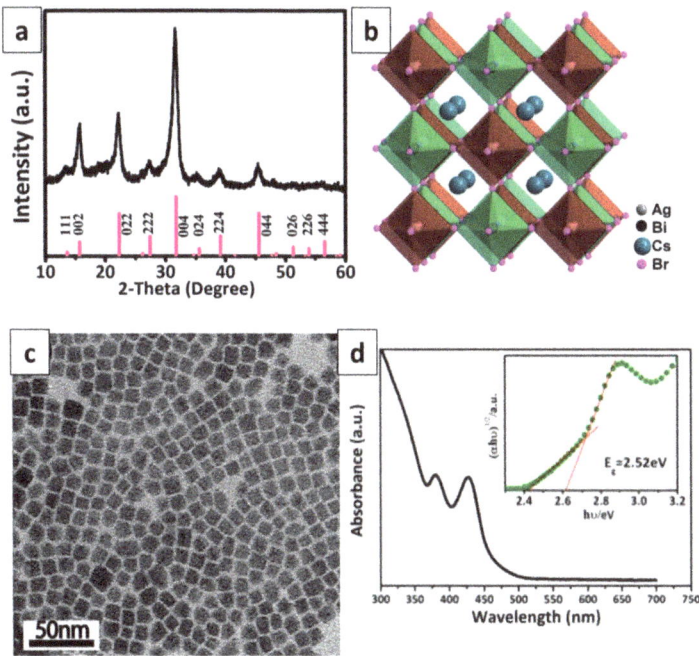

Figure 9. (**a**) XRD pattern of $Cs_2AgBiBr_6$ NCs synthesized at 200 °C. (**b**) Crystal structure of cubic $Cs_2AgBiBr_6$. (**c**) TEM image of $Cs_2AgBiBr_6$ NCs, (**d**) UV-Visible spectra of colloidal $Cs_2AgBiBr_6$ NCs, and the inset shows the Tauc plots. Reproduced with permission from Ref. [64], Copyright (2018) John Wiley and Sons.

Despite mounting interest in and remarkable achievements on colloidal synthesis of lead and lead-free halide perovskite nanocrystal, an important issue is their instability during the purification process upon exposure to the polar solvents and under ambient atmospheric conditions that consequently limit their use for several potential applications. Lead-free halide perovskite nanocrystals are suffering from poor PLQY compared to lead-based halide perovskite NCs. So, overall, it is indispensable for the scientific community to synthesize heterostructures or core shell NCs that allow their dispersion into a variety of solvents while maintaining their colloidal integrity and eventual incorporation into various devices.

3. The Technologies and Mechanism of the Growth of Large Crystal Grains of Perovskite Films

Perovskite materials have an ABX_3 structure, where A usually represents the monovalent cation including organic small cation MA, FA and inorganic Cs, Ag, B is the transition metal (Pb, Sn, Sb and so on), and X is the halogen anion (I, Cl, Br, O) or mixed. The preparation of these materials should mix two participants following which solution-process methods are adopted including one-step and two-step [65], with the advantages of a low-cost and simple fabrication process. On the other hand, physical methods including thermal evaporation, sputtering and pulsed laser deposition are widely used to grow materials with low solubility and high quality such as perovskite oxides. Unlike bulk crystal growth methods such as the solution temperature-lowering (STL) method [66,67], inverse temperature crystallization (ITC) method [4,68], anti-solvent vapor-assisted crystallization (AVC) method [69], and melt crystallization method [70–72], thin films with a large size and high-quality crystals for high power conversion-efficiency devices are more difficult to synthesize. Therefore, advanced technologies are required. Herein, we discuss the approaches and the fabrication processes which have been recently developed.

3.1. Anti-Solvent Method

The anti-solvent method is one of the most widely used methods and is often applied during the one-step method [73–76]. The crystal growth is mainly divided into two states: firstly, the seeds are formed on the surface of the substrates during spin-coating; and then the growth of crystals during thermal depositing. The crystal size and density of seeds increase in two stages in the process of the synthesis. To form dense large crystal film, dense seeds should be formed before their size growth. The anti-solvent method is used to optimize as-prepared films. Figure 10 shows the details of this method [74]. The spin-coating of precursor solution is divided into two time-steps. The first 10 s (10 s) is for the solution dispersion and coverage on the whole surface of the substrate. After that, antisolvent is casted in one shot to form super-saturated solution during the spinning for the next 20 s. Then, samples are thermally heated to grow crystals and solidify films. Compared to traditional one-step method, this induces intermediate complex film with long-time stability and does not need extra mobilized solvent except the complex solvent. During the following heat treatment, the super-saturated film enables the formation of dense uniform seeds. It should be noted that the density and shape of the seeds strongly depends on the selected anti-solvent. Generally, the chosen anti-solvent should not dissolve the target material including the participants, meanwhile it should have sufficient ability to extract the solvent of the precursor solution. There are a variety of organic solvents including chlorobenzene, toluene and esters. Initial studies have demonstrated that a high boiling point and miscibility of the selective solvent could play an important role in extracting the solvent inside in order to avoid fast evaporation at room temperature on the surface of the substrate, which usually produces a non-uniform surface. Ether and dichloromethane (DCM) with a boiling point of lower than 50 °C leads to uncovered needle-like seeds. According to kinetic studies, precise control of competition between the formed seeds and the growth of the size of crystals during the antisolvent treatment would affect the density of the seeds. The fast evaporation of antisolvent (low boiling point) also accelerates the heterogeneous nucleation with a quick solvent extraction. The challenge is that homogeneous nucleation needs a lower evaporation rate, which meanwhile lengthens the time required to grow the crystal size.

The antisolvent method should consider the balance between the extraction time and crystal growth time at this stage. After all, the choice of antisolvent should depend on the solvent of the precursor solution and the specific perovskite material. The most promising antisolvents for lead-based perovskite in recent studies are TFT and some esters [74].

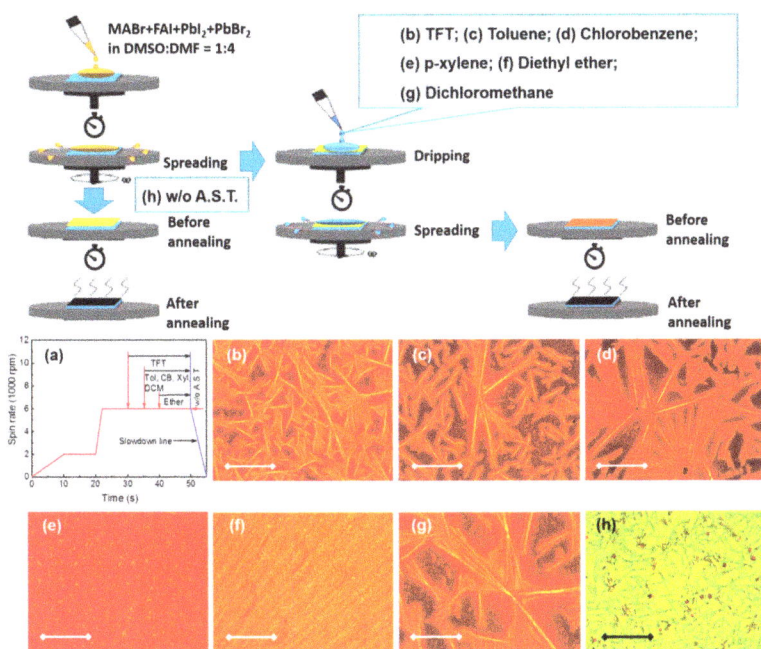

Figure 10. Schematic of fabrication of large crystals by antisolvent method. (**a**) The speed-time profile for the spin coating process with and without the various anti-solvent (A.S.T.) Optical micrographs (OM) illustrating oriented growth of needle shaped crystals with distinct size variation before thermal annealing as a function of various antisolvents with (**b–g**) and without A.S.T. (**h**) Scale bar: 300 μm. Reproduced with permission from Ref. [74].

3.2. Solvent Vapor Annealing Technology

Solvent vapor annealing technology (SVA) is a frequently used method to improve the surface morphology of the final film [77]. Unlike the antisolvent method, which improves the quality of as-prepared film to form uniform dense seeds, SVA focuses on the crystal growth step after the seeds formed. SVA utilizes the additional force of the solvent vapor, not only leading to a long diffusion length of molecules or ions, but also controlling the speed of removing the solvent, which improves the crystal size and surface morphology and overcomes the pin holes always formed inside the film because of the ultrafast evaporation of the solvent by conventional thermal annealing (TA). More importantly, TA leads to an unequal evaporation speed of the individual components, especially for a cosovlent solution because of the varied boiling points, such as the reported DMF and DMSO with 4:1 cosolvent in MASnI$_3$ solution [78,79]. Moreover, SVA can efficiently remove all the inside solvents at the same time for their excellent miscibility. An ideal solvent for SVA should have high evaporation pressure and a proper boiling point, while the choice of chamber size, position of samples and optional heat temperature should depend on the boiling point of the selected solvent.

The kinetic energy of solvent that is molecular in the vapor state is larger than that in the liquid state, which enables the nucleation and enlarges the crystals to micrometers. Meanwhile, solvent has the potential to dissolve seeds. Therefore, the evaporation of solvent of samples should be faster than the supplement to make sure that crystals are grown rather than being dissolved over time. Recent experiments establish an incomplete closed space with random holes inside the cap of the chamber for SVA, which avoids the over accumulation of solvent on the surface. The time taken for the SVA approach is also a factor affecting the growth of grains. Figure 11 shows a recent reported

SVA method [80], in which double layers are spin-coated. After the nano-polycrystals are formed by inter-diffusing, SVA with DMF solvent merges the nanocrystals and leads to uniform and dense miro-size crystals.

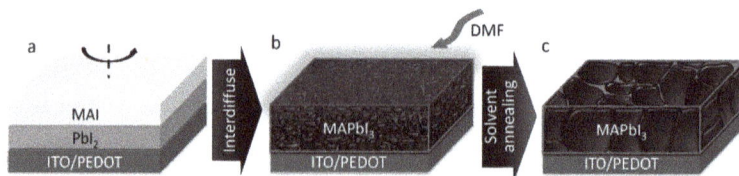

Figure 11. Schematics of the inter-diffusion approach and solvent vapor annealing to increase grain size. (a) The deposed PbI$_2$ and MAI two layers by spin-coating. (b) Solvent vapor annealing using DMF vapor. (c) The formed high quality film with large grains. Reproduced with permission from Ref. [80].

By contrast, anti-solvent can also be used. Compared to solvent of perovskite, which always slow down the evaporation of solvent, the added antisolvent speeds up the removal of the inside solvent of the immediate film like the heating method. In fact, the removal speed mainly depends on the miscibility rather than the boiling point in the heating method.

3.3. Ostwald Ripening

Ostwald ripening is another method to obtain large-size crystals, in which the large particles are more energetically favoured than smaller ones. The molecules are less stable on the surface of grains than inside because of the additional surface energy, which motivates the movement of the particles. Therefore, this gives the possibility of healing the pin holes formed between the crystals after spin-coating of the precursor solutions, especially in the one-step method. The movement of grains is proved to be the slowest process and the size growth is expressed as the following equation,

$$R^3 - R_0{}^3 = \frac{6\gamma c_\infty v^2 D}{9 R_g T} t$$

where R is the average radius of all the particles, γ is particle surface tension or surface energy, c_∞ is solubility of the particle material, v is molar volume of the particle, D is material diffusion coefficient of the particle, R_g is a constant related to the ideal gas, T is constant absolute temperature, and t is time. According to Lifshitz and Slyozov's equation, the growth of R strongly depends on the solubility known as liquid–liquid systems. In addition, additional force is needed to speed up its process. The initial studies use ion exchange as the additional force [74], and salt solution; for example methylammonium iodide (MAI) in IPA solvent is used to connect MAPbI$_3$ particles. Note that this method always introduces foreign material or changes the origin ratio in the final film. However, this can be ignored if the target films include such foreign material, for example, in the case of hybrid halide CH$_3$NH$_3$PbI$_{3-x}$Br$_x$ (Figure 12) [81]. MABr solution is introduced to exchange the MAI in the as-prepared MAPbI$_3$ film, which enhances the connection between crystals and heal pin holes. Additionally, the diffusion of molecules in solution plays an important role. To enhance the diffusion length of salt molecules, a low concentration is needed, which is consistent with the limited influence of the high salt solution, while high-quality and large crystal grains can be obtained by low concentration. After the treatment, the redundant matter should be removed and the disadvantage is obviously that the healing process is too slow, while the ratio is difficult to control.

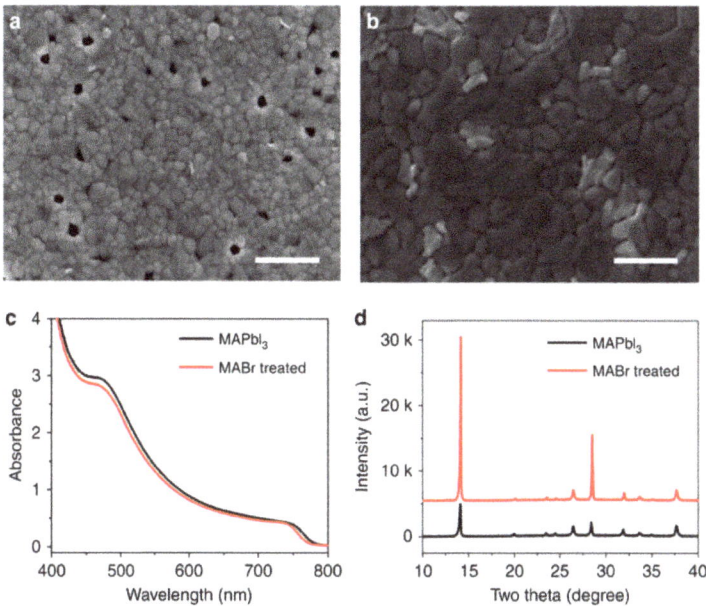

Figure 12. Top view of the scanning electron microscope (SEM) images of (**a**) MAPbI3 and (**b**) MABr-treated MAPbI$_3$ films. Scale bars, 1 μm. Comparison of (**c**) ultraviolet–visible (UV-vis) absorption spectra and (**d**) XRD patterns of MAPbI$_3$ films with and without MABr treatment. Reproduced with permission from Ref. [81].

3.4. Geometrically-Confined Lateral Crystal Growth

The methods mentioned above mainly focus on the fabrication of large-size polycrsytals, and in these methods the surfaces of substrate almost have no influence to the final films. A geometrically-confined lateral crystal growth method is utilized to form well-arranged single crystals, of which the substrate, unlike conventional glass or fluorine doped tin oxide (FTO), is also single crystal substrate and prepared to patented wettable surface by molecules, which have hydrophobic groups such as Triethoxy-1H,1H,2H,2H-tridecafluoro-n-octylsilane. This surface will confine the dispersed solution and guide the growth direction of perovskite crystals. As with the self-assembled organic molecular technology [82], hybrid perovskite molecules are favorable to assembly along the crystal surface of the substrate. This lateral crystal growth is a common method via epitaxial lateral overgrowth (ELO) and the vertical direction is restricted by geometrical confinement. The dispersed perovskite solution can only exist on the wettable region. The formed seeds in this restricted region grow and connect with each other during the removal of solvent. Then, single crystal belts are formed, filled and arranged. The reported method introduces an inner printer to control the amount of solution on the surface, which is heated at a certain temperature (Figure 13) [83].

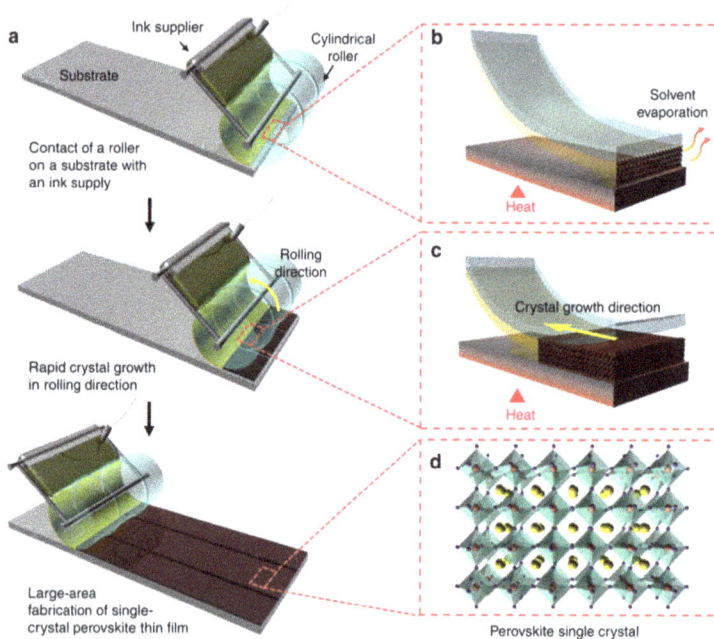

Figure 13. Procedure of single-crystal perovskite thin films using geometrically confined lateral crystal growth with a rolling module. (**a**) General process of the film fabrication. (**b**) Schematic of solvent evaporation by thermal heating. (**c**) Schematic of crystal growth direction. (**d**) The perovskite single crystal structure. Reproduced with permission from Ref. [83].

Due to the special requirement of substrate such as silicon or silicon nitride substrates, a conventional structure FTO/compact-TiO_2/meso-TiO_2 with perovskite/HTM/gold of devices is impossible to realize in this situation. Two electrodes are located on the same surface. The obtained power conversion efficiency is very low at about 4%, with low fill factor and dense current compared to other solar cells. Moreover, the electron selective layer (ETL) and hole transport material (HTM) material are not introduced, while the cost of the substrate is high and the incoming light is week.

3.5. Additives

Additives are usually applied to interconnect nanocrystals, which enhance the performance [84,85]. Unlike the crystal growth method, additives almost have no influence on the fabrication of large crystals but connect them by using chemical bonds. A hydrogen bond N–H···I– as well as P–OH···I– hydrogen at two ends of butylphosphonic acid 4-ammonium chloride (4-ABPACl) (Figure 14) reported by the Grätzel group connect neighboring perovskite crystals [86]. The use of this additive after a one-step solution-processing strategy results in stable and high-performance perovskite solar cells. 4-ABPACl molecules acted as crosslinking agents between neighbouring perovskite grains leading to a smooth and uniform perovskite layer rather than discontinuous crystal grains without additive. It is worth noting that the carbon length, which determines the distance between two grains, should be proper. A distance that is too long will lead to a loose interaction, while a short one cannot connect two grains. Other additives including MAI, ethylammonium iodide (EAI), bulky aromatic cations, tetraphenylphosphonium iodide (TPPI) and chloride (TPPCl) successfully improve the morphology and make dense, smooth film by utilizing the incorporation of one end of I– or Cl– into the crystals, while the strong hydrogen bond of N–H···I– forms at another end of the neighbouring crystals. Likewise, a small amount of inorganic molecular H_2O, HI/HCl were added into PbI_2/MAI in DMF/DMSO solution to make a homogenous precursor solution and give rise to high-quality full coverage film without pin holes. The difference is that these inorganic additive molecules having an important influence during the film fabrication for the supplement of extra hydrogen bonds will evaporate and not exist in the final film.

Additives always lead to an extra high humidity resistance for the tight films, which has great significance for applications. However, compared to the inside chemical bonds, the relative weak interaction of hydrogen bonds induces instability under high temperature. The transport of carriers between two grains is also limited. This may give the option for future additives to establish covalent bond with the excellent extraction of carriers between grains.

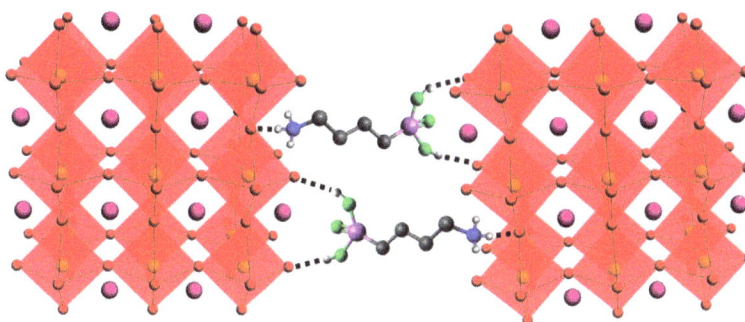

Figure 14. Schematic illustration of inter-connected two neighbouring grain structures by additives. Reproduced with permission from Ref. [86].

3.6. Physical Method–Vapor Assisted Growth

Solution-processed methods enable low-cost, high reproducible approaches for large size, high-quality film growth for perovskite. Compared to them, physical methods are seldom used because of the requirement of special conditions including a vacuum or high temperature. The following vapor-assisted growth method utilizes the vapor state of participants. Take $MAPbI_3$ as an example (Figure 15) [87], double layers PbI_2/MAI after two times spin-coating are put inside oven, which is heated at more than 200 °C surrounded with MAI vapor. Because of the nano polycrystals of $MAPbI_3$ formed, the growth of large crystals needs the movement of the crystals formed by utilizing the thermal energy. As reported by Kawamura et al., the equivalent lattice constant of the perovskite crystal increases with the temperature due to the wide path and movement of the ion inside perovskite crystals that is caused, which rearranges orientation and fuses the grains to grow larger. This means a lower temperature cannot change the size of final $MAPbI_3$ crystals. As the organic molecules degrade under certain temperature, the surrounding temperature should not be too high. As with Ostwald ripening, this growth is also slow and the movement of ions and molecules needs a longer time, usually 1–4 h is suggested. The final ratio of MAI and PbI_2 is difficult to control and excessive MAI will limit the final performance. Moreover, this reported method was limited to special substrate, and still no available devices are based on it. Another vapor-based method uses a double vapor source, whereby both MAI and PbI_2 vapor are introduced in the reaction chamber, in which samples are immersed. This modified vapor-assisted approach utilizes high-vacuum equipment to speed up the evaporation and reduce the process temperature. The enlarged diffusion length in a high-vacuum environment greatly leads to the movement of ions and molecules, which improves the crystal quality [88].

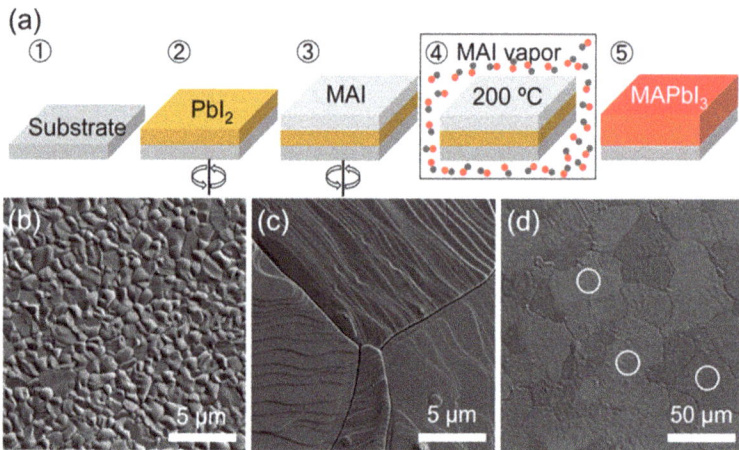

Figure 15. (a) Schematic of process of vapor assisted large crystal growth and (**b**–**d**) SEM of the film after treatment with different time. Reproduced with permission from Ref. [87].

4. Summary and Outlook

In summary, we have reviewed the recent frequently used approaches to realize different types of perovskite nanocrystals (NCs) as well as high-quality, large-scale and dense perovskite films for quantum dots (QD) and planar perovskite devices. Ligand-assisted reprecipitation and hot injection are the two dominant methods being applied for the synthesis of perovskite NCs. The most striking feature of these NCs is their extremely high PLQY. To further tune the optical properties of the perovskite NCs, the doping of NCs is realized by adding salts containing the doping metal ions into the precursor solutions for hot-injection synthesis. Anion exchange reactions provide an efficient way for compositional and optical property tuning. Given the major concerns of lead-based perovskites, lead-free perovskite NCs are also intensively studied and reviewed here.

Large and high-quality perovskite crystal films for planar devices possess many excellent properties including long diffusion length, high carrier life time and mobility. In addition, devices with higher power conversion efficiency for their devices have been obtained than those of traditional nano-polycrystals films. We divided the recent high-quality crystal growth methods into solution-proceed methods (including the anti-solvent method, SVA, Ostwald ripening, additives and geometrically-confined lateral crystal growth) and physical methods (vapor-assisted crystal growth). Moreover, the underlying mechanism of crystal growth in these treatments were investigated. For practical applications in devices, solution-proceed methods are more useful due to their low cost. Among the several solution-proceed methods, the anti-solvent method which only involves spin-coating and needs the lowest treatment time is the best option. In addition, additives are frequently used to interconnect neighoring grains as cross-linker agents, and geometrically-confined lateral crystal growth paves the way to single-crystal film growth in which the need for special substrates limits the power conversion efficiency.Future methods may focus on the fabrication of single crystals on cheap frequently used glass substrates.

Acknowledgments: Haijiao Harsan Ma acknowledges financial support of Xidian University. Zhiya Dang and Muhammad Imran acknowledges funding from the European Union under grant agreement no. 614897 (ERC Grant TRANS-NANO).

Conflicts of Interest: The authors declare no conflict of interest.

References

1. Hu, Z.; Kapil, G.; Shimazaki, H.; Pandey, S.S.; Ma, T.; Hayase, S. Transparent Conductive Oxide Layer and Hole Selective Layer Free Back-Contacted Hybrid Perovskite Solar Cell. *J. Phys. Chem. C* **2017**, *121*, 4214–4219. [CrossRef]
2. Todorov, T.K.; Reuter, K.B.; Mitzi, D.B. High-efficiency solar cell with earth-abundant liquid-processed absorber. *Adv. Mater.* **2010**, *22*, E156–E159. [CrossRef] [PubMed]
3. Meloni, S.; Moehl, T.; Tress, W.; Franckevičius, M.; Saliba, M.; Lee, Y.H.; Gao, P.; Nazeeruddin, M.K.; Zakeeruddin, S.M.; Rothlisberger, U. Ionic polarization-induced current–voltage hysteresis in CH3NH3PbX3 perovskite solar cells. *Nat. Commun.* **2016**, *7*, 10334. [CrossRef] [PubMed]
4. Saidaminov, M.I.; Abdelhady, A.L.; Murali, B.; Alarousu, E.; Burlakov, V.M.; Peng, W.; Dursun, I.; Wang, L.; He, Y.; Maculan, G. High-quality bulk hybrid perovskite single crystals within minutes by inverse temperature crystallization. *Nat. Commun.* **2015**, *6*, 7586. [CrossRef] [PubMed]
5. Setter, N.; Cross, L. The role of B-site cation disorder in diffuse phase transition behavior of perovskite ferroelectrics. *J. Appl. Phys.* **1980**, *51*, 4356–4360. [CrossRef]
6. Baikie, T.; Fang, Y.; Kadro, J.M.; Schreyer, M.; Wei, F.; Mhaisalkar, S.G.; Graetzel, M.; White, T.J. Synthesis and crystal chemistry of the hybrid perovskite (CH 3 NH 3) PbI 3 for solid-state sensitised solar cell applications. *J. Mater. Chem. A* **2013**, *1*, 5628–5641. [CrossRef]
7. Huang, H.; Polavarapu, L.; Sichert, J.A.; Susha, A.S.; Urban, A.S.; Rogach, A.L. Colloidal lead halide perovskite nanocrystals: Synthesis, optical properties and applications. *NPG Asia Mater.* **2016**, *8*, e328. [CrossRef]
8. Amgar, D.; Aharon, S.; Etgar, L. Inorganic and Hybrid Organo-Metal Perovskite Nanostructures: Synthesis, Properties, and Applications. *Adv. Funct. Mater.* **2016**, *26*, 8576–8593. [CrossRef]
9. Gonzalez-Carrero, S.; Galian, R.E.; Pérez-Prieto, J. Organic-inorganic and all-inorganic lead halide nanoparticles [Invited]. *Opt. Express* **2016**, *24*, A285–A301. [CrossRef] [PubMed]
10. Li, X.; Cao, F.; Yu, D.; Chen, J.; Sun, Z.; Shen, Y.; Zhu, Y.; Wang, L.; Wei, Y.; Wu, Y.; et al. All Inorganic Halide Perovskites Nanosystem: Synthesis, Structural Features, Optical Properties and Optoelectronic Applications. *Small* **2017**, *13*, 1603996. [CrossRef] [PubMed]
11. González-Carrero, S.; Galian, R.E.; Pérez-Prieto, J. Organometal Halide Perovskites: Bulk Low-Dimension Materials and Nanoparticles. *Part. Part. Syst. Charact.* **2015**, *32*, 709–720. [CrossRef]
12. Shamsi, J.; Dang, Z.Y.; Bianchini, P.; Canale, C.; Di Stasio, F.; Brescia, R.; Prato, M.; Manna, L. Colloidal Synthesis of Quantum Confined Single Crystal CsPbBr3 Nanosheets with Lateral Size Control up to the Micrometer Range. *J. Am. Chem. Soc.* **2016**, *138*, 7240–7243. [CrossRef] [PubMed]
13. Kojima, A.; Teshima, K.; Shirai, Y.; Miyasaka, T. Organometal Halide Perovskites as Visible-Light Sensitizers for Photovoltaic Cells. *J. Am. Chem. Soc.* **2009**, *131*, 6050–6051. [CrossRef] [PubMed]
14. Zhang, F.; Zhong, H.; Chen, C.; Wu, X.-G.; Hu, X.; Huang, H.; Han, J.; Zou, B.; Dong, Y. Brightly Luminescent and Color-Tunable Colloidal CH3NH3PbX3 (X = Br, I, Cl) Quantum Dots: Potential Alternatives for Display Technology. *ACS Nano* **2015**, *9*, 4533–4542. [CrossRef] [PubMed]
15. Schmidt, L.C.; Pertegás, A.; González-Carrero, S.; Malinkiewicz, O.; Agouram, S.; Mínguez Espallargas, G.; Bolink, H.J.; Galian, R.E.; Pérez-Prieto, J. Nontemplate Synthesis of CH3NH3PbBr3 Perovskite Nanoparticles. *J. Am. Chem. Soc.* **2014**, *136*, 850–853. [CrossRef] [PubMed]
16. Imran, M.; Caligiuri, V.; Wang, M.; Goldoni, L.; Prato, M.; Krahne, R.; De Trizio, L.; Manna, L. Benzoyl Halides as Alternative Precursors for the Colloidal Synthesis of Lead-Based Halide Perovskite Nanocrystals. *J. Am. Chem. Soc.* **2018**. [CrossRef] [PubMed]
17. Vybornyi, O.; Yakunin, S.; Kovalenko, M.V. Polar-solvent-free colloidal synthesis of highly luminescent alkylammonium lead halide perovskite nanocrystals. *Nanoscale* **2016**, *8*, 6278–6283. [CrossRef] [PubMed]
18. Cohen, S. Biological reactions of carbonyl halides. In *Acyl Halides (1972)*; John Wiley & Sons, Ltd.: Hoboken, NJ, USA, 1972; pp. 313–348. [CrossRef]
19. Sun, S.; Yuan, D.; Xu, Y.; Wang, A.; Deng, Z. Ligand-Mediated Synthesis of Shape-Controlled Cesium Lead Halide Perovskite Nanocrystals via Reprecipitation Process at Room Temperature. *ACS Nano* **2016**, *10*, 3648–3657. [CrossRef] [PubMed]
20. Akkerman, Q.A.; Motti, S.G.; Kandada, A.R.S.; Mosconi, E.; D'Innocenzo, V.; Bertoni, G.; Marras, S.; Kamino, B.A.; Miranda, L.; De Angelis, F.; et al. Solution Synthesis Approach to Colloidal Cesium Lead

Halide Perovskite Nanoplatelets with Monolayer-Level Thickness Control. *J. Am. Chem. Soc.* **2016**, *138*, 1010–1016. [CrossRef] [PubMed]

21. Protesescu, L.; Yakunin, S.; Bodnarchuk, M.I.; Krieg, F.; Caputo, R.; Hendon, C.H.; Yang, R.X.; Walsh, A.; Kovalenko, M.V. Nanocrystals of Cesium Lead Halide Perovskites (CsPbX3, X = Cl, Br, and I): Novel Optoelectronic Materials Showing Bright Emission with Wide Color Gamut. *Nano Lett.* **2015**, *15*, 3692–3696. [CrossRef] [PubMed]

22. Kim, Y.; Yassitepe, E.; Voznyy, O.; Comin, R.; Walters, G.; Gong, X.; Kanjanaboos, P.; Nogueira, A.F.; Sargent, E.H. Efficient Luminescence from Perovskite Quantum Dot Solids. *ACS Appl. Mater. Interfaces* **2015**, *7*, 25007–25013. [CrossRef] [PubMed]

23. Tong, Y.; Bladt, E.; Aygüler, M.F.; Manzi, A.; Milowska, K.Z.; Hintermayr, V.A.; Docampo, P.; Bals, S.; Urban, A.S.; Polavarapu, L.; et al. Highly Luminescent Cesium Lead Halide Perovskite Nanocrystals with Tunable Composition and Thickness by Ultrasonication. *Angew. Chem. Int. Ed.* **2016**, *55*, 13887–13892. [CrossRef] [PubMed]

24. Cottingham, P.; Brutchey, R.L. On the crystal structure of colloidally prepared CsPbBr3 quantum dots. *Chem. Commun.* **2016**, *52*, 5246–5249. [CrossRef] [PubMed]

25. Dang, Z.; Shamsi, J.; Palazon, F.; Imran, M.; Akkerman, Q.A.; Park, S.; Bertoni, G.; Prato, M.; Brescia, R.; Manna, L. In Situ Transmission Electron Microscopy Study of Electron Beam-Induced Transformations in Colloidal Cesium Lead Halide Perovskite Nanocrystals. *ACS Nano* **2017**, *11*, 2124–2132. [CrossRef] [PubMed]

26. Bertolotti, F.; Protesescu, L.; Kovalenko, M.V.; Yakunin, S.; Cervellino, A.; Billinge, S.J.L.; Terban, M.W.; Pedersen, J.S.; Masciocchi, N.; Guagliardi, A. Coherent Nanotwins and Dynamic Disorder in Cesium Lead Halide Perovskite Nanocrystals. *ACS Nano* **2017**, *11*, 3819–3831. [CrossRef] [PubMed]

27. Protesescu, L.; Yakunin, S.; Kumar, S.; Bär, J.; Bertolotti, F.; Masciocchi, N.; Guagliardi, A.; Grotevent, M.; Shorubalko, I.; Bodnarchuk, M.I.; et al. Dismantling the "Red Wall" of Colloidal Perovskites: Highly Luminescent Formamidinium and Formamidinium–Cesium Lead Iodide Nanocrystals. *ACS Nano* **2017**, *11*, 3119–3134. [CrossRef] [PubMed]

28. Liu, F.; Zhang, Y.; Ding, C.; Kobayashi, S.; Izuishi, T.; Nakazawa, N.; Toyoda, T.; Ohta, T.; Hayase, S.; Minemoto, T.; et al. Highly Luminescent Phase-Stable CsPbI3 Perovskite Quantum Dots Achieving Near 100% Absolute Photoluminescence Quantum Yield. *ACS Nano* **2017**, *11*, 10373–10383. [CrossRef] [PubMed]

29. Akkerman, Q.A.; Meggiolaro, D.; Dang, Z.Y.; De Angelis, F.; Manna, L. Fluorescent Alloy CsPbxMn1-x,I-3 Perovskite Nanocrystals with High Structural and Optica Stability. *ACS Energy Lett.* **2017**, *2*, 2183–2186. [CrossRef] [PubMed]

30. Di Stasio, F.; Imran, M.; Akkerman, Q.A.; Prato, M.; Manna, L.; Krahne, R. Reversible Concentration-Dependent Photoluminescence Quenching and Change of Emission Color in CsPbBr3 Nanowires and Nanoplatelets. *J. Phys. Chem. Lett.* **2017**, *8*, 2725–2729. [CrossRef] [PubMed]

31. He, X.H.; Qiu, Y.C.; Yang, S.H. Fully-Inorganic Trihalide Perovskite Nanocrystals: A New Research Frontier of Optoelectronic Materials. *Adv. Mater.* **2017**, *29*, 1700775. [CrossRef] [PubMed]

32. Pan, A.Z.; He, B.; Fan, X.Y.; Liu, Z.K.; Urban, J.J.; Alivisatos, A.P.; He, L.; Liu, Y. Insight into the Ligand-Mediated Synthesis of Colloidal CsPbBr3 Perovskite Nanocrystals: The Role of Organic Acid, Base, and Cesium Precursors. *ACS Nano* **2016**, *10*, 7943–7954. [CrossRef] [PubMed]

33. Imran, M.; Di Stasio, F.; Dang, Z.Y.; Canale, C.; Khan, A.H.; Shamsi, J.; Brescia, R.; Prato, M.; Manna, L. Colloidal Synthesis of Strongly Fluorescent CsPbBr3 Nanowires with Width Tunable down to the Quantum Confinement Regime. *Chem. Mater.* **2016**, *28*, 6450–6454. [CrossRef] [PubMed]

34. Bekenstein, Y.; Koscher, B.A.; Eaton, S.W.; Yang, P.D.; Alivisatos, A.P. Highly Luminescent Colloidal Nanoplates of Perovskite Cesium Lead Halide and Their Oriented Assemblies. *J. Am. Chem. Soc.* **2015**, *137*, 16008–16011. [CrossRef] [PubMed]

35. Pellet, N.; Gao, P.; Gregori, G.; Yang, T.-Y.; Nazeeruddin, M.K.; Maier, J.; Grätzel, M. Mixed-Organic-Cation Perovskite Photovoltaics for Enhanced Solar-Light Harvesting. *Angew. Chem. Int. Ed.* **2014**, *53*, 3151–3157. [CrossRef] [PubMed]

36. Eperon, G.E.; Stranks, S.D.; Menelaou, C.; Johnston, M.B.; Herz, L.M.; Snaith, H.J. Formamidinium lead trihalide: A broadly tunable perovskite for efficient planar heterojunction solar cells. *Energy Environ. Sci.* **2014**, *7*, 982–988. [CrossRef]

37. Amat, A.; Mosconi, E.; Ronca, E.; Quarti, C.; Umari, P.; Nazeeruddin, M.K.; Grätzel, M.; De Angelis, F. Cation-Induced Band-Gap Tuning in Organohalide Perovskites: Interplay of Spin–Orbit Coupling and Octahedra Tilting. *Nano Lett.* **2014**, *14*, 3608–3616. [CrossRef] [PubMed]

38. Levchuk, I.; Osvet, A.; Tang, X.; Brandl, M.; Perea, J.D.; Hoegl, F.; Matt, G.J.; Hock, R.; Batentschuk, M.; Brabec, C.J. Brightly Luminescent and Color-Tunable Formamidinium Lead Halide Perovskite FAPbX3 (X = Cl, Br, I) Colloidal Nanocrystals. *Nano Lett.* **2017**, *17*, 2765–2770. [CrossRef] [PubMed]

39. Protesescu, L.; Yakunin, S.; Bodnarchuk, M.I.; Bertolotti, F.; Masciocchi, N.; Guagliardi, A.; Kovalenko, M.V. Monodisperse Formamidinium Lead Bromide Nanocrystals with Bright and Stable Green Photoluminescence. *J. Am. Chem. Soc.* **2016**, *138*, 14202–14205. [CrossRef] [PubMed]

40. Minh, D.N.; Kim, J.; Hyon, J.; Sim, J.H.; Sowlih, H.H.; Seo, C.; Nam, J.; Eom, S.; Suk, S.; Lee, S.; et al. Room-Temperature Synthesis of Widely Tunable Formamidinium Lead Halide Perovskite Nanocrystals. *Chem. Mater.* **2017**, *29*, 5713–5719. [CrossRef]

41. Zhumekenov, A.A.; Saidaminov, M.I.; Haque, M.A.; Alarousu, E.; Sarmah, S.P.; Murali, B.; Dursun, I.; Miao, X.-H.; Abdelhady, A.L.; Wu, T.; et al. Formamidinium Lead Halide Perovskite Crystals with Unprecedented Long Carrier Dynamics and Diffusion Length. *ACS Energy Lett.* **2016**, *1*, 32–37. [CrossRef]

42. Swarnkar, A.; Ravi, V.K.; Nag, A. Beyond Colloidal Cesium Lead Halide Perovskite Nanocrystals: Analogous Metal Halides and Doping. *ACS Energy Lett.* **2017**, *2*, 1089–1098. [CrossRef]

43. Meinardi, F.; Akkerrnan, Q.A.; Bruni, F.; Park, S.; Mauri, M.; Dang, Z.Y.; Manna, L.; Brovelli, S. Doped Halide Perovskite Nanocrystals for Reabsorption-Free Luminescent Solar Concentrators. *Acs Energy Lett.* **2017**, *2*, 2368–2377. [CrossRef]

44. Parobek, D.; Roman, B.J.; Dong, Y.T.; Jin, H.; Lee, E.; Sheldon, M.; Son, D.H. Exciton-to-Dopant Energy Transfer in Mn-Doped Cesium Lead Halide Perovskite Nanocrystals. *Nano Lett.* **2016**, *16*, 7376–7380. [CrossRef] [PubMed]

45. Begum, R.; Parida, M.R.; Abdelhady, A.L.; Murali, B.; Alyami, N.M.; Ahmed, G.H.; Hedhili, M.N.; Bakr, O.M.; Mohammed, O.F. Engineering Interfacial Charge Transfer in CsPbBr3 Perovskite Nanocrystals by Heterovalent Doping. *J. Am. Chem. Soc.* **2017**, *139*, 731–737. [CrossRef] [PubMed]

46. Xu, K.Y.; Lin, C.C.; Xie, X.B.; Meijerink, A. Efficient and Stable Luminescence from Mn2+ in Core and Core-Isocrystalline Shell CsPbCl3 Perovskite Nanocrystals. *Chem. Mater.* **2017**, *29*, 4265–4272. [CrossRef] [PubMed]

47. Zou, S.H.; Liu, Y.S.; Li, J.H.; Liu, C.P.; Feng, R.; Jiang, F.L.; Li, Y.X.; Song, J.Z.; Zeng, H.B.; Hong, M.C.; et al. Stabilizing Cesium Lead Halide Perovskite Lattice through Mn(II) Substitution for Air-Stable Light-Emitting Diodes. *J. Am. Chem. Soc.* **2017**, *139*, 11443–11450. [CrossRef] [PubMed]

48. Akkerman, Q.A.; D'Innocenzo, V.; Accornero, S.; Scarpellini, A.; Petrozza, A.; Prato, M.; Manna, L. Tuning the Optical Properties of Cesium Lead Halide Perovskite Nanocrystals by Anion Exchange Reactions. *J. Am. Chem. Soc.* **2015**, *137*, 10276–10281. [CrossRef] [PubMed]

49. Zhang, T.; Li, G.; Chang, Y.; Wang, X.; Zhang, B.; Mou, H.; Jiang, Y. Full-spectra hyperfluorescence cesium lead halide perovskite nanocrystals obtained by efficient halogen anion exchange using zinc halogenide salts. *CrystEngComm* **2017**, *19*, 1165–1171. [CrossRef]

50. Parobek, D.; Dong, Y.; Qiao, T.; Rossi, D.; Son, D.H. Photoinduced Anion Exchange in Cesium Lead Halide Perovskite Nanocrystals. *J. Am. Chem. Soc.* **2017**, *139*, 4358–4361. [CrossRef] [PubMed]

51. Guhrenz, C.; Benad, A.; Ziegler, C.; Haubold, D.; Gaponik, N.; Eychmüller, A. Solid-State Anion Exchange Reactions for Color Tuning of CsPbX3 Perovskite Nanocrystals. *Chem. Mater.* **2016**, *28*, 9033–9040. [CrossRef]

52. Nedelcu, G.; Protesescu, L.; Yakunin, S.; Bodnarchuk, M.I.; Grotevent, M.J.; Kovalenko, M.V. Fast Anion-Exchange in Highly Luminescent Nanocrystals of Cesium Lead Halide Perovskites (CsPbX3, X = Cl, Br, I). *Nano Lett.* **2015**, *15*, 5635–5640. [CrossRef] [PubMed]

53. Liu, H.W.; Wu, Z.N.; Shao, J.R.; Yao, D.; Gao, H.; Liu, Y.; Yu, W.L.; Zhang, H.; Yang, B. CsPbxMn1-xCl3 Perovskite Quantum Dots with High Mn Substitution Ratio. *ACS Nano* **2017**, *11*, 2239–2247. [CrossRef] [PubMed]

54. Momma, K.; Izumi, F. VESTA 3 for three-dimensional visualization of crystal, volumetric and morphology data. *J. Appl. Crystallogr.* **2011**, *44*, 1272–1276. [CrossRef]

55. Jellicoe, T.C.; Richter, J.M.; Glass, H.F.J.; Tabachnyk, M.; Brady, R.; Dutton, S.E.; Rao, A.; Friend, R.H.; Credgington, D.; Greenham, N.C.; et al. Synthesis and Optical Properties of Lead-Free Cesium Tin Halide Perovskite Nanocrystals. *J. Am. Chem. Soc.* **2016**, *138*, 2941–2944. [CrossRef] [PubMed]

56. Wang, A.F.; Guo, Y.Y.; Muhammad, F.; Deng, Z.T. Controlled Synthesis of Lead-Free Cesium Tin Halide Perovskite Cubic Nanocages with High Stability. *Chem. Mater.* **2017**, *29*, 6493–6501. [CrossRef]

57. Wang, A.F.; Yan, X.G.; Zhang, M.; Sun, S.B.; Yang, M.; Shen, W.; Pan, X.Q.; Wang, P.; Deng, Z.T. Controlled Synthesis of Lead-Free and Stable Perovskite Derivative Cs2SnI6 Nanocrystals via a Facile Hot-Injection Process. *Chem. Mater.* **2016**, *28*, 8132–8140. [CrossRef]

58. Dolzhnikov, D.S.; Wang, C.; Xu, Y.D.; Kanatzidis, M.G.; Weiss, E.A. Ligand-Free, Quantum-Confined Cs2SnI6 Perovskite Nanocrystals. *Chem. Mater.* **2017**, *29*, 7901–7907. [CrossRef]

59. Leng, M.Y.; Chen, Z.W.; Yang, Y.; Li, Z.; Zeng, K.; Li, K.H.; Niu, G.D.; He, Y.S.; Zhou, Q.C.; Tang, J. Lead-Free, Blue Emitting Bismuth Halide Perovskite Quantum Dots. *Angew. Chem. Int. Ed.* **2016**, *55*, 15012–15016. [CrossRef] [PubMed]

60. Leng, M.; Yang, Y.; Zeng, K.; Chen, Z.; Tan, Z.; Li, S.; Li, J.; Xu, B.; Li, D.; Hautzinger, M.P.; et al. All-Inorganic Bismuth-Based Perovskite Quantum Dots with Bright Blue Photoluminescence and Excellent Stability. *Adv. Funct. Mater.* **2018**, *28*, 1704446. [CrossRef]

61. Yang, B.; Chen, J.S.; Hong, F.; Mao, X.; Zheng, K.B.; Yang, S.Q.; Li, Y.J.; Pullerits, T.; Deng, W.Q.; Han, K.L. Lead-Free, Air-Stable All-Inorganic Cesium Bismuth Halide Perovskite Nanocrystals. *Angew. Chem. Int. Ed.* **2017**, *56*, 12471–12475. [CrossRef] [PubMed]

62. Zhang, J.; Yang, Y.; Deng, H.; Farooq, U.; Yang, X.K.; Khan, J.; Tang, J.; Song, H.S. High Quantum Yield Blue Emission from Lead Free Inorganic Antimony Halide Perovskite Colloidal Quantum Dots. *ACS Nano* **2017**, *11*, 9294–9302. [CrossRef] [PubMed]

63. Creutz, S.E.; Crites, E.N.; De Siena, M.C.; Gamelin, D.R. Colloidal Nanocrystals of Lead-Free Double-Perovskite (Elpasolite) Semiconductors: Synthesis and Anion Exchange To Access New Materials. *Nano Lett.* **2018**. [CrossRef] [PubMed]

64. Zhou, L.; Xu, Y.-F.; Chen, B.-X.; Kuang, D.-B.; Su, C.-Y. Synthesis and Photocatalytic Application of Stable Lead-Free Cs2AgBiBr6 Perovskite Nanocrystals. *Small* **2018**. [CrossRef]

65. Burschka, J.; Pellet, N.; Moon, S.-J.; Humphry-Baker, R.; Gao, P.; Nazeeruddin, M.K.; Grätzel, M. Sequential deposition as a route to high-performance perovskite-sensitized solar cells. *Nature* **2013**, *499*, 316–319. [CrossRef] [PubMed]

66. Aggarwal, M.; Choi, J.; Wang, W.; Bhat, K.; Lal, R.; Shields, A.D.; Penn, B.G.; Frazier, D.O. Solution growth of a novel nonlinear optical material: L-histidine tetrafluoroborate. *J. Cryst. Growth* **1999**, *204*, 179–182. [CrossRef]

67. Owens, C.; Bhat, K.; Wang, W.; Tan, A.; Aggarwal, M.; Penn, B.G.; Frazier, D.O. Bulk growth of high quality nonlinear optical crystals of L-arginine tetrafluoroborate (L-AFB). *J. Cryst. Growth* **2001**, *225*, 465–469. [CrossRef]

68. Maculan, G.; Sheikh, A.D.; Abdelhady, A.L.; Saidaminov, M.I.; Haque, M.A.; Murali, B.; Alarousu, E.; Mohammed, O.F.; Wu, T.; Bakr, O.M. CH3NH3PbCl3 single crystals: Inverse temperature crystallization and visible-blind UV-photodetector. *J. Phys. Chem. Lett.* **2015**, *6*, 3781–3786. [CrossRef] [PubMed]

69. Park, N.-G. Crystal growth engineering for high efficiency perovskite solar cells. *CrystEngComm* **2016**, *18*, 5977–5985. [CrossRef]

70. Suwa, T.; Takehisa, M.; Machi, S. Melting and crystallization behavior of poly(tetrafluoroethylene). New method for molecular weight measurement of poly(tetrafluoroethylene) using a differential scanning calorimeter. *J. Appl. Polym. Sci.* **1973**, *17*, 3253–3257. [CrossRef]

71. Lofgren, G.; Hargraves, R. Experimental studies on the dynamic crystallization of silicate melts. In *Physics of Magmatic Processes*; Princeton University Press: Princeton, NJ, USA, 1980; Volume 487-503.

72. König, A.; Stepanski, M.; Kuszlik, A.; Keil, P.; Weller, C. Ultra-purification of ionic liquids by melt crystallization. *Chem. Eng. Res. Des.* **2008**, *86*, 775–780. [CrossRef]

73. Cohen, B.-E.; Aharon, S.; Dymshits, A.; Etgar, L. Impact of antisolvent treatment on carrier density in efficient hole-conductor-free perovskite-based solar cells. *J. Phys. Chem. C* **2015**, *120*, 142–147. [CrossRef]

74. Paek, S.; Schouwink, P.; Athanasopoulou, E.N.; Cho, K.T.; Grancini, G.; Lee, Y.; Zhang, Y.; Stellacci, F.; Nazeeruddin, M.K.; Gao, P. From Nano- to Micrometer Scale: The Role of Antisolvent Treatment on High Performance Perovskite Solar Cells. *Chem. Mater.* **2017**, *29*, 3490–3498. [CrossRef]

75. Park, N.-G.; Grätzel, M.; Miyasaka, T.; Zhu, K.; Emery, K. Towards stable and commercially available perovskite solar cells. *Nat. Energy* **2016**, *1*, 16152. [CrossRef]

76. Konstantakou, M.; Perganti, D.; Falaras, P.; Stergiopoulos, T. Anti-solvent crystallization strategies for highly efficient perovskite solar cells. *Crystals* **2017**, *7*, 291. [CrossRef]

77. Zuo, L.; Dong, S.; De Marco, N.; Hsieh, Y.-T.; Bae, S.-H.; Sun, P.; Yang, Y. Morphology evolution of high efficiency perovskite solar cells via vapor induced intermediate phases. *J. Am. Chem. Soc.* **2016**, *138*, 15710–15716. [CrossRef] [PubMed]

78. Fujihara, T.; Terakawa, S.; Matsushima, T.; Qin, C.; Yahiro, M.; Adachi, C. Fabrication of high coverage MASnI 3 perovskite films for stable, planar heterojunction solar cells. *J. Mater. Chem. C* **2017**, *5*, 1121–1127. [CrossRef]

79. Yokoyama, T.; Cao, D.H.; Stoumpos, C.C.; Song, T.-B.; Sato, Y.; Aramaki, S.; Kanatzidis, M.G. Overcoming short-circuit in lead-free CH3NH3SnI3 perovskite solar cells via kinetically controlled gas–solid reaction film fabrication process. *J. Phys. Chem. Lett.* **2016**, *7*, 776–782. [CrossRef] [PubMed]

80. Xiao, Z.; Dong, Q.; Bi, C.; Shao, Y.; Yuan, Y.; Huang, J. Solvent Annealing of Perovskite-Induced Crystal Growth for Photovoltaic-Device Efficiency Enhancement. *Adv. Mater.* **2014**, *26*, 6503–6509. [CrossRef] [PubMed]

81. Yang, M.; Zhang, T.; Schulz, P.; Li, Z.; Li, G.; Kim, D.H.; Guo, N.; Berry, J.J.; Zhu, K.; Zhao, Y. Facile fabrication of large-grain CH3NH3PbI3−xBrx films for high-efficiency solar cells via CH3NH3Br-selective Ostwald ripening. *Nat. Commun.* **2016**, *7*, 12305. [CrossRef] [PubMed]

82. Chinwangso, P.; Lee, H.J.; Jamison, A.C.; Marquez, M.D.; Park, C.S.; Lee, T.R. Structure, Wettability, and Thermal Stability of Organic Thin-Films on Gold Generated from the Molecular Self-Assembly of Unsymmetrical Oligo(ethylene glycol) Spiroalkanedithiols. *Langmuir* **2017**, *33*, 1751–1762. [CrossRef] [PubMed]

83. Lee, L.; Baek, J.; Park, K.S.; Lee, Y.-E.; Shrestha, N.K.; Sung, M.M. Wafer-scale single-crystal perovskite patterned thin films based on geometrically-confined lateral crystal growth. *Nat. Commun.* **2017**, *8*, 15882. [CrossRef] [PubMed]

84. Li, T.; Pan, Y.; Wang, Z.; Xia, Y.; Chen, Y.; Huang, W. Additive engineering for highly efficient organic–inorganic halide perovskite solar cells: Recent advances and perspectives. *J. Mater. Chem. A* **2017**, *5*, 12602–12652. [CrossRef]

85. Zuo, L.; Guo, H.; Jariwala, S.; De Marco, N.; Dong, S.; DeBlock, R.; Ginger, D.S.; Dunn, B.; Wang, M.; Yang, Y. Polymer-modified halide perovskite films for efficient and stable planar heterojunction solar cells. *Sci. Adv.* **2017**, *3*, e1700106. [CrossRef] [PubMed]

86. Li, X.; Dar, M.I.; Yi, C.; Luo, J.; Tschumi, M.; Zakeeruddin, S.M.; Nazeeruddin, M.; Han, H.; Graetzel, M. Improved performance and stability of perovskite solar cells by crystal crosslinking with alkylphosphonic acid ?ω-ammonium chlorides. *Nat. Chem.* **2015**, *7*, 703–711. [CrossRef] [PubMed]

87. Ma, T.; Zhang, Q.; Tadaki, D.; Hirano-Iwata, A.; Niwano, M. Fabrication and Characterization of High-Quality Perovskite Films with Large Crystal Grains. *J. Phys. Chem. Lett.* **2017**, *8*, 720–726. [CrossRef] [PubMed]

88. Sessolo, M.; Momblona, C.; Gil-Escrig, L.; Bolink, H.J. Photovoltaic devices employing vacuum-deposited perovskite layers. *MRS Bull.* **2015**, *40*, 660–666. [CrossRef]

Review

Anti-Solvent Crystallization Strategies for Highly Efficient Perovskite Solar Cells

Maria Konstantakou [1], Dorothea Perganti [1,2,3], Polycarpos Falaras [2] and Thomas Stergiopoulos [1,*]

[1]   Laboratory of Physical Chemistry, Department of Chemistry, Aristotle University of Thessaloniki, Thessaloniki 54124, Greece; marykon21@hotmail.com (M.K.); d.perganti@inn.demokritos.gr (D.P.)
[2]   Institute of Nanoscience and Nanotechnology, National Centre for Scientific Research Demokritos, Athens 15310, Greece; p.falaras@inn.demokritos.gr
[3]   School of Chemical Engineering, National Technical University of Athens, Athens 15780, Greece
*   Correspondence: stergt@chem.auth.gr

Academic Editor: Wei Zhang
Received: 3 September 2017; Accepted: 26 September 2017; Published: 28 September 2017

Abstract: Solution-processed organic-inorganic halide perovskites are currently established as the hottest area of interest in the world of photovoltaics, ensuring low manufacturing cost and high conversion efficiencies. Even though various fabrication/deposition approaches and device architectures have been tested, researchers quickly realized that the key for the excellent solar cell operation was the quality of the crystallization of the perovskite film, employed to assure efficient photogeneration of carriers, charge separation and transport of the separated carriers at the contacts. One of the most typical methods in chemistry to crystallize a material is anti-solvent precipitation. Indeed, this classical precipitation method worked really well for the growth of single crystals of perovskite. Fortunately, the method was also effective for the preparation of perovskite films by adopting an anti-solvent dripping technique during spin-coating the perovskite precursor solution on the substrate. With this, polycrystalline perovskite films with pure and stable crystal phases accompanied with excellent surface coverage were prepared, leading to highly reproducible efficiencies close to 22%. In this review, we discuss recent results on highly efficient solar cells, obtained by the anti-solvent dripping method, always in the presence of Lewis base adducts of lead(II) iodide. We present all the anti-solvents that can be used and what is the impact of them on device efficiencies. Finally, we analyze the critical challenges that currently limit the efficacy/reproducibility of this crystallization method and propose prospects for future directions.

Keywords: perovskite; solar cell; anti-solvent; efficiency

---

1. Introduction

Halide perovskite solar cells were introduced for the first time in 2009 [1] and it then took three years for devices to attain power conversion efficiencies (PCEs) higher than ~10% under standard 1 sun AM1.5G illumination [2–4]. Within less than a year, the efficiencies went up to 15% using a so-called two-step solution [5] or a vapor deposition [6] processes. However, the most influential publications in the field (in our opinion) came with the introduction of anti-solvent crystallization in 2014; a poor solvent is poured onto the precursor perovskite film during spin coating, causing the salts to precipitate out of solution into a smooth, compact film [7,8]. Notably, in one of the publications, anti-solvent crystallization was combined with evidences that the perovskite forms through an intermediate that includes methylammonium iodide (MAI), PbI_2 and dimethylsulfoxide (DMSO). The presence of this intermediate (or adduct) was indispensable in order to deliver hysteresis-free efficiencies over 16% [8]. Then a scientific "full moon" hit the field, motivating independent laboratories to adopt intermediate/anti-solvent crystallization in order to further enhance the efficiencies. Luckily, PCEs went up to close to 22% [9] within three years of experiments, utilizing various device architectures, passivation strategies and mixed perovskites. These PCEs are equivalent to others attained from emerging thin-film technologies such as CdTE and CIGS solar cells [10]. Notably, this success story found also application in other optoelectronic devices such as photodetectors and LEDs [11].

There are already literature reviews on perovskite film formation where the adduct/anti-solvent crystallization process is included [12–17]. However, at least according to our knowledge, there is no work describing in detail results adopting solely this method. More specifically here, we attempt to describe the method in comparison with typical anti-solvent precipitation that grows perovskite single crystals. We discuss the role of the stable MAI·PbI$_2$·DMSO adduct and how we can achieve this intermediate phase through control of the precursor solution stoichiometry. We demonstrate all the recent results based on record PCEs attained from various types of devices based on typical MAPbI$_3$ or mixed FA$_{1-x}$MA$_x$PbI$_{3-y}$Br$_y$ perovskites (where FA stands for formamidinium cation). Emphasis is given on novel approaches based on MAI or PbI$_2$-excessive perovskite precursor solutions as well as on the addition of inorganic cations (Cs$^+$ and Rb$^+$) in order to fabricate optimal films. We also review very recent results where dripping an anti-solvent is accompanied by the deposition of an active compound at the very top surface of the perovskite. Finally, we provide all the range of anti-solvents used thus far, suggesting the five more promising systems.

Reading this review, one will always meet the terms "stabilized" (or steady-state) and/or "reverse scan" efficiency. We would like to clarify here that we refer to the scan-rate dependent hysteresis in the J-V curves of the perovskite solar cells. When current is measured from reverse-to-forward bias sweep direction (called reverse scan by us), an overestimation of the photovoltaic performance is taking place. In turn, if voltage is swept oppositely, one finds lower performances. As a consequence, recommendations are suggested so as to show photovoltaic behavior without masking the detrimental hysteresis effect. This is often realized by the "stabilized efficiency" experiment; thus, with the term "stabilized" efficiency, we refer to the determined value of the efficiency after holding the tested device at a constant voltage around the maximum power-point and track the power-output until it reaches a constant value [18].

2. Anti-solvent Precipitation Method to Grow Perovskite Single Crystals

2.1. Solution Processed Growth of Perovskite Single Crystals: The Conventional Way

The typical method to grow perovskite single crystals of high quality out of solution employs a large excess of HI/H$_2$O as the reaction solvent. Precipitation in a solution occurs rapidly when the concentration of a compound exceeds its solubility, e.g., from a supersaturated solution. Supersaturation can be easily achieved for instance by cooling. As a general rule, the more heat is added to a system, the more soluble a substance becomes (this is not the case for various perovskites-in-solvent systems [19]). Therefore, at high temperatures, more solute can be dissolved than at room temperature. If this solution was to be suddenly cooled at a rate faster than the rate of precipitation, the solution will become supersaturated until the solute precipitates to the temperature-determined saturation point. With this, pure stoichiometric compounds are isolated with a very low carrier concentration (being nearly intrinsic) [20].

2.2. Anti-Solvent Precipitation: A Short Description

In another approach, supersaturation can be simply realized by exposing a solution of the product to another solvent (or multiple ones) in which the product is sparingly soluble (thus called anti-solvent). Again, precipitation will occur since the solubility of the desired product will be drastically reduced (Figure 1a). The efficacy of the approach (or otherwise stated, the quality of the crystals) depends on several parameters such as the nature of the anti-solvent being used, the volume ratio between solvent and anti-solvent, the exact time of the diffusion of the anti-solvent, the diffusion rate etc.

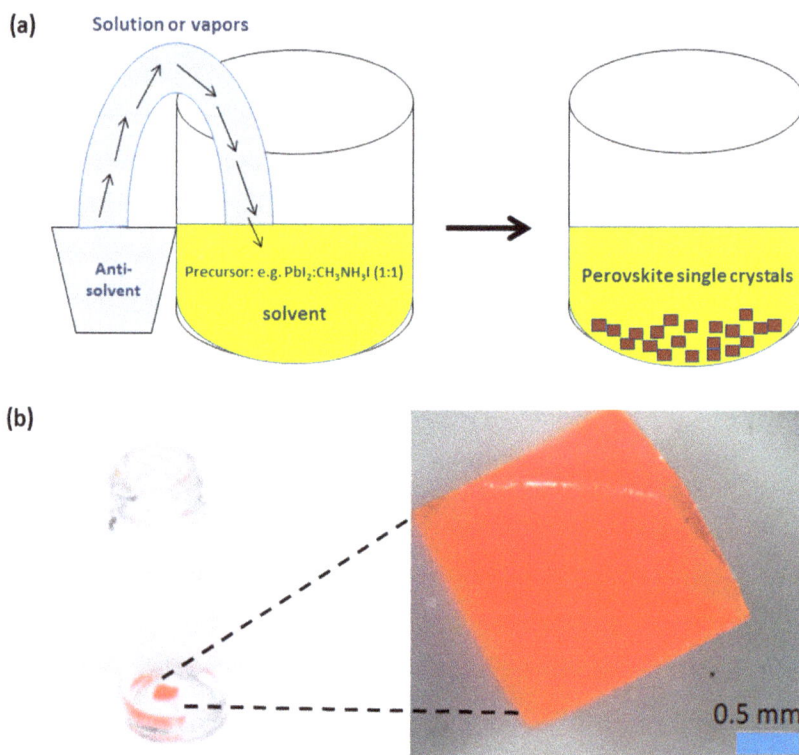

Figure 1. (**a**) A schematic diagram of a typical, anti-solvent crystallization process for MAPbI$_3$ perovskite formation. (**b**) A vial containing an orange single crystal of MAPbBr$_3$, grown by the anti-solvent (IPA) crystallization method and optical microscope image of the crystal. Reprinted with permission from [21]. Copyright (2014) American Chemical Society.

2.3. MAPbBr$_3$ Single Crystals Grown Via Anti-Solvent (IPA) Precipitation

The first work on the anti-solvent precipitation method came out on June 2014. Tidhar et al. have grown MAPbBr$_3$ single crystals by slowly adding isopropanol (IPA) vapors in the precursors dissolved in pure dimethylformamide (DMF); cubic perovskite crystals of large size were formed within 10–12 h (Figure 1b). The authors could not apply the same method to grow crystals for the MAPbI$_3$ perovskite [21].

2.4. MAPbBr$_3$ and MAPbI$_3$ Single Crystals Grown Via Anti-Solvent (DCM) Precipitation

In a seminal work, submitted on January 2015 in Science, Shi et al. took the approach in a step forward. They replaced IPA by dichloromethane (DCM), which is a much poorer solvent of the perovskite (IPA can dissolve one of the constituents of the precursor, i.e., MAI or MABr). Vapors of DCM were allowed to diffuse into the precursor solutions that employed highly polar but non-highly coordinating solvents. Slow diffusion was chosen in order to develop adequate surface area to primarily achieve growth (vs. nucleation). With this, the ionic building blocks of the perovskite were co-precipitated from solution stoichiometrically, producing high-quality, millimeter-sized single crystals. Interestingly, these crystals exhibited exceptionally low trap densities (10^9–10^{10} cm^{-3}) and impressively long diffusion lengths exceeding 10 μm [22]. We should state here that this was the first work ever to grow single crystals of MAPbI$_3$ following anti-solvent crystallization; to achieve that, the authors have used an excess of MAI (PbI$_2$:MAI = 1:3) in gamma-butyrolactone, gBL (instead of a stoichiometric 1:1 molar ratio of PbBr$_2$:MABr in DMF, being employed for the crystals of MAPbBr$_3$).

3. Anti-Solvent Dripping Technique to Prepare Polycrystalline Perovskite Films

3.1. Anti-Solvent Dripping: An Introduction

Normally, to replicate the anti-solvent precipitation procedure, one should prepare yellow films of MAI:PbI$_2$ (1:1) out of solution in an aprotic solvent (recall that this is a yellow solution) via spin-coating, and then inject the anti-solvent to immediately form a dark brown/red perovskite film; this film has to be purified (separation from mother liquor is not a prerequisite here as in the case of anti-solvent precipitation) by removing any additional phases, impurities (usually by copious washing with ethanol [20]) and the anti-solvent itself. There are inherently two main issues here: first crystallization of the film during spin-coating is quite fast, producing a perovskite film with bad morphology and very low surface coverage; if one injects the anti-solvent on top after the end of spinning (or if one simply immerses the "bad" film in a bath of anti-solvent), nothing will change. This implies that the anti-solvent needs to be dripped at a certain point (and quite quickly) during spinning. The second issue is that after forming a perovskite film (at the end of spin-coating), a proper procedure has to be adopted in order to get rid of all the stuff that are unnecessary (anything that is not MAPbI$_3$). This suggests that perovskite films should be washed with an appropriate solvent and dried adequately so that impurities evaporate out of the film.

3.2. Anti-Solvent (CB) Dripping Using Precursor Solution Dissolved Solely in DMF

Simultaneously with the first work on anti-solvent precipitation for the growth of single crystals, two papers on anti-solvent crystallized perovskite films appeared in literature. First, Xiao et al. fabricated, inside a N$_2$-filled glovebox, compact films with full surface coverage by employing a fast deposition-crystallization (FDC) procedure when dripping an anti-solvent during the first seconds of spin-coating a precursor solution dissolved in DMF [7]. When an anti-solvent was not involved in the process, an uncontrollable, needle-like structure was formed (Figure 2a). A large variety of anti-solvents was tested such as chlorobenzene (CB or CBZ), benzene, xylene (XYL), toluene (TL), methanol, ethanol, ethylene glycol, 2-propanol (IPA), chloroform (CF), tetrahydrofuran, acetonitrile and benzonitrile (some of these solvents, like ethanol or ethylene glycol, could dissolve the ammonium halide salt and destroy the final perovskite film). Due to the centrifugal forces during spinning, the anti-solvent should be dripped at the center of the film. The morphology of the film was uniform over the entire substrate only in the case of chlorobenzene, benzene, xylene and toluene; mainly the central area of the films was non-uniform when other anti-solvents were adopted.

To fabricate films with the desirable morphology and structure, CB as an anti-solvent along with a specific delay time of 4–6 s were proven to be the most adequate choices. The authors rationalized this as follows: in the first 3 s after spinning at 5000 rpm was commenced, removal of excess precursor solution is a dominant process. Introduction of CB at this stage does not lead to full surface coverage possibly because the perovskite solution is far from supersaturation. Then at 4th to 6th s, evaporation of the residue solvent occurs significantly, concentrating the perovskite solution from which a dense and uniform film is formed when the CB is introduced. After 7 s, the liquid film starts to dry and addition of CB does not help (Figure 2b).

After dripping, the film instantly darkened (from yellow to light brown), as perovskite constituents precipitate out and the material crystallized. Then, spinning lasted for additional 24–26 s. To purify and fully dry the material, the films were subsequently subjected to annealing at 100 °C for 10 min. The obtained crystalline grains spanned the thickness of the film (~micron size), being large and free of boundaries in the perpendicular direction, allowing for beneficial charge transport/recombination dynamics.

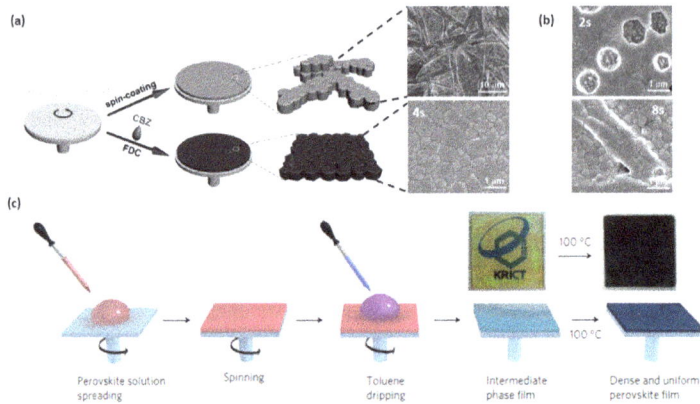

Figure 2. (a) Schematic illustration of anti-solvent dripping (denoted as FDC) and conventional spin-coating process for fabricating perovskite films. Conventional spin-coating (top) results in a gray film (on top of a compact TiO$_2$ layer) composed of rod-like crystals. In the FDC process (bottom), a light brown film is formed, consisting of uniformly sized perovskite grains. (b) SEM images of the surface morphology of films prepared by adding CBZ at different delay times from the start of the spin-coating process (after 2 s and 8 s) Note that a delay time between 4 and 6 s produces the optimum morphology. Reproduced from Ref. [7] with permission from Wiley-VCH. (c) Toluene dripping during precursor perovskite solution (gBL-DMSO) spin-casting onto m-TiO$_2$ substrate for preparing a stable intermediate which easily transforms into the perovskite after annealing. Reproduced from ref. [8] with permission from Nature Publishing Group.

3.3. Anti-Solvent (TL) Dripping Using Precursor Solution Dissolved In DMSO-gBL Mixtures: Lewis Acid-Base Interactions

The second work came from the group of Sang Il Seok who employed a "solvent engineering" technique (Figure 2b) to deposit MAPb(I$_{(1-x)}$Br$_x$)$_3$ onto thin mesoporous TiO$_2$ (m-TiO$_2$) scaffolds [8]. They used a mixture of gBL and DMSO (7:3 v/v) as a co-solvent for the precursors, which was coated onto the substrate by a consecutive two-step spin-coating process at 1000 and 5000 rpm for 10 and 20 s, respectively. During the second spin-coating step, TL was dripped onto the film. When gBL was used alone, dark hazy brown films, presenting very poor surface coverage, were obtained, with or without the anti-solvent treatment. In the presence of DMSO, the authors concluded that during anti-solvent dripping, the salts rapidly precipitated out of solution into a smooth, pre-crystalized, transparent yellowish film, potentially consisting of a MAI-PbI$_2$-DMSO intermediate phase (or otherwise called, an adduct). The role of DMSO in the MAI-PbI$_2$-DMSO phase is to retard the rapid reaction between PbI$_2$ and MAI during the evaporation of solvent in the spin-coating process. This is realized due to the interaction between Lewis base DMSO, iodide (I$^-$) and Lewis acid PbI$_2$. As a final step, annealing at 100 °C for 10 min leads to formation of the Br-containing MAPbI$_3$ perovskite (dark brown film) by driving out the entrapped coordinated DMSO. This procedure results finally in compact and uniform capping layers with grain sizes in the range of 100 to 500 nm with a 100% surface coverage of the substrate.

The same group adopted the toluene dripping technique in order to fabricate highly efficient MAPbI$_3$ solar cells by replacing m-TiO$_2$ with mesoporous La–doped BaSnO$_3$, which presents better electron mobility [23]. This time, the authors used a bromide-free MAPbI$_3$ perovskite dissolved in a gBL/DMSO co-solvent system (4:3 v/v) but 2-methoxyethanol was added as well. Additionally, every fabrication step was conducted in air under relative humidity below 25% at 25 °C. The solar cells, having a meso n-i-p structure (FTO/compact TiO$_2$/mesoporous BaSnO$_3$ (or m-TiO$_2$)/doped polytriarylamine (PTAA)/Au), have shown a steady-state PCE of 21.2%, under AM1.5G full sun illumination. The viability of the concept of TL dripping along with a gBL/DMSO system was confirmed by other groups [24].

3.4. Anti-Solvent (DE) Dripping Using Precursor Solution Dissolved in DMSO-DMF Mixtures: The Role of the Stable MAI·PbI$_2$·DMSO (1:1:1 mol%) Adduct

The existence of the intermediate MAI·PbI$_2$·DMSO phase was justified by Jeon et al. [8], however it was unclear whether this was indeed a 1:1:1 stoichiometric compound; first because this specific crystalline structure could not be identified by X-ray diffraction, and then because the anti-solvent (TL) is fully miscible with gBL and DMSO, thus some of the DMSO could be washed away during dripping.

To do so, the group of Nam-Gyu Park suggested a novel Lewis acid–base adduct approach. The authors added an equimolar mixture of PbI$_2$, MAI and DMSO (1:1:1 mol%) in DMF. The precursor solution was spin-coated onto the m-TiO$_2$ substrate at 4000 rpm for 25 s. 10 s before the surface changed to be turbid, caused by rapid vaporization of DMF, diethyl ether (DE) was slowly dripped on the rotating substrate. This eventually results in a transparent film that is directly indicative of the formation of the adduct. The purpose of using DE (au lieu of toluene), which is immiscible with DMSO, is to remove only DMF to form the 1:1:1 adduct film; FTIR confirmed the adduct formation. The transparent film is then converted to a dark brown film upon heating at low temperature of 65 °C for 1 min due to removal of the volatile DMSO from the adduct (Figure 3). Average PCEs of 18.3% with a record value of 19.7% were attained via the adduct approach [25].

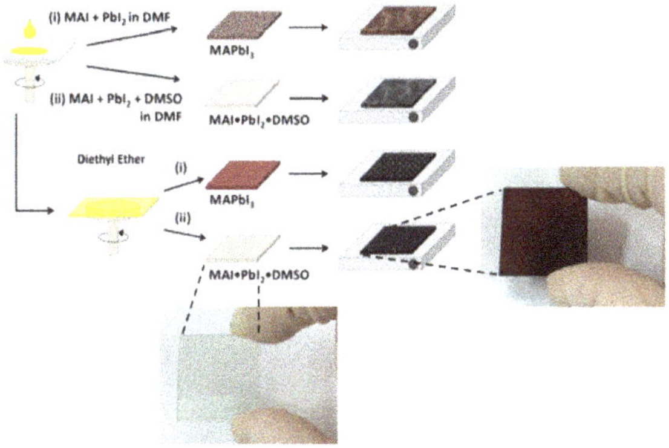

Figure 3. Schematic representation of the anti-solvent (DE) dripping method when incorporating or not DMSO as a co-solvent in the perovskite precursor solution. Macroscopical photos of the films in the intermediate (transparent) and perovskite phase (dark brown) are also shown. Perovskite is formed after annealing at 65 °C for 1 min and then at 100 °C for 2 min. [25]. Copyright (2015) American Chemical Society.

The approach was adopted by the same group in order to fabricate p-i-n solar cells. For this, perovskite films were deposited on poly(3,4-ethylenedioxythiophene)-poly(styrenesulfonate) or simply called PEDOT:PSS. A best PCE of 18.8% was attained [26]. Interestingly, the method worked also well for typical n-i-p solar cells and, most importantly, for a perovskite with a slightly lower bandgap (than the conventional MAPbI$_3$ perovskite). Wang et al. were able to deposit high quality MA$_{0.7}$FA$_{0.3}$PbI$_3$ (where FA stands for formamidinium cation) films on C$_{60}$ SAM/SnO$_2$ substrates via DE dripping (SAM being a self-assembled monolayer); 2% Pb(SCN)$_2$ was added in order to enhance the crystal grain size. With these, a stabilized PCE of 20.3% was recorded [27]. Likewise, a stabilized PCE of 19% was obtained by depositing the MAPbI$_3$ perovskite via the adduct method onto low temperature UV-treated Nb-doped compact TiO$_2$ substrate [28].

3.5. The Presence of Intermediate Phases: Is 1:1:1 the Correct Molar Ratio for MAI:PbI₂:DMSO to Make a Precursor Solution?

Despite the excellent results attained, the exact crystal structure and chemical composition of adduct were in doubt since X-ray diffraction data were missing. Simultaneously with the work of Ahn et al. [25], a very important paper was published, identifying the structure of the intermediate phase as $MA_2Pb_3I_8(DMSO)_2$, using a mixed DMF/DMSO (1:3 v/v) solvent and TL as the anti-solvent (Figure 1a) [29]. It was then suggested that the presence of this specific intermediate was a key point for the formation of a perovskite layer of the desirable morphology [30].

From the above works, an open question unveiled concerning whether the MAI:PbI₂:DMSO ratio should remain exactly at 1:1:1 in order to form the $MA_2Pb_3I_8(DMSO)_2$ intermediate. Indeed, Bai et al. systematically manipulated the composition of intermediate perovskite films aiming to optimize the perovskite layer for highly efficient solar cells. By carefully controlling the ratio of DMSO:PbI₂ in the perovskite precursor, a pure $MA_2Pb_3I_8(DMSO)_2$ intermediate phase was obtained at the DMSO:PbI₂ ratio of 10.0:1 (Figure 4b,c) [31]. In other works, perovskite solar cells prepared by the precursor solution of PbI₂:MAI:xDMSO (x = 2.0 [32] or x = 1.5 [33]) exhibited the highest reproducibility and efficiency (compared to the standard 1:1:1 ratio). Quite importantly, the presence of the intermediate was also confirmed during DMSO solvent annealing of a MAPbI₃ perovskite film; DMSO vapors react with MAPbI₃ to form a $MA_2Pb_3I_8(DMSO)_2$ intermediate at grain boundaries between perovskite crystalline domains. The decomposition of the intermediate can facilitate the grain boundary migration. As a result, the overall crystal size of MAPbI₃ crystalline domains increases substantially [34].

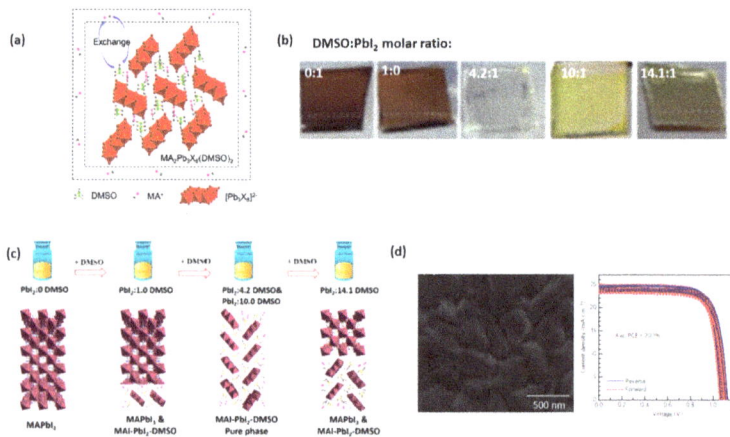

Figure 4. (a) The schematic crystal structure and composition of the intermediate $MA_2Pb_3I_8(DMSO)_2$ phase. Reprinted from [30] with permission from Royal Society of Chemistry. (b) Photos of different intermediate films on NiO substrates based on precursor solutions varying the DMF/DMSO volume ratio. Toluene is adopted as an anti-solvent here. Note that a transparent film appears in the case of DMSO:PbI₂ 4.2:1 (while the transparent film of Figure 3 shows up at a 1:1 molar ratio when DE is dripped during spinning), highlighting the importance of the anti-solvent being used (the substrate also should play a role here). (c) Schematic illustrating the relation between the amount of DMSO and the component of intermediate film. Note that the best results (photovoltaic efficiencies) were obtained after conversion of the yellow film (DMSO:PbI₂ = 10:1) into a perovskite through annealing at 95 °C for 30 min. Reprinted from [31] with permission from Elsevier. (d) By adding excess of CH_3NH_3I in the precursor solution, a yellowish intermediate film was prepared which was later converted to a perovskite film with self-formed grain boundaries (top SEM view image). These films led to champion photovoltaic efficiencies (average PCEs over 20%) for a $CH_3NH_3PbI_3$ perovskite. Reproduced from Ref. [35] with permission from Nature Publishing Group.

3.6. The Unique Case of MAI-Excessive MAPbI₃ Perovskite Films

Interestingly, the group of Nam-Gyu Park published a recent work proving that a novel precursor solution containing excess of MAI (e.g., $(1 + x)$MAI:PbI$_2$:DMSO) was able to introduce an effective passivation layer on the MAPbI$_3$ perovskite grain boundaries (Figure 4d). This grain boundary healing was found to play a crucial role in carrier lifetime improvement, suppression of non-radiative recombination at grain boundaries, and effective extraction of charge carriers at the interface between perovskite and selective contacts. Consequently, the photovoltaic performance was improved, leading to average power conversion efficiencies exceeding 20% at $x = 0.06$ (recall that average PCEs were 18.3% for $x = 0$) [35].

3.7. The Importance of the Presence of the Intermediate Phase: A New Era for the Future of Highly-Efficient Perovskite Solar Cells

The discovery of this intermediate species along with the possibility to extend the MAI:PbI$_2$:DMSO ratio into a much wider region motivated lots of researchers to work more closely on anti-solvent crystallization, attempting the fabrication of highly efficient solar cells with a more easily controlled and reproducible manner. Fortunately, this fact has dramatically accelerated the development of perovskite solar cells. In the following figure, we plot PCEs for a large range of highly efficient perovskite solar cells. In the case of pure lead perovskites, we can see that the results are highly reproducible for the solar cells using perovskite films prepared via the intermediate/anti-solvent method. Fourteen independent works have been published, reporting PCEs over 20% (Figure 5a). Other techniques can be similarly efficient; however it is evident that researchers do not prefer to change from anti-solvent treatment. Certainly, we should note here that the current record efficiency (a certified PCE of 22.1%) belongs to a two-step solution process via a triiodide-assisted (through FAI/I$_2$ reaction in IPA) FA$^+$/MA$^+$ intercalation process [36].

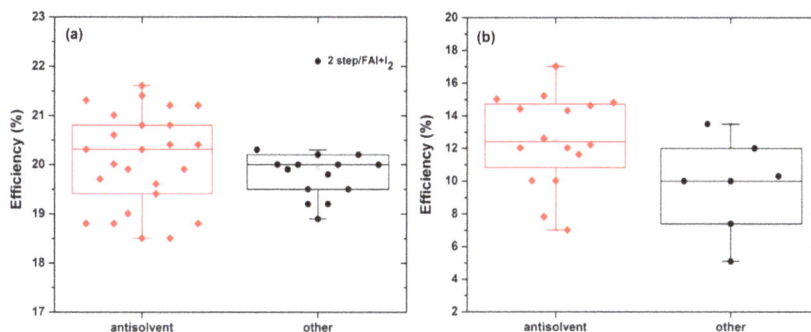

Figure 5. Overall photovoltaic efficiencies of (various types of) perovskite solar cells with perovskite films fabricated via a combined intermediate/anti-solvent crystallization method: (**a**) solar cells based on pure Pb^{2+} perovskite and (**b**) solar cells based on mixed Pb/Sn perovskites. Note that in the case of pure lead perovskite solar cells, only efficiencies higher than 18.5% were introduced in the statistical analysis. Data were taken from the references existing in the present manuscript.

Even more importantly, in the case of pure tin or mixed lead/tin perovskite solar cells, anti-solvent crystallization was the sole manner to fabricate films of high quality and excellent surface coverage (Figure 5b). Quite notably, the current record efficiencies (8.1% for pure tin [37] and a certified 17% for 60% tin perovskite [38]) are attained by this technique. We do not intend to comment further on the tin perovskite solar cells here, as we have already published a review paper on the topic [39].

3.8. The Case of PbI₂-Excessive Perovskite Films

Getting back to the previous discussion, in full contradiction to the work of Park [35], the groups of Grätzel/Nazeeruddin and Sang Il Seok have published (almost simultaneously) two papers on highly efficient MAPbI₃ solar cells based on a non-stoichiometric ratio between PbI₂ and MAI precursors (x:1, being x ≥ 1). In the first case, pure DMSO was used as a solvent and CB was dripped during spinning. 10% excessive PbI₂ led to efficiencies of 19% under 1 sun illumination [40]. Likewise, Kim et al. adopted the usual gBL/DMSO co-solvent system and TL was utilized as the anti-solvent. 5.7% excess of PbI₂ led to optimum results. Interestingly, the approach could be employed for the (FAPbI₃)₀.₈₅(MAPbBr₃)₀.₁₅ perovskite as well. This time a DMF/DMSO (6:1 v/v) system dissolved the perovskite constituents and DE was poured onto the substrate. Quite notably, the excess of PbI₂ led to slightly smaller grains when compared to pure stoichiometric compound; there was no such effect in the case of MAPbI₃ perovskite. The excessive PbI₂ perovskite gave a stabilized efficiency of 20.1%, certified at 19.8% [41].

Even though multiple explanations have been suggested in order to interpret the beneficial role of extra PbI₂, the most probable cause (in our opinion) should be passivation of grain boundaries (stoichiometric perovskites seem to possess grain boundaries with enriched organic species, impeding charge transport) [42]. Even though these works look completely contradictory to the work of Park [33] (excess of MAI or PbI₂ passivates the grain boundaries after all?), we have to notice that, in these works, crystallization occurs via a different road to form the MA₂Pb₃I₈(DMSO)₂ intermediate. Therefore, we do not believe that a fair comparison between these two approaches is possible here.

3.9. PbI₂-Excessive Perovskite Films: The Key to Attain Record Efficiencies for Mixed Perovskites Grown on a Mesoporous Substrate?

We have previously seen that a PbI₂-excessive (FAPbI₃)₀.₈₅(MAPbBr₃)₀.₁₅ perovskite could attain PCEs over 20% [41]. This was a motivation for researchers to adopt the approach for various mixed perovskites and enhance further the efficiencies. First, the ~20% efficiencies were confirmed by others [43]. Then, the Grätzel group was able to enhance the efficiency up to 20.8% (with a record open-circuit potential of 1.18 V) by simply optimizing the PbI₂:FAI molar ratio at 1.05 [44]. The PCE was further increased up to 21.3% when a thin compositional gradient layer of FAPbBr₃₋ₓIₓ was introduced at the rear interface between the (FAPbI₃)₀.₈₅(MAPbBr₃)₀.₁₅ perovskite and the hole transporting material, severely reducing recombination.

Introduction of a small inorganic cation (Cs⁺) into the perovskite lattice resulted in a triple cation perovskite composition that contained less (yellow) phase impurities and was less sensitive to processing conditions. In our opinion, the key to achieve this was that, after anti-solvent (CB) treatment, films with Cs⁺ turned dark immediately (i.e., crystallized) after spin coating [45]. Thus, the situation is similar to the classical anti-solvent precipitation; a perovskite is formed when the anti-solvent is injected into the precursor solution (see Section 2.1). Drying is realized then only to drive out the impurities (e.g., CB) and coarsening the film. This was also observed in the case of the intercalation of FAI/MABr/I₂ into the initial PbI₂/PbBr₂ film [36]. We are not aware whether this is happening also due to the involvement of the extra PbI₂ in the process. For instance, in the case of the Cs$_x$FA$_{(1-x)}$PbI₃ perovskite, Cs incorporation led to small grain sizes, which limited the device performance due to short carrier lifetimes. Addition of 0.5 mol% excess of Pb(SCN)₂ was a prerequisite to enlarge the grain size and significantly decrease recombination within the crystal [46]. With this in mind, PCEs of 20.6% were recorded by triple cation perovskites grown on Cs₂CO₃-treated m-TiO₂ substrate [47]. In a seminal work, Saliba et al. managed to fabricate 5 mol% Cs⁺-doped FAMA films with highly uniform perovskite grains (Figure 6a), extending from the electron to the hole collecting layer, consistent with seed-assisted crystal growth (Figure 6b). Devices could attain record PCEs of 21.2% (under reverse scan,), stabilized at 21.1% (Figure 6c). Interestingly, the devices that incorporated cesium cation in the perovskite film presented a PCE of 18% (at maximum power point) even after 250 h under operational conditions (Figure 6d) [45].

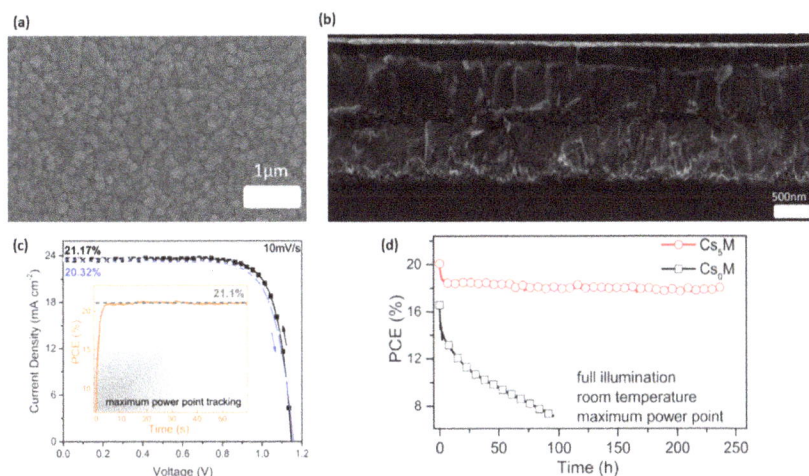

Figure 6. (**a**) Top (**b**) and cross-sectional view scanning electron microscope (SEM) images of the Cs-doped FAMA perovskite films and the whole meso n-i-p devices, respectively. (**c**) Forward and reverse scanned J-V curves under 1 sun illumination and stabilized output power characteristics recorded for 60 s. (**d**) Aging for 250 h of devices with (Cs$_5$M) or without Cs$^+$ (Cs$_0$M) in a nitrogen atmosphere held at room temperature under constant illumination and maximum power point tracking [45]. Published by The Royal Society of Chemistry.

Similar or even better results were drawn when a fourth cation was incorporated into the Cs$_y$FA$_{(1-x)}$MA$_{(1-x-y)}$PbI$_3$ perovskite lattice. Surprisingly, the Rb$^+$-doped perovskite film had a narrow photoluminescence peak at 770 nm attributable to perovskite, without any further annealing. Thus, the addition of Rb$^+$ (5 mol%) enforces a controllable crystallization and avoids any additional insulating (yellow) phases. With these films, Saliba et al. achieved stabilized efficiencies of up to 21.6% (average value: 20.2%) on small areas (and a stabilized 19.0% on a 0.5 cm^2 cell) as well as an electroluminescence of 3.8% [9]. Saliba's results were independently confirmed by Peng and coworkers who reported a PCE of 20% attained by a meso n-i-p device using a Cs$_{0.07}$Rb$_{0.03}$FA$_{0.765}$MA$_{0.135}$PbI$_{2.55}$Br$_{0.45}$ perovskite. To achieve these large values, the authors incorporated an ultrathin passivation layer, consisting of a mixture of polymethyl(methacrylate), PMMA, and [6,6]-phenyl C$_{61}$ butyric acid methyl ester (PCBM), on top of the perovskite, effectively passivating defects at or near to the perovskite/TiO$_2$ interface, significantly suppressing interfacial recombination [48].

3.10. Mixed Perovskites in n-i-p and p-i-n Solar Cells: Is the Excess of PbI$_2$ a Prerequisite?

Thus, far, almost every work, employing excess of PbI$_2$ in mixed perovskites, referred to meso n-i-p solar cells. There are only two reports on record devices without this concept [23,48]. However, what happens when PbI$_2$-excess concept is applied on n-i-p or p-i-n solar cells?

In the case of the n-i-p devices, stabilized efficiencies of 18% could be attained when a mixed FA$_{(1-x)}$MA$_x$I$_{(3-y)}$Br$_y$ perovskite was grown onto Mg-doped TiO$_2$ compact layer [49]. By replacing TiO$_2$ with a low temperature SnO$_2$ with higher charge selectivity for efficient electron extraction, a stabilized PCE of 20.8% could be obtained by a Cs$^+$-doped FAMA perovskite [50]. For p-i-n solar cells, (CsPbI$_3$)$_{0.05}$[(FAPbI$_3$)$_{0.83}$(MAPbBr$_3$)$_{0.17}$]$_{0.95}$ films on top of PTAA led to PCEs of 20% with a fill factor (close to 0.8) approaching the Shockley-Queisser limit [51]. Despite these results, the question remained whether one can fabricate highly efficient, mixed perovskite planar solar cells without an excess of PbI$_2$.

The answer came recently with two breakthrough results. The group of Jinsong Huang published 21% (certified at 20.6%) efficient p-i-n solar cells employing a $FA_{0.85}MA_{0.15}Pb(I_{0.85}Br_{0.15})_3$ perovskite grown on PTAA substrate via TL dripping/DMSO adduct method. The trick to enhance the efficiencies (from 19.2 up to 21.0%) was to passivate ionic defects at the top surface of the perovskite by choline chloride. With this, the density of states was severely reduced over the whole trap depth region [52]. In the case of n-i-p solar cells, the group of Ted Sargent in Toronto used a chlorine-capped TiO_2 layer (au lieu of typical TiO_2) as the crystallization substrate, passivating traps and reducing recombination at the interface which forms upon contact of the n-type layer with the $Cs_{0.05}FA_{0.81}MA_{0.14}PbI_{2.55}Br_{0.45}$ perovskite. Efficiencies of 21.4% were recorded for the best cell, certified at 20.1% [53].

In the following Figure 7a, we compare the efficiencies of various types of solar cells based on MAPbI$_3$ and mixed perovskite films. It is evident that, using a thin mesoporous n-type layer, the efficiencies are very reproducible. However, we should note that record planar (both n-i-p and p-i-n) perovskite solar cells have now reached the values of PCE attained by the meso n-i-p devices. Accidentally or not, these two formulations (FAMA for p-i-n and CsFAMA for n-i-p devices) did not contain an excess of PbI$_2$ (or MAI).

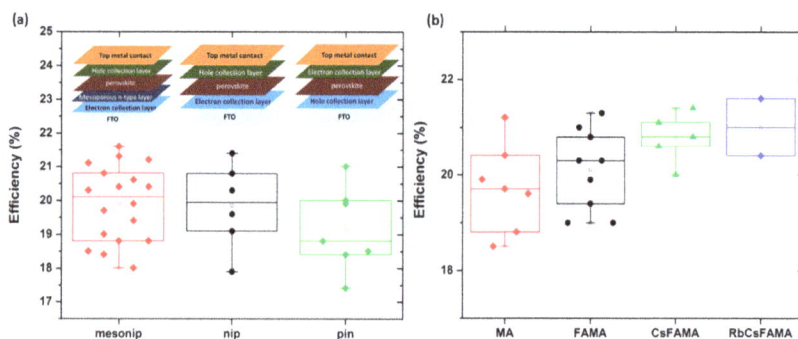

Figure 7. Overall photovoltaic efficiencies of perovskite solar cells with various mixed (or not) perovskite films fabricated via a combined intermediate/anti-solvent crystallization method: (**a**) mesonip, nip and pin cells refer to a specific architecture depicted in the figure as an inset. See text for further details. (**b**) MA, FAMA, CsFAMA and RbCsFAMA refer to compounds consisting of pure MAPbI$_3$, mixed FAPbI$_3$-MAPbI$_3$, Cs$^+$-doped mixed FAPbI$_3$-MAPbI$_3$ and Rb$^+$/Cs$^+$-co-doped mixed FAPbI$_3$-MAPbI$_3$, respectively. Note that various types (meso n-i-p, n-i-p and p-i-n) of highly efficient (over 18.5%) solar cells are included in the statistical analysis. Data were taken from the references existing in the present manuscript.

Then, in Figure 7b we show the evolution of the efficiency by simply adding one, two or three cations in the archetypal MAPbI$_3$ lattice. In the case of FAMA perovskites, the bandgap lowers, hence more photocurrent is extracted. For the other two cations (Cs$^+$ and Rb$^+$), a small amount of them (solely 5 mol%), introduced in the lattice, stabilizes the system so that only the pure perovskite phase precipitates out.

Additionally, in Table 1, we summarize all the electrical parameters of the most efficient solar cells, providing a more comprehensive picture of the current status of the literature, concerning various perovskite films (mixed or not) and device structures. With this, one can also have a clearer view on the hysteresis (or not) of the devices (simply, when average PCE is close to stabilized PCE within 1–2%, hysteresis is not an issue).

Table 1. Parameters for solar cells (measured at 1 sun-AM1.5G) based on record-efficient perovskite solar cells (adopting various n- or p-type contacts and under various cell structures). J_{sc}, V_{oc} and FF values are taken from the best device, reported in every reference.

Perovskite	Device Structure	J_{sc} (mA cm^{-2})	V_{oc} (V)	FF	Average PCE (%)	Stabilized PCE (%)	Certified PCE (%)	Reference
MAPbI$_3$	Meso n-i-p	23.7	1.12	0.78	20.1	>20	-	[35]
FAMA	Meso n-i-p	23.7	1.14	0.78	21.0	-	21.0	[54]
FAMA	Meso n-i-p	23.4	1.12	0.81	20.3	-	21.2	[23]
FAMA	p-i-n	23.7	1.14	0.78	19.4	19.6	20.6	[52]
CsFAMA	n-i-p	22.3	1.19	0.81	19.8	20.9	20.1	[53]
RbCsFAMA	Meso n-i-p	22.8	1.18	0.81	20.2	21.6	-	[9]

3.11. Dripping Active Compounds at the Top Surface of the Perovskite with the Help of the Anti-Solvent

A year ago, anti-solvent dripping was adopted in order to incorporate active compounds into the perovskite film during its formation. Two papers were published at the same day in Nature Energy. First, a molecular fullerene, PCBM, dissolved in toluene, was dripped onto the MAPbI$_3$ perovskite precursor (dissolved in pure DMSO) film during spinning (Figure 8a). This led to the formation of a mixed interlayer with a gradient of electron acceptors in the perovskite light absorption layer, which was denoted as a graded heterojunction structure (Figure 8b). This structure was found capable of enhancing the PCE of inverted-structured PSCs as it improved the photoelectron collection and reduced recombination loss. With this, certified PCEs exceeding 18%, based on a cell with an aperture area greater than 1 cm^2, were obtained [55]. The approach worked really well also in the case of meso n-i-p devices; using a mixed PbI$_2$-excessive FAMA perovskite, efficiencies of 19.9% were recorded [56]. The authors also prepared a fullerene derivative (α-bis-PCBM) through purification of the as-produced PCBM isomer mixture. Quite notably, when α-bis-PCBM replaced standard PCBM in the process, PCE was further increased up to 20.8%. The rationale behind this was that the α-bis-PCBM fulfills (in a better way, compared to PCBM) the vacancies and grain boundaries of the perovskite film, enhancing the crystallization of perovskites and addressing the issue of slow electron extraction. Luckily, α-bis-PCBM also resists the ingression of moisture and passivates voids or pinholes generated in the hole-transporting layer [56].

The second work, published at the same day in Nature Energy, originated from the group of Michael Grätzel at EPFL. The authors adopted the previously reported concept to use an insulating polymer as a template to control nucleation and crystal growth of the perovskite [57]. For this, a PMMA solution in CB/TL (9:1 v/v) was pipetted onto the substrate, 15 s before the end of spin-coating step. Spin-coating was realized at 6500 rpm for 30 s using a typical PbI$_2$-excessive FAMA precursor dissolved in a mixed solvent of DMF, NMP and DMSO (where the molar ratio of DMF/DMSO is 5:1 and the molar ratio of Pb^{2+}/[(DMSO)$_{0.8}$(NMP)$_{0.2}$] is 1:1). With this, shiny, smooth perovskite films of excellent electronic quality were obtained, manifested by a remarkably long photoluminescence lifetime. Stable meso n i p solar cells with excellent reproducibility were fabricated with record PCEs of up to 21.6% (certified at 21.0%) under standard AM 1.5G illumination conditions [54]. This very interesting concept (polymer-assisted crystallization) gave efficiencies over 20% also in the case of n-i-p MAPbI$_3$ perovskite solar cells (adopting the two step deposition method though) [58].

A unique approach was applied during anti-solvent crystallization by Bai and coworkers [59]. The authors slowly dripped a TL solution of phenethylammonium iodide (PEAI) during spin-coating a MAPbI$_3$ solution in DMF:DMSO (2:7 v/v) co-solvent onto NiO substrate (Figure 8c). PEA$^+$ is a large cation that cannot preserve a 3D perovskite structure; instead, when reacting with PbI$_2$, it forms a 2D PEA$_2$PbI$_4$ perovskite [60].

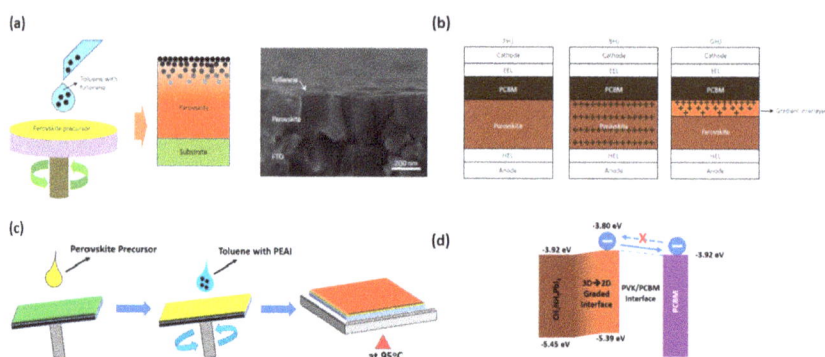

Figure 8. (**a**) A schematic of the method of depositing fullerene molecules at the very top surface of the perovskite along with a cross-sectional SEM image of the dripping-induced perovskite–PCBM stacking. (**b**) Schematic of the p-i-n solar cell depicting the formation of a graded heterojunction, GHJ (perovskite with a gradient distribution of PCBM), in comparison with typical planar (PHJ) and bulk heterojunction (BHJ) devices. Reproduced from Ref. [55] with permission from Nature Publishing Group. (**c**) The schematic of the 3D-2D perovskite deposition. (**d**) Schematic of energy level alignment at the 3D-2D perovskite and PCBM interface. Reproduced from Ref [59] with permission from Wiley-VCH.

With this, a 3D-2D (MAPbI$_3$-PEA$_2$Pb$_2$I$_4$) graded perovskite interface is formed, modifying the interface energy levels in such a way that reduces interface charge recombination and simultaneously favors electron extraction (into PCBM electron acceptor) with a marginal resistance (Figure 8d). These properties were translated to PCEs of 19.9% for p-i-n solar cell devices, accompanied with an ultrahigh V$_{oc}$ of 1.17 V. Moreover, benefiting from the large hydrophobic groups at the interface as well as the grain boundaries, the device presented enhanced moisture stability. Additionally, the devices proved to be thermally stable due to suppression of cross-layer ion migration [59]. The fabrication of efficient, ambient-air solar cells with 3D-2D perovskites is a recent trend in literature [61–63].

In another approach, electron-deficient aromatic compounds were incorporated at the top surface of perovskite films in order to enhance their photovoltaic performance. More specifically, various nitrogen-containing polycyclic aromatic hydrocarbons, dissolved in DE or CB anti-solvents, were dripped during spin-coating a MAPbI$_3$ solution in DMF:DMSO (10.5:1 v/v) co-solvent onto m-TiO$_2$ substrate. Photoluminescence measurements proved that, upon dripping, non-radiative recombination was severely reduced due to the fact that the organic molecules mainly reside at the grain boundaries of the perovskite films, effectively passivating them [64].

3.12. Various Anti-Solvents Dripping Combined with Adduct Method

Reaching the end of this review, a question evidently comes into the reader's mind: Why should dripping always involve chlorobenzene, toluene or diethyl ether? What's wrong with other solvents? In the last part of the review, we will not provide a clear answer but we will briefly discuss the impact of the nature of the anti-solvent on the crystallization process. As far as we are aware of, there are only three reports that compare a whole bunch of poor solvents (for perovskites) that can be used for the process. In two of them, two novel anti-solvents seem to deliver very promising results.

In the first comparative study, Li et al. compared TL, DE and DCM in the crystallization of MAPbI$_3$ (dissolved in pure DMSO) on m-TiO$_2$ substrates. The as-prepared films had very different appearance between each other (Figure 9a), due to the different ratios of MAPbI$_3$ perovskite to MA$_2$Pb$_3$I$_8$(DMSO)$_2$ intermediate phase (recall Figure 4b,c). After annealing, a dense, compact and pinhole free film was obtained only for DE anti-solvent; in both other cases, the films had large, round (DCM) or small sized pinholes (Figure 9b). The authors rationalized the different film formation as follows: when DCM and

TL were used as the anti-solvents, DMSO was probably over-extracted and thereby a high content perovskite phase was formed in the as-prepared thin films with more pinholes. On the contrary, in the case of DE, which is immiscible with DMSO, a purer (or more homogeneous) intermediate phase was formed which effectively retards the rapid nucleation. Annealing is not capable of altering the situation when the intermediate phase gradually transforms into the perovskite phase through the intramolecular exchange between DMSO and MAI (Figure 9c). In any event, we have to point out here that all anti-solvents were dripped in an identical way, namely similar volume and dripping rate/time [65].

Figure 9. (**a**) Photographs of the as-prepared perovskite films using different anti-solvents. (**b**) SEM images of the same films after annealing. (**c**) Schematic illustration for the formation process of the perovskite films via the intermediate/anti-solvent approach. Reproduced from Ref. [65] with permission from Royal Society of Chemistry.

In the second work, a much larger variety of anti-solvents were dripped on a PbI$_2$-excessive perovskite precursor solution, spin-cast on m-TiO$_2$ substrates (Figure 10a). Interestingly, alcohols such as isobutyl alcohol (IBA) and IPA worked quite efficiently, producing PCEs of 15.8% and 14.8%, respectively. However, the most amazing feature of the work was that ethyl acetate (EA) worked even better than CB, producing pinhole-free films composed of big crystalline grains and reduced number of grain boundaries (Figure 10b). Devices fabricated using the above films, exhibited record PCEs of 19.4%. Again, we should also emphasize the fact that all anti-solvents were dripped in an exactly similar manner [66]. The viability of EA as an anti-solvent was already confirmed by others [67–69].

Figure 10. (**a**) Photographs of perovskite films fabricated using different anti-solvents before (the top brown one) and after annealing (the bottom black one) (**b**) Top view SEM images of the EA-processed perovskite films after annealing. Reproduced from Ref. [66] with permission from Wiley-VCH.

A very concise comparison was recently published by the group of Nazeeruddin at EPFL in Sion. Again lots of solvents were tested, but this time the quantity (of the anti-solvent) and the time of dripping were optimized for every anti-solvent separately. Four of the tested anti-solvents of high boiling point formed perovskite films with a similar grain density and coverage on a nanometer scale; DE-treated films (or not treated at all) have shown pinholes at their surface (Figure 11a).

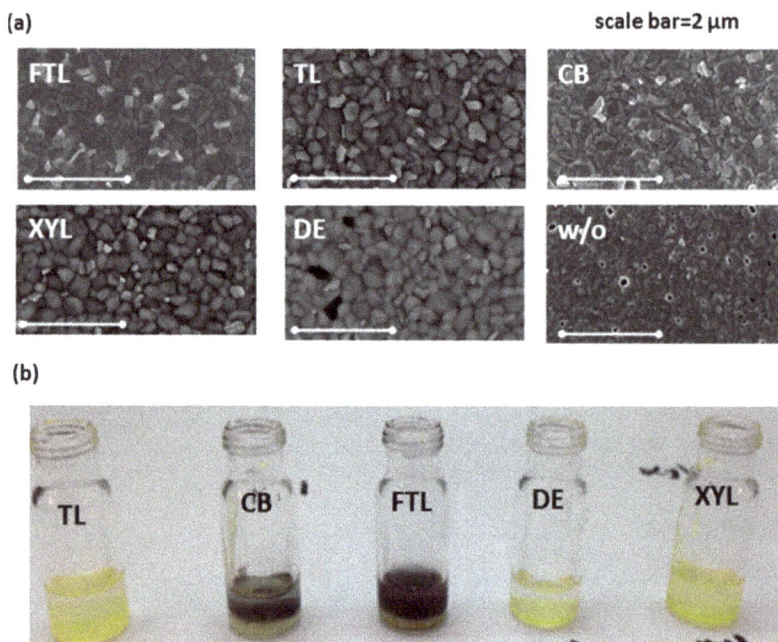

Figure 11. (a) SEM images of perovskite films treated (or not) by various anti-solvents (trifluorotoluene: FTL, toluene:TL, chlorobenzene: CB, xylene: XYL, diethyl ether: DE); (b) Photograph of adding anti-solvents into precursor solutions. Reproduced from Ref. [70] with permission from American Chemical Society.

Interestingly, xylene, which is immiscible with both DMF and DMSO, produced PCEs of 17.8%. However, even better efficiencies were obtained by a new anti-solvent; quite notably, trifluorotoluene (TFL) was proven to give higher PCEs (best at 20.3%) than TL and CB, making it a good candidate for future experiments. We should also notice here that DE produced significantly lower PCEs when compared to rival CB and TL. Finally, the authors suggested a good trick to see whether an anti-solvent is good or bad: one injects the anti-solvent in a vial containing the precursor solution; the formation of the perovskite crystal (brown-blackish precipitate) can be a promising sign for high efficiencies (Figure 11b) [70].

Besides xylene, another solvent (hexane, HX), which is also immiscible with both DMF and DMSO, gave appreciable photovoltaic efficiencies (15.5%). Certainly, these values were increased up to 16.5% when a solvent miscible with DMF (DE) was added in a 50–50 ratio [71]. Another interesting co-solvent system (6% IPA in CB) assisted to get an even higher PCE (19.2%) when dripped on a CsFAMA precursor perovskite film [72]. In the following Figure 12, we summarize all the PCEs attained using various anti-solvents in literature thus far.

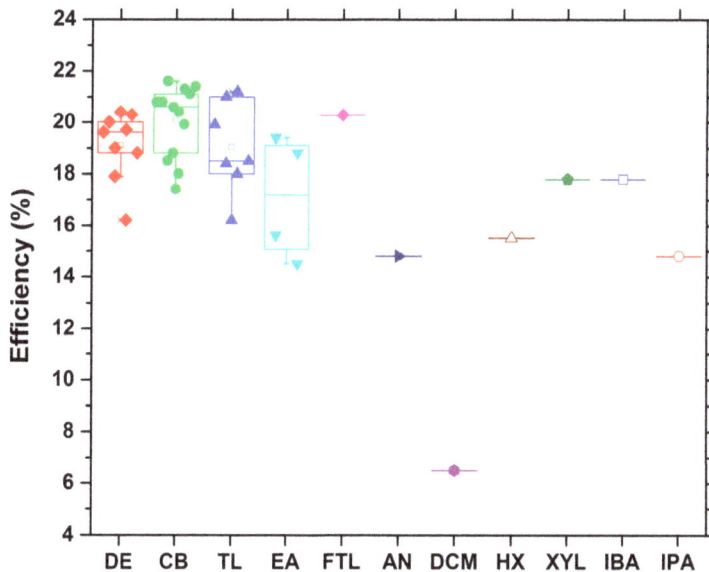

Figure 12. Overall photovoltaic efficiencies of various types of solar cells which use perovskite films fabricated by the application of different anti-solvents in the combined intermediate/anti-solvent crystallization method. Note that in the case of anisole (AN), a highly efficient solar cell based on a Pb/Sn mixed perovskite [73] was included in the analysis.

Apart from high crystallinity and controllable nucleation/growth kinetics, which result in a polycrystalline film with low defect density, anti-solvent was found to affect the electronic properties of the perovskite. Nawaz et al. evidenced a downward shift of the conduction and valence bands of the MAPbI$_3$ perovskite via toluene-assisted crystallization, leading to favored interfacial charge transfer and consequently to higher open-circuit potential [74]. In turn, Cohen et al. demonstrated, through combined conductive atomic force microscopy and surface photovoltage measurements, that the perovskite film became slightly more intrinsic during TL anti-solvent treatment. They attributed this behavior to the net positive charge on the Pb atoms that is left when halide and methylammonium ions are removed from the surface, resulting in a more conductive surface of the perovskite [75].

4. Summary and Outlook

In summary, we have reviewed all the literature data concerning the intermediate (or adduct)/anti-solvent crystallization method in order to grow (MAPbI$_3$ or mixed) perovskite films of high (and pure) crystallinity and rich substrate coverage; DMSO intermediate with MAI and PbI$_2$ plays a key role to achieve this. Most of the solvents that are poor for perovskites seem to work quite efficiently, irrespectively of miscibility (with DMF, DMSO or both) and boiling point (and vapor pressure) properties. CB, TL and DE are definitely the winning anti-solvents even though TFL and EA have recently shown very interesting results. Meso n-i-p architectures are preferred due to much higher reproducibility and reduced hysteresis, however recent results prove that planar n-i-p and p-i-n devices can reach certified efficiencies of more than 20% as well. Whether a PbI$_2$ or MAI excess in the perovskite precursor solution is a prerequisite or not is still an open question. However, it was clear that adding a second, third or fourth cation into the typical MAPbI$_3$ perovskite, the bandgap lowers delivering more photocurrent (with FA) or a purer perovskite precipitates (with Cs and/or Rb), producing films of excellent electronic quality and record efficiencies.

How can we get at more than 21.6%? Certainly an easy answer is, if we take into account that we are already preparing films of 100% surface coverage, by precipitating a 100% pure polycrystalline perovskite. A trick to this could be an additional washing step after spin-coating (and before annealing) in order to drive away any impurities [68,73,76].

Certainly, the key will be to fully understand the intermediate-to-perovskite transition. Our current understanding is that, in the case of MAPbI$_3$ perovskite, a yellow (e.g., MAI-excessive) intermediate film is enough to produce the perovskite via a simple annealing step for a few min. However, when FA is incorporated, a different lattice is formed and crystallization process could change completely; what is the intermediate that forms then? In any event, dark, well crystalline films should be produced after the end of spin casting.

To conclude, having in mind the recent results obtained by the group of Seok [36], we suspect that researchers will soon attempt to transform the I-deficient [(Pb$_3$I$_8$)$_n$]$^{2n-}$ intermediate into the perovskite via reaction with I$_2$-rich MAI. We argue that efficiency (and stability) increase of solar cell devices will prove to be beneficial for other optoelectronic applications, such as photodetectors, LEDs and X-Ray detectors [11].

Acknowledgments: The current work was funded by the "RESEARCH PROJECTS FOR EXCELLENCE IKY/SIEMENS" and the "IKY FELLOWSHIPS OF EXCELLENCE FOR POSTGRADUATE STUDIES IN GREECE-SIEMENS PROGRAMME" Programmes.

Author Contributions: Thomas Stergiopoulos conceived the topic of this review. Maria Konstantakou wrote the manuscript under the supervision of Thomas Stergiopoulos. Dorothea Perganti and Polycarpos Falaras made comments and revised the manuscript.

Conflicts of Interest: The authors declare no conflict of interest.

References

1. Kojima, A.; Teshima, K.; Shirai, Y.; Miyasaka, T. Organometal Halide Perovskites as Visible-Light Sensitizers for Photovoltaic Cells. *J. Am. Chem. Soc.* **2009**, *131*, 6050–6051. [CrossRef] [PubMed]
2. Lee, M.M.; Teuscher, J.; Miyasaka, T.; Murakami, T.N.; Snaith, H.J. Efficient hybrid solar cells based on meso-superstructured organometal halide perovskites. *Science* **2012**, *338*, 643–647. [CrossRef] [PubMed]
3. Kim, H.S.; Lee, C.R.; Im, J.H.; Lee, K.B.; Moehl, T.; Marchioro, A.; Moon, S.J.; Humphry-Baker, R.; Yum, J.H.; Moser, J.E.; et al. Lead iodide perovskite sensitized all-solid-state submicron thin film mesoscopic solar cell with efficiency exceeding 9%. *Sci. Rep.* **2012**, *2*, 591. [CrossRef] [PubMed]
4. Heo, J.H.; Im, S.H.; Noh, J.H.; Mandal, T.N.; Lim, C.S.; Chang, J.A.; Lee, Y.H.; Kim, H.J.; Sarkar, A.; Nazeeruddin, M.K.; et al. Efficient inorganic-organic hybrid heterojunction solar cells containing perovskite compound and polymeric hole conductors. *Nat. Photonics* **2013**, *7*, 487–492. [CrossRef]
5. Burschka, J.; Pellet, N.; Moon, S.J.; Humphry-Baker, R.; Gao, P.; Nazeeruddin, M.K.; Grätzel, M. Sequential deposition as a route to high-performance perovskite-sensitized solar cells. *Nature* **2013**, *499*, 316–319. [CrossRef] [PubMed]
6. Liu, M.; Johnston, M.B.; Snaith, H.J. Efficient planar heterojunction perovskite solar cells by vapour deposition. *Nature* **2013**, *501*, 395–398. [CrossRef] [PubMed]
7. Xiao, M.; Huang, F.; Huang, W.; Dkhissi, Y.; Zhu, Y.; Etheridge, J.; Gray-Weale, A.; Bach, U.; Cheng, Y.B.; Spiccia, L.A. Fast deposition-crystallization procedure for highly efficient lead iodide perovskite thin-film solar cells. *Angew. Chem. Int. Ed. Engl.* **2014**, *53*, 9898–9903. [CrossRef] [PubMed]
8. Jeon, N.J.; Noh, J.H.; Kim, Y.C.; Yang, W.S.; Ryu, S.; Seok, S.I. Solvent engineering for high-performance inorganic-organic hybrid perovskite solar cells. *Nat. Mater.* **2014**, *13*, 897–903. [CrossRef] [PubMed]
9. Saliba, M.; Matsui, T.; Domanski, K.; Seo, J.Y.; Ummadisingu, A.; Zakeeruddin, S.M.; Correa-Baena, J.P.; Tress, W.R.; Abate, A.; Hagfeldt, A.; et al. Incorporation of rubidium cations into perovskite solar cells improves photovoltaic performance. *Science* **2016**, *354*, 206–209. [CrossRef] [PubMed]
10. Best Research—Cell Efficiencies. Available online: https://www.nrel.gov/pv/assets/images/efficiency-chart.png (accessed on 27 September 2017).
11. Zhang, W.; Eperon, G.E.; Snaith, H.J. Metal halide perovskites for energy applications. *Nat. Energy* **2016**, *1*, 16048. [CrossRef]

12. Stranks, S.D.; Nayak, P.K.; Zhang, W.; Stergiopoulos, T.; Snaith, H.J. Formation of thin films of organic-inorganic perovskites for high-efficiency solar cells. *Angew. Chem. Int. Ed. Engl.* **2015**, *54*, 3240–3248. [CrossRef] [PubMed]

13. Cohen, B.E.; Etgar, L. Parameters that control and influence the organo-metal halide perovskite crystallization and morphology. *Front. Optoelectron.* **2016**, *9*, 44–52. [CrossRef]

14. Lee, J.W.; Kim, H.S.; Park, N.G. Lewis acid–base adduct approach for high efficiency perovskite solar cells. *Acc. Chem. Res.* **2016**, *49*, 311–319. [CrossRef] [PubMed]

15. Seo, J.; Noh, J.H.; Il Seok, S. Rational strategies for efficient perovskite solar cells. *Acc. Chem. Res.* **2016**, *49*, 562–572. [CrossRef] [PubMed]

16. Chen, Y.; He, M.; Peng, J.; Sun, Y.; Liang, Z. Structure and growth control of organic–inorganic halide perovskites for optoelectronics: From polycrystalline films to single crystals. *Adv. Sci.* **2016**, *3*, 1500392. [CrossRef] [PubMed]

17. Fakharuddin, A.; Schmidt-Mende, L.; Garcia-Belmonte, G.; Jose, R.; Mora-Sero, I. Interfaces in perovskite solar cells. *Adv. Energy Mater.* **2017**, *65*, 1700623. [CrossRef]

18. Snaith, H.J.; Abate, A.; Ball, J.M.; Eperon, G.E.; Leijtens, T.; Noel, N.K.; Stranks, S.D.; Wang, J.T.W.; Wojciechowski, K.; Zhang, W. Anomalous hysteresis in perovskite solar cells. *J. Phys. Chem. Lett.* **2014**, *5*, 1511–1515. [CrossRef] [PubMed]

19. Saidaminov, M.I.; Abdelhady, A.L.; Murali, B.; Alarousu, E.; Burlakov, V.M.; Peng, W.; Dursun, I.; Wang, L.; He, Y.; Maculan, G.; et al. High-quality bulk hybrid perovskite single crystals within minutes by inverse temperature crystallization. *Nat. Commun.* **2015**, *6*, 7586. [CrossRef] [PubMed]

20. Stoumpos, C.C.; Malliakas, C.D.; Kanatzidis, M.G. Semiconducting tin and lead iodide perovskites with organic cations: Phase transitions, high mobilities, and near-infrared photoluminescent properties. *Inorg. Chem.* **2013**, *52*, 9019–9038. [CrossRef] [PubMed]

21. Tidhar, Y.; Edri, E.; Weissman, H.; Zohar, D.; Hodes, G.; Cahen, D.; Rybtchinski, B.; Kirmayer, S. Crystallization of methyl ammonium lead halide perovskites: Implications for photovoltaic applications. *J. Am. Chem. Soc.* **2014**, *136*, 13249–13256. [CrossRef] [PubMed]

22. Shi, D.; Adinolfi, V.; Comin, R.; Yuan, M.; Alarousu, E.; Buin, A.; Chen, Y.; Hoogland, S.; Rothenberger, A.; Katsiev, K.; et al. Low trap-state density and long carrier diffusion in organolead trihalide perovskite single crystals. *Science* **2015**, *347*, 519–522. [CrossRef] [PubMed]

23. Shin, S.S.; Yeom, E.J.; Yang, W.S.; Hur, S.; Kim, M.G.; Im, J.; Seo, J.; Noh, J.H.; Il Seok, S. Colloidally prepared La-doped BaSnO₃ electrodes for efficient, photostable perovskite solar cells. *Science* **2017**, *356*, 167–171. [CrossRef] [PubMed]

24. Yang, M.; Zhang, T.; Schulz, P.; Li, Z.; Li, G.; Kim, D.H.; Guo, N.; Berry, J.J.; Zhu, K.; Zhao, Y. Facile fabrication of large-grain CH₃NH₃PbI₃₋ₓBrₓ films for high-efficiency solar cells via CH₃NH₃Br selective Ostwald ripening. *Nat. Commun.* **2016**, *7*, 12305. [CrossRef] [PubMed]

25. Ahn, N.; Son, D.Y.; Jang, I.H.; Min, K.S.; Choi, M.; Park, N.G. Highly reproducible perovskite solar cells with average efficiency of 18.3% and best efficiency of 19.7% fabricated via lewis base adduct of lead(ii) iodide. *J. Am. Chem. Soc.* **2015**, *137*, 8696–8699. [CrossRef] [PubMed]

26. Sung, H.; Ahn, N.; Jang, M.S.; Lee, J.K.; Yoon, H.; Park, N.G.; Choi, M. Transparent conductive oxide-free graphene-based perovskite solar cells with over 17% efficiency. *Adv. Energy Mater.* **2015**, *6*, 1501873. [CrossRef]

27. Wang, C.; Xiao, C.; Yu, Y.; Zhao, D.; Awni, R.A.; Grice, C.R.; Ghimire, K.; Constantinou, D.; Liao, W.; Cimaroli, A.J.; et al. Understanding and eliminating hysteresis for highly efficient planar perovskite solar cells. *Adv. Energy Mater.* **2017**, 1700414. [CrossRef]

28. Jeong, I.; Jung, H.; Park, M.; Park, J.S.; Hae, J.; Joo, J.; Lee, J.; Ko, M.J. A tailored TiO₂ electron selective layer for high-performance flexible perovskite solar cells via low temperature UV process. *Nano Energy* **2016**, *28*, 380–389. [CrossRef]

29. Rong, Y.; Tang, Z.; Zhao, Y.; Zhong, X.; Venkatesan, S.; Graham, H.; Patton, M.; Jing, Y.; Guloy, A.M.; Yao, Y. Solvent engineering towards controlled grain growth in perovskite planar heterojunction solar cells. *Nanoscale* **2015**, *7*, 10595–10599. [CrossRef] [PubMed]

30. Rong, Y.; Venkatesan, S.; Guo, R.; Wang, Y.; Bao, J.; Li, W.; Fan, Z.; Yao, Y. Critical kinetic control of non-stoichiometric intermediate phase transformation for efficient perovskite solar cells. *Nanoscale* **2016**, *8*, 12892–12899. [CrossRef] [PubMed]

31. Bai, Y.; Xiao, S.; Hu, C.; Zhang, T.; Meng, X.; Li, Q.; Yang, Y.; Wong, K.S.; Chen, H.; Yang, S. A pure and stable intermediate phase is key to growing aligned and vertically monolithic perovskite crystals for efficient PIN planar perovskite solar cells with high processibility and stability. *Nano Energy* **2017**, *34*, 58–68. [CrossRef]

32. Tu, Y.; Wu, J.; He, X.; Guo, P.; Luo, H.; Liu, Q.; Lin, J.; Huang, M.; Huang, Y.; Fan, L.; et al. Controlled growth of $CH_3NH_3PbI_3$ films towards efficient perovskite solar cells by varied-stoichiometric intermediate adduct. *Appl. Surf. Sci.* **2017**, *403*, 572–577. [CrossRef]

33. Ren, Y.K.; Liu, S.D.; Duan, B.; Xu, Y.F.; Li, Z.Q.; Huang, Y.; Hu, L.H.; Zhu, J.; Dai, S.Y. Controllable intermediates by molecular self-assembly for optimizing the fabrication of large-grain perovskite films via one-step spin-coating. *J. Alloy. Compd.* **2017**, *705*, 205–210. [CrossRef]

34. Xiao, S.; Bai, Y.; Meng, X.; Zhang, T.; Chen, H.; Zheng, X.; Hu, C.; Qu, Y.; Yang, S. Unveiling a key intermediate in solvent vapor post-annealing to enlarge crystalline domains of organometal halide perovskite films. *Adv. Funct. Mater.* **2017**, *27*, 1604944. [CrossRef]

35. Son, D.Y.; Lee, J.W.; Choi, Y.J.; Jang, I.H.; Lee, S.; Yoo, P.J.; Shin, H.; Ahn, N.; Choi, M.; Kim, D.; et al. Self-formed grain boundary healing layer for highly-efficient $CH_3NH_3PbI_3$ perovskite solar cells. *Nat. Energy* **2016**, *1*, 16081. [CrossRef]

36. Yang, W.S.; Park, B.W.; Jung, E.H.; Jeon, N.J.; Kim, Y.C.; Lee, D.U.; Shin, S.S.; Seo, J.; Kim, E.K.; Noh, J.H.; et al. Iodide management in formamidinium-lead-halide–based perovskite layers for efficient solar cells. *Science* **2017**, *356*, 1376. [CrossRef] [PubMed]

37. Zhao, Z.; Gu, F.; Li, Y.; Sun, W.; Ye, S.; Rao, H.; Liu, Z.; Bian, Z.; Huang, C. Mixed-Organic-Cation tin iodide for lead-free perovskite solar cells with an efficiency of 8.12%. *Adv. Sci.* **2017**, *131*, 1700204. [CrossRef]

38. Zhao, D.; Yu, Y.; Wang, C.; Liao, W.; Shrestha, N.; Grice, C.R.; Cimaroli, A.J.; Lei, G.; Ellingson, R.J.; Zhu, K.; et al. Low-bandgap mixed tin–lead iodide perovskite absorbers with long carrier lifetimes for all-perovskite tandem solar cells. *Nat. Energy* **2017**, *2*, 17018. [CrossRef]

39. Konstantakou, M.; Stergiopoulos, T. A critical review on tin halide perovskite solar cells. *J. Mater. Chem. A* **2017**, *5*, 11518–11549. [CrossRef]

40. Roldán-Carmona, C.; Gratia, P.; Zimmermann, I.; Grancini, G.; Gao, P.; Graetzel, M.; Nazeeruddin, M.K. High efficiency methylammonium lead triiodide perovskite solar cells: The relevance of non-stoichiometric precursors. *Energy Environ. Sci.* **2015**, *8*, 3550–3556. [CrossRef]

41. Kim, Y.C.; Jeon, N.J.; Noh, J.H.; Yang, W.S.; Seo, J.; Yun, J.; Ho-Baillie, A.; Huang, S.; Green, M.A.; Seidel, J.; et al. Beneficial effects of PBI_2 incorporated in organo-lead halide perovskite solar cells. *Adv. Energy Mater.* **2016**, *6*, 1502104. [CrossRef]

42. Jacobsson, T.J.; Correa-Baena, J.P.; Anaraki, E.H.; Philippe, B.; Stranks, S.D.; Bouduban, M.E.F.; Tress, W.; Schenk, K.; Teuscher, J.; Moser, J.E.; et al. Unreacted PbI_2 as a double-edged sword for enhancing the performance of perovskite solar cells. *J. Am. Chem. Soc.* **2016**, *138*, 10331–10343. [CrossRef] [PubMed]

43. Xie, L.Q.; Chen, L.; Nan, Z.A.; Lin, H.X.; Wang, T.; Zhan, D.; Yan, J.W.; Mao, B.W.; Tian, Z.Q. Understanding the cubic phase stabilization and crystallization kinetics in mixed cations and halides perovskite single crystals. *J. Am. Chem. Soc.* **2017**, *139*, 3320–3323. [CrossRef] [PubMed]

44. Bi, D.; Tress, W.M.; Dar, I.; Gao, P.; Luo, J.; Renevier, C.; Schenk, K.; Abate, A.; Giordano, F.; Correa Baena, J.P.; et al. Efficient luminescent solar cells based on tailored mixed-cation perovskites. *Sci. Adv.* **2016**, *2*, 1501170. [CrossRef] [PubMed]

45. Saliba, M.; Matsui, T.; Seo, J.Y.; Domanski, K.; Correa-Baena, J.P.; Nazeeruddin, M.K.; Zakeeruddin, S.M.; Tress, W.; Abate, A.; Hagfeldt, A.; et al. Cesium-containing triple cation perovskite solar cells: Improved stability, reproducibility and high efficiency. *Energy Environ. Sci.* **2016**, *9*, 1989–1997. [CrossRef] [PubMed]

46. Yu, Y.; Wang, C.; Grice, C.R.; Shrestha, N.; Chen, J.; Zhao, D.; Liao, W.; Cimaroli, A.J.; Roland, P.J.; Ellingson, R.J.; et al. Improving the performance of formamidinium and cesium lead triiodide perovskite solar cells using lead thiocyanate additives. *ChemSusChem* **2016**, *9*, 3288–3297. [CrossRef] [PubMed]

47. Ye, T.; Petrović, M.; Peng, S.; Yoong, J.L.K.; Vijila, C.; Ramakrishna, S. Enhanced charge carrier transport and device performance through dual-cesium doping in mixed-cation perovskite solar cells with near unity free carrier ratios. *ACS Appl. Mater. Interfaces* **2017**, *9*, 2358–2368. [CrossRef] [PubMed]

48. Peng, J.; Wu, Y.; Ye, W.; Jacobs, D.A.; Shen, H.; Fu, X.; Wan, Y.; Duong, T.; Wu, N.; Barugkin, C.; et al. Interface passivation using ultrathin polymer–fullerene films for high-efficiency perovskite solar cells with negligible hysteresis. *Energy Environ. Sci.* **2017**, *10*, 1792–1800. [CrossRef]

49. Zhang, H.; Shi, J.; Xu, X.; Zhu, L.; Luo, Y.; Li, D.; Meng, Q. Mg-doped TiO_2 boosts the efficiency of planar perovskite solar cells to exceed 19%. *J. Mater. Chem. A* **2016**, *4*, 15383–15389. [CrossRef]

50. Anaraki, E.H.; Kermanpur, A.; Steier, L.; Domanski, K.; Matsui, T.; Tress, W.; Saliba, M.; Abate, A.; Grätzel, M.; Hagfeldt, A.; et al. Highly efficient and stable planar perovskite solar cells by solution-processed tin oxide. *Energy Environ. Sci.* **2016**, *9*, 3128–3134. [CrossRef]

51. Stolterfoht, M.; Wolff, C.M.; Amir, Y.; Paulke, A.; Perdigón-Toro, L.; Caprioglio, P.; Neher, D. Approaching the fill factor Shockley–Queisser limit in stable, dopant-free triple cation perovskite solar cells. *Energy Environ. Sci.* **2017**, *10*, 1530–1539. [CrossRef]

52. Zheng, X.; Chen, B.; Dai, J.; Fang, Y.; Bai, Y.; Lin, Y.; Wei, H.; Zeng, X.C.; Huang, J. Defect passivation in hybrid perovskite solar cells using quaternary ammonium halide anions and cations. *Nat. Energy* **2017**, *2*, 17102. [CrossRef]

53. Tan, H.; Jain, A.; Voznyy, O.; Lan, X.; de Arquer, F.P.G.; Fan, J.Z.; Quintero-Bermudez, R.; Yuan, M.; Zhang, B.; Zhao, Y.; et al. Efficient and stable solution-processed planar perovskite solar cells via contact passivation. *Science* **2017**, *355*, 722–726. [CrossRef] [PubMed]

54. Bi, D.; Yi, C.; Luo, J.; Décoppet, J.D.; Zhang, F.; Zakeeruddin, S.M.; Li, X.; Hagfeldt, A.; Grätzel, M. Polymer-templated nucleation and crystal growth of perovskite films for solar cells with efficiency greater than 21%. *Nat. Energy* **2016**, *1*, 16142. [CrossRef]

55. Wu, Y.; Yang, X.; Chen, W.; Yue, Y.; Cai, M.; Xie, F.; Bi, E.; Islam, A.; Han, L. Perovskite solar cells with 18.21% efficiency and area over 1 cm^2 fabricated by heterojunction engineering. *Nat. Energy* **2016**, *1*, 16148. [CrossRef]

56. Zhang, F.; Shi, W.; Luo, J.; Pellet, N.; Yi, C.; Li, X.; Zhao, X.; Dennis, T.J.S.; Li, X.; Wang, S.; et al. Isomer-pure bis-pcbm-assisted crystal engineering of perovskite solar cells showing excellent efficiency and stability. *Adv. Mater.* **2017**, *29*, 1606806. [CrossRef] [PubMed]

57. Zhao, Y.; Wei, J.; Li, H.; Yan, Y.; Zhou, W.; Yu, D.; Zhao, Q. A polymer scaffold for self-healing perovskite solar cells. *Nat. Commun.* **2016**, *7*, 10228. [CrossRef] [PubMed]

58. Zuo, L.; Guo, H.; de Quilettes, D.W.; Jariwala, S.; de Marco, N.; Dong, S.; de Block, R.; Ginger, D.S.; Dunn, B.; Wang, M.; et al. Polymer-modified halide perovskite films for efficient and stable planar heterojunction solar cells. *Sci. Adv.* **2017**, *3*. [CrossRef] [PubMed]

59. Bai, Y.; Xiao, S.; Hu, C.; Zhang, T.; Meng, X.; Lin, H.; Yang, Y.; Yang, S. Dimensional engineering of a graded 3d–2d halide perovskite interface enables ultrahigh v_{oc} enhanced stability in the p-i-n photovoltaics. *Adv. Energy Mater.* **2017**, 1701038. [CrossRef]

60. Mitzi, D.B.; Field, C.A.; Harrison, W.T.A.; Guloy, A.M. Conducting tin halides with a layered organic-based perovskite structure. *Nature* **1994**, *369*, 467. [CrossRef]

61. Grancini, G.; Roldán-Carmona, C.; Zimmermann, I.; Mosconi, E.; Lee, X.; Martineau, D.; Narbey, S.; Oswald, F.; de Angelis, F.; Graetzel, M.; et al. One-Year stable perovskite solar cells by 2D/3D interface engineering. *Nat. Commun.* **2017**, *8*, 15684. [CrossRef] [PubMed]

62. Li, N.; Zhu, Z.; Chueh, C.C.; Liu, H.; Peng, B.; Petrone, A.; Li, X.; Wang, L.; Jen, A.K.Y. Mixed cation $FA_xPEA_{1-x}PbI_3$ with enhanced phase and ambient stability toward high-performance perovskite solar cells. *Adv. Energy Mater.* **2017**, *7*, 1601307. [CrossRef]

63. Wang, Z.; Lin, Q.; Chmiel, F.P.; Sakai, N.; Herz, L.M.; Snaith, H.J. Efficient ambient-air-stable solar cells with 2D–3D heterostructured butylammonium-caesium formamidinium lead halide perovskites. *Nat. Energy* **2017**, *6*, 17135. [CrossRef]

64. Ngo, T.T.; Suarez, I.; Antonicelli, G.; Cortizo-Lacalle, D.; Martinez-Pastor, J.P.; Mateo-Alonso, A.; Mora-Sero, I. Enhancement of the performance of perovskite solar cells, LEDs, and optical amplifiers by anti-solvent additive deposition. *Adv. Mater.* **2017**, *29*, 1604056. [CrossRef] [PubMed]

65. Li, Y.; Wang, J.; Yuan, Y.; Dong, X.; Wang, P. Anti-solvent dependent device performance in $CH_3NH_3PbI_3$ solar cells: The role of intermediate phase content in the as-prepared thin films. *Sustain. Energy Fuels* **2017**, *1*, 1041–1048. [CrossRef]

66. Bu, T.; Wu, L.; Liu, X.; Yang, X.; Zhou, P.; Yu, X.; Qin, T.; Shi, J.; Wang, S.; Li, S.; et al. Synergic interface optimization with green solvent engineering in mixed perovskite solar cells. *Adv. Energy Mater.* **2017**, 1700576. [CrossRef]

67. Yin, M.; Xie, F.; Chen, H.; Yang, X.; Ye, F.; Bi, E.; Wu, Y.; Cai, M.; Han, L. Annealing-free perovskite films by instant crystallization for efficient solar cells. *J. Mater. Chem. A* **2016**, *4*, 8548–8553. [CrossRef]

68. Fei, C.; Li, B.; Zhang, R.; Fu, H.; Tian, J.; Cao, G. Highly efficient and stable perovskite solar cells based on monolithically grained ch₃nh₃pbi₃ film. *Adv. Energy Mater.* **2017**, 1602017. [CrossRef]
69. Troughton, J.; Hooper, K.; Watson, T.M. Humidity resistant fabrication of CH₃NH₃PbI₃ perovskite solar cells and modules. *Nano Energy* **2017**, *39*, 60–68. [CrossRef]
70. Paek, S.; Schouwink, P.; Athanasopoulou, E.N.; Cho, K.T.; Grancini, G.; Lee, Y.; Zhang, Y.; Stellacci, F.; Nazeeruddin, M.K.; Gao, P. From nano to micrometer scale: The role of anti-solvent treatment on the high-performance perovskite solar cells. *Chem. Mater.* **2017**, *29*, 3490–3498. [CrossRef]
71. Yu, Y.; Yang, S.; Lei, L.; Cao, Q.; Shao, J.; Zhang, S.; Liu, Y. Ultrasmooth perovskite film via mixed anti-solvent strategy with improved efficiency. *ACS Appl. Mater. Interfaces* **2017**, *9*, 3667–3676. [CrossRef] [PubMed]
72. Wang, Y.; Wu, J.; Zhang, P.; Liu, D.; Zhang, T.; Ji, L.; Gu, X.; Chen, Z.D.; Li, S. Stitching triple cation perovskite by a mixed anti-solvent process for high performance perovskite solar cells. *Nano Energy* **2017**, *39*, 616–625. [CrossRef]
73. Eperon, G.E.; Leijtens, T.; Bush, K.A.; Prasanna, R.; Green, T.; Wang, J.T.W.; McMeekin, D.P.; Volonakis, G.; Milot, R.L.; May, R.; et al. Perovskite-perovskite tandem photovoltaics with optimized band gaps. *Science* **2016**, *354*, 861–865. [CrossRef] [PubMed]
74. Nawaz, A.; Erdinc, A.K.; Gultekin, B.; Tayyib, M.; Zafer, C.; Wang, K.; Akram, M.N.; Wong, K.K.; Hussain, S.; Schmidt-Mende, L.; et al. Insights into optoelectronic properties of anti-solvent treated perovskite films. *J. Mater. Sci. Mater. Electron.* **2017**, 1–7. [CrossRef]
75. Cohen, B.E.; Aharon, S.; Dymshits, A.; Etgar, L. Impact of antisolvent treatment on carrier density in efficient hole-conductor-free perovskite-based solar cells. *J. Phys. Chem. C* **2016**, *120*, 142–147. [CrossRef]
76. Cao, J.; Jing, X.; Yan, J.; Hu, C.; Chen, R.; Yin, J.; Li, J.; Zheng, N. Identifying the molecular structures of intermediates for optimizing the fabrication of high-quality perovskite films. *J. Am. Chem. Soc.* **2016**, *138*, 9919–9926. [CrossRef] [PubMed]

MDPI

St. Alban-Anlage 66

4052 Basel

Switzerland

Tel. +41 61 683 77 34

Fax +41 61 302 89 18

www.mdpi.com

Crystals Editorial Office

E-mail: crystals@mdpi.com

www.mdpi.com/journal/crystals

www.ingramcontent.com/pod-product-compliance
Lightning Source LLC
Chambersburg PA
CBHW051845210326
41597CB00033B/5781